5 3 $\frac{00}{70}$ C

SEMICONDUCTORS AND SEMIMETALS

VOLUME 15

Contacts, Junctions, Emitters

SEMICONDUCTORS AND SEMIMETALS

Edited by   ROBERT K. WILLARDSON
COMINCO AMERICAN INC.
SPOKANE, WASHINGTON

ALBERT C. BEER
BATTELLE COLUMBUS LABORATORIES
COLUMBUS, OHIO

VOLUME 15
Contacts, Junctions, Emitters

ACADEMIC PRESS
A Subsidiary of Harcourt Brace Jovanovich, Publishers
New York   London   Toronto   Sydney   San Francisco

COPYRIGHT © 1981, BY ACADEMIC PRESS, INC.
ALL RIGHTS RESERVED.
NO PART OF THIS PUBLICATION MAY BE REPRODUCED OR
TRANSMITTED IN ANY FORM OR BY ANY MEANS, ELECTRONIC
OR MECHANICAL, INCLUDING PHOTOCOPY, RECORDING, OR ANY
INFORMATION STORAGE AND RETRIEVAL SYSTEM, WITHOUT
PERMISSION IN WRITING FROM THE PUBLISHER.

ACADEMIC PRESS, INC.
111 Fifth Avenue, New York, New York 10003

United Kingdom Edition published by
ACADEMIC PRESS, INC. (LONDON) LTD.
24/28 Oval Road, London NW1 7DX

Library of Congress Cataloging in Publication Data
Main entry under title:

Semiconductors and semimetals.

   Includes bibliographical references.
   Contents: v. 1-2. Physics of III-V compounds.--
v. 3. Optical properties of III-V compounds.--[etc.]--
v. 15. Contacts, junctions, emitters.
   1. Semiconductors--Collected works. 2. Semimetals--
Collected works. I. Willardson, Robert K.
II. Beer, Albert C.
QC610.9.S47   537.6'22   65-26048
ISBN 0-12-752115-1 (v. 15)

PRINTED IN THE UNITED STATES OF AMERICA

81 82 83 84   9 8 7 6 5 4 3 2 1

Contents

LIST OF CONTRIBUTORS ... vii
PREFACE ... ix
CONTENTS OF PREVIOUS VOLUMES ... xi

Chapter 1 Ohmic Contacts to III–V Compound Semiconductors
B. L. Sharma

I. Introduction ... 1
II. General Ohmic Contact Concepts ... 2
III. Methods for Making Ohmic Contacts ... 9
IV. Methods for Evaluating Contact Resistance ... 18
V. Structural Characteristics of Ohmic Contacts ... 24
VI. Conclusions ... 32
References ... 32

Chapter 2 The Theory of Semiconducting Junctions
Allen Nussbaum

List of Symbols ... 39
I. Introduction ... 41
II. A Description of the *PN* Junction prior to Injection ... 41
III. The *PN* Junction in the Low-Injection Regime ... 59
IV. The *PN* Junction under High Injection ... 104
V. Heterojunctions and Contacts ... 135
VI. Summary ... 189
References ... 190

Chapter 3 NEA Semiconductor Photoemitters
John S. Escher

List of Symbols ... 195
I. Introduction ... 197
II. Photoemission from Metals ... 201
III. Photoemission from Semiconductors ... 205
IV. Work Function Lowering with Cesium and Oxygen ... 211
V. The NEA Semiconductor Photoemitter ... 213
VI. Basic NEA-Related Surface Studies ... 216
VII. The Three-Step Model of Photoemission ... 233
VIII. Photocathode Sensitivity Measurements ... 258

IX. NEA Devices 260
　 X. Dark Current Emission from NEA Cathodes 270
　XI. Shelf Life and Operating Life of NEA Devices 274
　XII. Bias-Assisted Photoemitters 276
XIII. Summary 289
　　　References 290

INDEX 301

List of Contributors

Numbers in parentheses indicate the pages on which the authors' contributions begin.

JOHN S. ESCHER, *Corporate Solid State Laboratory, Varian Associates, Inc., Palo Alto, California 94303* (195)

ALLEN NUSSBAUM, *Electrical Engineering Department, University of Minnesota, Minneapolis, Minnesota 55455* (39)

B. L. SHARMA, *Solid State Physics Laboratory, Lucknow Road, Delhi-110007, India* (1)

Preface

All of the articles in the present volume deal in some way with phenomena involving interfaces. The first chapter concerns ohmic contacts to III–V compounds. Here, after an appropriate presentation giving the reader the basic concepts, the emphasis is placed on practical aspects of value to persons making devices from III–V materials. Detailed information is presented on the various metals and alloys and on associated fabrication techniques that have been found effective in producing good ohmic contacts on the major III–V semiconductors and mixed crystals thereof. Included also is a variety of methods for measurement of contact resistance and for structural characterization of the surface. Important here are the metallurgy of the contact system, the semiconductor surface character, and the compositional and structural changes accompanying the fabrication process. Such considerations have an important effect on device performance, reliability, product yield, and aging.

A review of the fundamental physics of semiconductor junctions occupies the second chapter. This article concentrates on the various approximations and associated boundary conditions used to obtain information from the set of differential equations describing charge, current, potentials, and fields in the interfacial regions. The complications of multicomponent systems, such as the heterojunction, are discussed in detail, with attention drawn to a number of controversial areas still in need of resolution.

The final chapter reviews in detail the physics and technology of electron emission into vacuum from semiconductor surfaces. Emphasis is on the significant achievements of the past decade, involving surface activation by cesium plus oxygen of p-type GaAs and related III–V semiconductors (and also silicon) to produce sufficient work-function lowering to permit the development of photoemissive devices sensitive in the visible and near-infrared regions. Such *negative-electron-affinity* surfaces have demonstrated outstanding performance in developmental and commercial devices, including photoemitters, secondary-electron emitters, and cold-cathode emitters.

The editors are indebted to the many contributors and their employers who make this treatise possible. They wish to express their appreciation to Cominco American Incorporated and Battelle Memorial Institute for providing the facilities and environment necessary for such an endeavor. Special thanks are also due the editors' wives for their patience and understanding.

R. K. WILLARDSON
ALBERT C. BEER

Semiconductors and Semimetals

Volume 1 Physics of III–V Compounds
C. Hilsum, Some Key Features of III–V Compounds
Franco Bassani, Methods of Band Calculations Applicable to III–V Compounds
E. O. Kane, The $k \cdot p$ Method
V. L. Bonch-Bruevich, Effect of Heavy Doping on the Semiconductor Band Structure
Donald Long, Energy Band Structures of Mixed Crystals of III–V Compounds
Laura M. Roth and Petros N. Argyres, Magnetic Quantum Effects
S. M. Puri and T. H. Geballe, Thermomagnetic Effects in the Quantum Region
W. M. Becker, Band Characteristics near Principal Minima from Magetoresistance
E. H. Putley, Freeze-Out Effects, Hot Electron Effects, and Submillimeter Photoconductivity in InSb
H. Weiss, Magnetoresistance
Betsy Ancker-Johnson, Plasmas in Semiconductors and Semimetals

Volume 2 Physics of III–V Compounds
M. G. Holland, Thermal Conductivity
S. I. Novkova, Thermal Expansion
U. Piesbergen, Heat Capacity and Debye Temperatures
G. Giesecke, Lattice Constants
J. R. Drabble, Elastic Properties
A. U. Mac Rae and G. W. Gobeli, Low Energy Electron Diffraction Studies
Robert Lee Mieher, Nuclear Magnetic Resonance
Bernard Goldstein, Electron Paramagnetic Resonance
T. S. Moss, Photoconduction in III–V compounds
E. Antončik and J. Tauc, Quantum Efficiency of the Internal Photoelectric Effect in InSb
G. W. Gobeli and F. G. Allen, Photoelectric Threshold and Work Function
P. S. Pershan, Nonlinear Optics in III–V Compounds
M. Gershenzon, Radiative Recombination in the III–V Compounds
Frank Stern, Stimulated Emission in Semiconductors

Volume 3 Optical of Properties III–V Compounds
Marvin Hass, Lattice Reflection
William G. Spitzer, Multiphonon Lattice Absorption
D. L. Stierwalt and R. F. Potter, Emittance Studies
H. R. Philipp and H. Ehrenreich, Ultraviolet Optical Properties
Manuel Cardona, Optical Absorption above the Fundamental Edge
Earnest J. Johnson, Absorption near the Fundamental Edge
John O. Dimmock, Introduction to the Theory of Exciton States in Semiconductors
B. Lax and J. G. Mavroides, Interband Magnetooptical Effects
H. Y. Fan, Effects of Free Carriers on Optical Properties
Edward D. Palik and George B. Wright, Free-Carrier Magnetooptical Effects
Richard H. Bube, Photoelectronic Analysis
B. O. Seraphin and H. E. Bennett, Optical Constants

Volume 4 Physics of III–V Compounds

N. A. Goryunova, A. S. Borschevskii, and D. N. Tretiakov, Hardness
N. N. Sirota, Heats of Formation and Temperatures and Heats of Fusion of Compounds $A^{III}B^V$
Don L. Kendall, Diffusion
A. G. Chynoweth, Charge Multiplication Phenomena
Robert W. Keyes, The Effects of Hydrostatic Pressure on the Properties of III–V Semiconductors
L. W. Aukerman, Radiation Effects
N. A. Goryunova, F. P. Kesamanly, and D. N. Nasledov, Phenomena in Solid Solutions
R. T. Bate, Electrical Properties of Nonuniform Crystals

Volume 5 Infrared Detectors

Henry Levinstein, Characterization of Infrared Detectors
Paul W. Kruse, Indium Antimonide Photoconducutive and Photoelectromagnetic Detectors
M. B. Prince, Narrowband Self-Filtering Detectors
Ivars Melngailis and T. C. Harman, Single-Crystal Lead–Tin Chalcogenides
Donald Long and Joseph L. Schmit, Mercury–Cadmium Telluride and Closely Related Alloys
E. H. Putley, The Pyroelectric Detector
Norman B. Stevens, Radiation Thermopiles
R. J. Keyes and T. M. Quist, Low Level Coherent and Incoherent Detection in the Infrared
M. C. Teich, Coherent Detection in the Infrared
F. R. Arams, E. W. Sard, B. J. Peyton, and F. P. Pace, Infrared Heterodyne Detection with Gigahertz IF Response
H. S. Sommers, Jr., Microwave-Based Photoconductive Detector
Robert Sehr and Rainer Zuleeg, Imaging and Display

Volume 6 Injection Phenomena

Murray A. Lampert and Ronald B. Schilling, Current Injection in Solids: The Regional Approximation Method
Richard Williams, Injection by Internal Photoemission
Allen M. Barnett, Current Filament Formation
R. Baron and J. W. Mayer, Double Injection in Semiconductors
W. Ruppel, The Photoconductor–Metal Contact

Volume 7 Application and Devices: Part A

John A. Copeland and Stephen Knight, Applications Utilizing Bulk Negative Resistance
F. A. Padovani, The Voltage–Current Characteristics of Metal–Semiconductor Contacts
P. L. Hower, W. W. Hooper, B. R. Cairns, R. D. Fairman, and D. A. Tremere, The GaAs Field-Effect Transistor
Marvin H. White, MOS Transistors
G. R. Antell, Gallium Arsenide Transistors
T. L. Tansley, Heterojunction Properties

Volume 7 Application and Devices: Part B

T. Misawa, IMPATT Diodes
H. C. Okean, Tunnel Diodes
Robert B. Campbell and Hung-Chi Chang, Silicon Carbide Junction Devices
R. E. Enstrom, H. Kressel, and L. Krassner, High-Temperature Power Rectifiers of $GaAs_{1-x}P_x$

Volume 8 Transport and Optical Phenomena

Richard J. Stirn, Band Structure and Galvanomagnetic Effects in III–V Compounds with Indirect Band Gaps
Roland W. Ure, Jr., Thermoelectric Effects in III–V Compounds
Herbert Piller, Faraday Rotation
H. Barry Bebb and E. W. Williams, Photoluminescence I: Theory
E. W. Williams and H. Barry Bebb, Photoluminescence II: Gallium Arsenide

Volume 9 Modulation Techniques

B. O. Seraphin, Electroreflectance
R. L. Aggarwal, Modulated Interband Magnetooptics
Daniel F. Blossey and Paul Handler, Electroabsorption
Bruno Batz, Thermal and Wavelength Modulation Spectroscopy
Ivar Balslev, Piezooptical Effects
D. E. Aspnes and N. Bottka, Electric-Field Effects on the Dielectric Function of Semiconductors and Insulators

Volume 10 Transport Phenomena

R. L. Rode, Low-Field Electron Transport
J. D. Wiley, Mobility of Holes in III–V Compounds
C. M. Wolfe and G. E. Stillman, Apparent Mobility Enhancement in Inhomogeneous Crystals
Robert L. Peterson, The Magnetophonon Effect

Volume 11 Solar Cells

Harold J. Hovel, Introduction; Carrier Collection, Spectral Response, and Photocurrent; Solar Cell Electrical Characteristics; Efficiency; Thickness; Other Solar Cell Devices; Radiation Effects; Temperature and Intensity; Solar Cell Technology

Volume 12 Infrared Detectors (II)

W. L. Eiseman, J. D. Merriam, and R. F. Potter, Operational Characteristics of Infrared Photodetectors
Peter R. Bratt, Impurity Germanium and Silicon Infrared Detectors
E. H. Putley, InSb Submillimeter Photoconductive Detectors
G. E. Stillman, C. M. Wolfe, and J. O. Dimmock, Far-Infrared Photoconductivity in High Purity GaAs
G. E. Stillman and C. M. Wolfe, Avalanche Photodiodes
P. L. Richards, The Josephson Junction as a Detector of Microwave and Far-Infrared Radiation
E. H. Putley, The Pyroelectric Detector—An Update

Volume 13 Cadmium Telluride

Kenneth Zanio, Materials Prepartion; Physics; Defects; Applications

Volume 14 Lasers, Junctions, Transport

N. Holonyak, Jr. and M. H. Lee, Photopumped III–V Semiconductor Lasers
Henry Kressel and Jerome K. Butler, Heterojunction Laser Diodes
A. Van der Ziele, Space-Charge-Limited Solid-State Diodes
Peter J. Price, Monte Carlo Calculation of Electron Transport in Solids

CHAPTER 1

Ohmic Contacts to III–V Compound Semiconductors

B. L. Sharma

I. INTRODUCTION 1
II. GENERAL OHMIC CONTACT CONCEPTS 2
III. METHODS FOR MAKING OHMIC CONTACTS 9
IV. METHODS FOR EVALUATING CONTACT RESISTANCE 18
V. STRUCTURAL CHARACTERIZATION OF OHMIC CONTACTS . . . 24
VI. CONCLUSIONS 32
REFERENCES 32

I. Introduction

With the development of a wide variety of III–V compound semiconductor devices (e.g., injection lasers, light-emitting diodes, Gunn diodes, microwave Schottky-barrier and avalanche diodes, field-effect transistors, bistable resistors, photodiodes), the need for low resistance, reproducible, stable, and reliable ohmic contacts to these semiconducting materials is increasingly felt. In fact, a number of detailed investigations for getting suitable ohmic contacts to III–V compound semiconductors, especially to n-type GaAs, has been undertaken during the past decade. The empirical and experimental findings of these investigations are summarized in the literature (Milnes and Feucht, 1972; Sharma and Purohit, 1974; Rideout, 1975; Schwartz, 1969; Edwards *et al.*, 1972). It appears from most of these findings that achieving good, reliable, low resistance ohmic contact is still a "black" technical art. Lately, however, newer techniques have been utilized by some workers (Robinson, 1975; Sharma *et al.*, 1978; Yoneda *et al.*, 1975; Wittmer *et al.*, 1977; Hartnagel *et al.*, 1976; Vyas, 1976; Yokoyama *et al.*, 1975; Sebestyen *et al.*, 1976; Kim *et al.*, 1975; Chino and Wada, 1977) to study the metallurgy of contact regions, and it is hoped that in coming years with better insight this technology will no longer remain a technical art but will develop as a science.

Because it is necessary to consolidate and critically analyze the existing theoretical and experimental information for a better understanding of any system, it is deemed appropriate to review in this chapter the available theory and technology of ohmic contacts to III–V compound semiconductors. Various techniques used for evaluating and analyzing these contacts will also be summarized. Since it has been observed (Chino and Wada, 1977; Irie et al., 1976; Sebestyen et al., 1975a; Chino et al., 1975; Ohata and Ogawa, 1974; Cox and Hasty, 1969; Palmstrom et al., 1977; Nishitani et al., 1977; Schwartz and Sarace, 1966) that the gradual degradation of many III–V compound semiconductor devices is, to a large extent, related to contact problems, suitable criteria and composition of contacts for various devices will be discussed at the end of the chapter.

II. General Ohmic Contact Concepts

According to the simple Schottky model, when a metal is brought into intimate contact with a semiconductor, the ohmic or rectifying nature of the contact depends on the work functions ϕ_m and ϕ_s of the metal and the semiconductor, respectively. For instance, for a metal contact with an n-type semiconductor to be ohmic it is necessary that $\phi_m < \phi_s$, while for it to be rectifying $\phi_m > \phi_s$. The converse is true for a metal contact with a p-type semiconductor. In fact, if this model is applied to a rectifying metal–n-type semiconductor contact (see Fig. 1a), then the barrier for the movement of electrons from semiconductor to metal is $\phi_m - \phi_s$, while the barrier for the reverse flow of electrons from metal to semiconductor is $\phi_m - \chi_s$, where χ_s is the electron affinity of the semiconductor. Since to a first approximation, $\phi_m - \chi_s$ is a constant for a given metal–n-type-semiconductor

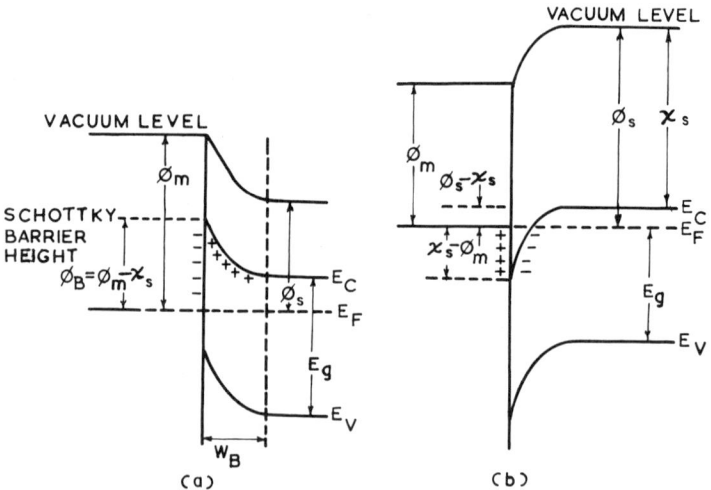

FIG. 1. Equilibrium energy-band diagrams of metal–(n-type)-semiconductor contacts. (a) Rectifying contact ($\phi_m > \phi_s$) and (b) ohmic contact ($\phi_m < \phi_s$).

contact and remains unaffected by applied voltage or doping level of the semiconductor, this barrier is considered to be the most important parameter and is defined as the barrier height ϕ_B of the rectifying (Schottky) contact. In the case of an ohmic metal–n-type-semiconductor contact, if $\phi_m - \chi_s$ turns out to be a negative quantity, then there exists a dipole surface charge barrier (see Fig. 1b). The relatively small barrier height $\phi_s - \chi_s$ implies a linear current–voltage characteristic with a flow of charge across the contact proceeding with equal ease in both directions. Actually, the observed barrier heights of III–V compound semiconductors do not follow these simple rules (see Table I).

In fact the relative independence of the barrier heights of Schottky contacts on metals and the lack of apparent correlation of ohmic contacts with work functions imply that the barrier heights are governed mainly by the large density of the surface states (Bardeen, 1947; Cowley and Sze, 1965; Sze, 1969; Kurtin et al., 1969) and that ohmic contacts are Schottky junctions having energy barriers that are narrow enough to permit extensive tunneling across them (Padovani, 1971).

Considering the effect of surface states for semiconductors (such as GaAs) having large density of surface states, the expression for ϕ_B can be written in modified form (Cowley and Sze, 1965) as

$$\phi_B = \gamma(\phi_m - \chi_s) + (1 - \gamma)(E_g - \phi_0), \quad (1)$$

where E_g is the bandgap of the semiconductor, ϕ_0 represents the position of the neutral level for the surface states from the top of the valence band, and γ is given by $\varepsilon_i/(\varepsilon_i + q\delta D_s)$, in which ε_i is the permittivity, δ the thickness of the interfacial layer, D_s is the density of the surface states, and q is the electronic charge. The assumption of this model, however, is the existence of a thin insulating interfacial barrier, separating the semiconductor surface from the metal. Although this model appears to describe the properties of Schottky barriers, the underlying physical mechanism responsible for stablization of the surface Fermi level has been questioned by some workers (Heine, 1965; Bennett and Duke, 1967; Phillips, 1974; Crowell, 1974). Various theoretical approaches for metal–semiconductor junctions are briefly discussed by Garcia-Moliner and Flores (1976) and Many et al. (1971).

Regarding the possible conduction mechanism in metal–semiconductor systems, various theories have been proposed that qualitatively describe the current in Schottky barriers [e.g., diffusion (Spenke, 1958), thermionic emission (Bethe, 1942), synthesis of thermionic emission and diffusion (Schultz, 1954; Crowell and Sze, 1966)].

Based on the most widely accepted thermionic theory, the current–voltage relation for an ideal Schottky contact can be expressed as

$$I = A^*ST^2 \exp(-\phi_B/kT)[\exp(qV/kT) - 1] = I_S[\exp(qV/kT) - 1], \quad (2)$$

TABLE I CALCULATED AND OBSERVED BARRIER HEIGHTS FOR METAL ON n-TYPE III-V COMPOUND SEMICONDUCTORS

Metal	Mean value[a] ϕ_m (eV)	Calculated $\phi_B = (\phi_m - \chi_s)$ (eV)[b]				Observed ϕ_B (eV, 300 K)				
		GaAs $\chi_s = 4.07$ eV	GaP $\chi_s = 4.0$ eV	InP $\chi_s = 4.40$ eV	InAs $\chi_s = 4.9$ eV	InSb $\chi_s = 4.59$ eV	GaAs	GaP	InP	InAs	InSb
Al	4.25	0.18	0.25	−0.15	−0.65	−0.34	0.80[c]	1.05[c]	0.52[n]	Ohmic[c]	—
Ag	4.36	0.29	0.36	−0.04	−0.54	−0.23	0.88[c]	1.20[c]	0.47[n]	Ohmic[c]	0.18[c,q] 0.079 (77 K)
Au	4.46	0.39	0.46	0.06	−0.44	−0.13	0.86[d]	1.30[c]	0.50[n]	Ohmic[c]	0.17[c,q] 0.076 (77 K)
Bi	4.17	0.10	0.17	−0.23	−0.73	−0.42	0.89[e]	—	—	—	—
Cr	4.38	0.31	0.38	−0.02	−0.52	−0.21	0.77[f]	—	—	—	—
Cu	4.46	0.39	0.46	0.06	−0.44	−0.13	0.82[c]	1.20[c]	0.49[n]	—	—
In	—	—	—	—	—	—	0.82[e]	—	Ohmic[n]	Ohmic[p]	—
Ni	4.64	0.57	0.64	0.24	−0.26	−0.05	0.83[g]	1.20[m]	—	—	—
Pt	4.94	0.87	0.94	0.54	0.04	0.35	0.86[c]	1.45[c]	—	—	—
Sb	4.14	0.07	0.14	−0.26	−0.76	−0.45	0.86[e]	—	—	—	—
Sn	4.36	0.29	0.36	−0.04	−0.54	−0.23	—	—	—	—	—
Ti	4.14	0.07	0.14	−0.26	−0.76	−0.45	0.82[f]	—	—	—	—
W	4.38	0.31	0.38	−0.02	−0.52	−0.21	0.64[h]	—	—	—	—
Au–Ga	—	—	—	—	—	—	0.71–0.75[d]	—	—	—	—
Au–Ge	—	—	—	—	—	—	0.27–0.35[i]	—	—	—	—
Au–Sn	—	—	—	—	—	—	0.72[j]	—	—	—	—
Au–Ti	—	—	—	—	—	—	—	—	0.53[o]	—	—
Pt–Ni	—	—	—	—	—	—	0.95[k]	—	—	—	—
Ti–Ag	—	—	—	—	—	—	0.80[l]	—	—	—	—

[a] Mean value of work functions obtained by contact potential method (Weast, 1972; Michaelson, 1950).
[b] χ_s values from Sharma and Purohit (1974, p. 24).
[c] Milnes and Feucht, 1972, p. 163. [d] Guha et al., 1977. [e] Tyagi, 1977. [f] Sato et al., 1970.
[g] Hackman and Harrop, 1972. [h] Linden, 1976. [i] Moroney and Anand, 1971. [j] Buene et al., 1976.
[k] Ogawa et al., 1971. [l] Irvin and Vanderwal, 1969. [m] Nannichi and Pearson, 1969. [n] Williams et al., 1977.
[o] Roberts and Pande, 1977. [p] Sharma, 1966, unpublished work. [q] Korwin-Pawlowski and Heasell, 1975.

where A^* is the effective Richardson constant, S the area of contact, T the absolute temperature, ϕ_B the barrier height, k the Boltzmann constant, q the electronic charge, V the applied bias, and I_S the saturation current. It is clear from this expression that the contact resistance at zero bias is kT/qI_S and that the plot of the logarithm of the current as a function of the forward applied bias for $V \gg kT/q$ is a straight line with slope q/kT. The intersection of this line with the current axis will give the value of I_S. In actual Schottky contact measurements, the straight line slope of the forward characteristics, instead of being equal to q/kT, is equal to q/nkT, where n is a dimensionless parameter often referred as an ideality factor of the contact.

If thermionic emission over the barrier were the only possible conduction mechanism in metal–semiconductor systems, then the number of ohmic contacts that could be made to III–V compound semiconductors would be very limited. In fact, there are two other conduction mechanisms, namely, thermionic-field (tunnel) emission (Crowell and Rideout, 1969a,b) and field emission (Rideout, 1975; Duke, 1970), that are possible in metal–semiconductor systems, depending on the temperature and the semiconductor doping concentration. For example, if an abrupt metallurgical transition between a metal and a nondegenerate semiconductor forms a rectifying Schottky barrier junction (see Fig. 1a) with thermionic emission as the dominant conduction mechanism, then by increasing the doping concentration, even though the barrier height ϕ_B remains essentially the same, the barrier width decreases with the square root of the doping concentration and thermionic-field emission starts to dominate the conduction mechanism. The current–voltage relation suitable for this thermionic-field emission can be expressed as (Rideout, 1975; Crowell and Rideout, 1969b)

$$I = I_S \{\exp(qV/nkT) - \exp[(n^{-1} - 1)qV/kT]\}; \tag{3}$$

when n equals unity, this equation reduces to Eq. (2). According to this expression, for $n > 2$ the Schottky contact conducts better in the reverse direction because of tunneling-dominated conduction. Finally, in the case of highly doped semiconductors, the barrier width becomes so small that field-emission tunneling becomes dominant, and the contact starts to behave like an ohmic contact with contact resistance reaching a sufficiently small value. This transition from rectifying to ohmic contact, accomplished by varying the doping concentration of the semiconductor, is illustrated schematically by Mead (1969) for Au contacts on n-type GaAs. As a matter of fact, it seems to be well established (Sebestyen et al., 1975a,b) that for ohmic contacts of superior quality, a highly doped semiconductor layer is deemed necessary between contact metal and the low-doped semiconductor. Recently, analysis of some properties of metal–semiconductor contacts that have a thin, heavily doped semiconductor layer between the metal and semiconductor have been reported by Shannon (1976) and Popovic (1978). Popovic (1978) has considered a case in which the heavily doped layer width is equal to or greater than the depletion region width, and he has compared qualitatively the experimental results with theory for metal–Si contacts. A typical energy-band diagram and current–voltage characteristic for such an arrangement is shown schematically in Fig. 2a and b. Gupta *et al.* (1971) have calculated the n^+–n junction resistance in n-type GaAs and have concluded that this resistance determines the lower limit of the contact resistance of a metal–n-type-GaAs alloyed junction.

It is clear from the above discussion that two general approaches can be applied to ohmic contacts to III–V compound semiconductors:

(a) The barrier height is reduced to such an extent that the thermally excited current over the barrier is large enough to give the required value of the contact resistance.

(b) The semiconductor at the interface is doped to such a high doping density that extensive tunneling takes place at the contact.

A unified model for both thermionic and tunneling ohmic contacts to semiconductors has been proposed by Pellegrini and Salardi (1975). The validity of this model, however, for ohmic contacts to III–V compounds has yet to be established.

These approaches, however, do not explain the behavior of many ohmic contacts (Rideout, 1975; Sebestyen, 1977; Jung, 1975; Gyulai *et al.*, 1971), e.g., contacts of such metals or their alloys as In, Au, Ag, Au–In, Ag–In, and Au–Ni to both types of III–V compound semiconductors. In these cases, the

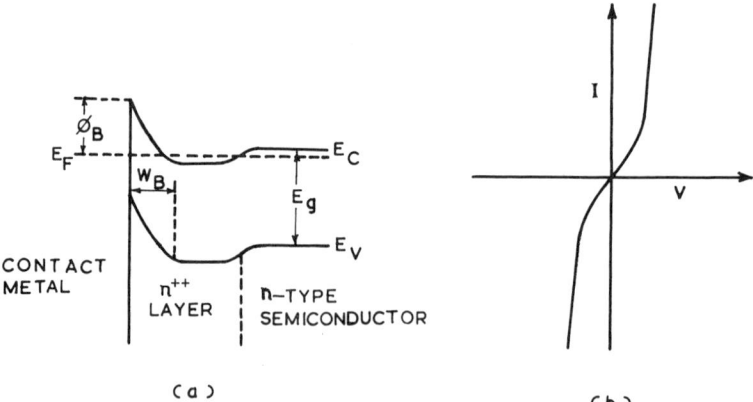

FIG. 2. Schematic representation of (a) energy-band diagram and (b) current–voltage characteristic of a metal–n^{++}–(n-type)-semiconductor ohmic contact.

formation of a highly doped semiconductor layer cannot be assumed since these metals are known to be unsuitable for high-level doping of III–V compounds. Even in cases for which epitaxial contact layers could be obtained from Au–Ge (Otsubo et al., 1977; Andrews and Holonyak, 1972), Au–Ge–Ni (Otsubo et al., 1977), and Au–Ge–Sb (Andrews and Holonyak, 1972) melts, the doping concentrations obtained are an order of magnitude less than those required for superior quality ohmic contacts (Sebestyen et al., 1975a; Chang et al., 1971). In order to explain the behavior of such ohmic contacts, Sebestyen (1977, 1978) has recently proposed a model based on the formation of a thin disordered semiconductor contact layer having a gradual decrease in the amount of disorder from the amorphous or highly disordered region (at the metal–contact layer interface) toward the crystalline semiconductor. The energy-band diagrams of an abrupt amorphous or highly disordered contact layer and of a graded disordered contact layer are shown schematically in Fig. 3a and b, respectively. In these diagrams, the localized states in the mobility gap of the amorphous or disordered contact layer are indicated by dashes.

It can be seen from Fig. 3a that the barrier height ϕ'_B at the metal–amorphous semiconductor interface does not limit or inhibit the current because hopping of electrons between the mobility gap states near the Fermi level is the dominant conduction mechanism in this case (Johnscher and Hill, 1975; Wey, 1976; Peterson and Adler, 1976). Such metal–amorphous semiconductor junctions are nonrectifying and thus the applied voltage drop is mainly in the bulk region of the amorphous semiconductor (Wey, 1976). Because of the barrier height ϕ_B (see Fig. 3a), the abrupt amorphous–crystalline semiconductor junction is generally rectifying and has some similarity to a Schottky barrier junction (Brodsky and Dohler, 1975). If the abrupt discontinuities between the mobility edges of the amorphous semiconductor and the band edges of the crystalline semiconductor are replaced by a smooth gradual variation (see Fig. 3b), then the barrier height ϕ_B of the abrupt case disappears and the graded amorphous–crystalline semiconductor heterojunction behaves like a metal–(n^+-n)-type ohmic contact. So far as the conduction mechanism of this model is concerned, it is similar to one proposed by Riben and Feucht (1966) and involves multistep tunneling and trap-assisted recombination through trap states. This model of ohmic contacts to III–V compounds is supported by the experimental findings of Gyulai et al. (1971) and Magee and Peng (1975), but the observed results of Farrow (1977) are contrary to the assumptions of this model.

With the development of techniques for growing graded semiconductor heterojunctions (Milnes and Feucht, 1972; Sharma and Purohit, 1974), another approach for improving the contactability in some special cases could be the formation of a graded crystalline layer of a mixed semiconductor

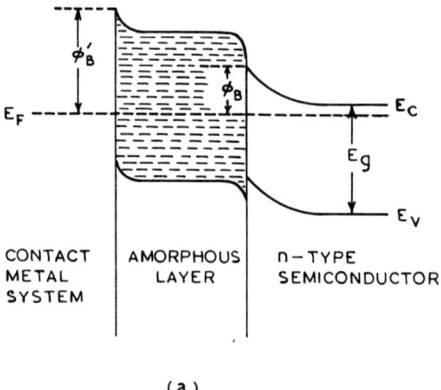

(a)

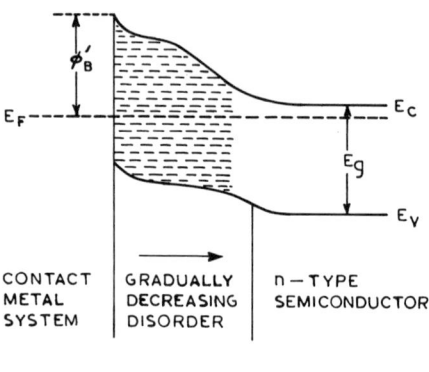

(b)

FIG. 3. Equilibrium energy-band diagrams of metal–(n-type)-semiconductor ohmic contacts having (a) an abrupt amorphous or highly disordered contact layer and (b) a graded disordered contact layer.

between the contact-metal system and the semiconductor. In fact, model calculations for graded heterojunctions have shown that for large graded layers the barrier height decreases to a negligible value and ohmic heterojunctions are formed (Sebestyen, 1978; Raymond and Hayes, 1977; Womac and Rediker, 1972; Cheung et al., 1975; Oldham and Milnes, 1963). Sebestyen (1978) has discussed theoretically the possibility of forming a graded InAs–GaAs-type heterojunction between In and n-type GaAs for good ohmic contact. He has shown that the length of the graded layer necessary for this purpose is greatly influenced by the interface space charge caused by a large lattice mismatch. In fact, the need for rather large graded layers and invariably large lattice mismatches between two III–V compound semiconductors (Sharma and Purohit, 1974) makes this model less probable and useful for ohmic contacts to III–V compounds.

III. Methods For Making Ohmic Contacts

As can be seen from Table II, various methods have been reported in the literature for making ohmic contracts to III–V compound semiconductors. In principle, nearly all methods depend on depositing a thin layer of a metal, a number of metals, or a mixture of metals on the relatively oxide-free clean surface of a semiconductor and heat-treating it during or after deposition in vacuum or in an inert atmosphere. Generally, it is preferred that the metal system deposited semiconductor be heat treated at a temperature higher than the alloying temperature because a heavily doped contact layer, required for good ohmic contact, is often formed between the metal system and the semiconductor during the cooling cycle.

The selection of a contact-metal system for ohmic contact to a particular III–V compound semiconductor depends on many factors. The foremost among them is that one of the components of the metal system should preferably be such that the semiconductors could be doped by it to produce a degenerate surface layer (e.g., for contacts to n-type semiconductor possible materials are Si, Ge, Sn, Se, or Te and for p-type Zn, Cd, Be, or Mg). In addition to this, a number of other factors (e.g., ease of deposition, good adherence, convenient alloying temperature, no interface reaction, minimum thermal mismatch, no surface tension effects during alloying, inertness to the ambient atmosphere, ease in defining contact geometry, good electrical and thermal behavior, and adaptability to thermocompression bonding) must also be considered before selecting a particular contact-metal system. Most of these factors, except for surface tension effects that cause "balling up" during alloying and ultimately lead to patchy nonuniform ohmic contacts, are well taken care of by using metal combinations or alloys having Ag, Au, In, or Sn as the major component and the doping material (mentioned above) as another component. In fact, it can be seen from Table II that Ag, Au, or In base alloys are the most widely used metal systems for ohmic contact to III–V compounds. The final choice of one particular system over another, however, depends on the eutectic temperature of the major component (i.e., Ag, Au, or In) with the semiconductor and its correlation with the temperature to which the semiconductor can safely be heated. For example, both Au–Ge and Ag–In–Ge alloys form good ohmic contacts with n-type GaAs, but because the eutectic temperature of Ag–GaAs (Bernstein, 1962) is higher than that of Au–GaAs (Panish, 1967), the former will be preferred for GaAs devices meant for operating at higher temperatures.

After suitable choice of contact-metal system, an appropriate technique has to be selected for depositing the contact-metal system onto the semiconductor surface. A number of techniques, e.g., evaporation (Maisel and Glang, 1970; Holland, 1961; Powell et al., 1966; Chopra, 1969), sputtering (Maisel and Glang, 1970; Campbell, 1976; Bessot, 1976; Weibmantel et al., 1972), and plating (electrolytic or electroless) (Hakki and Knight, 1966; Champman and Anderson, 1974; Hillegas and Schnable, 1963; Sullivan, 1976; Lawless, 1967; Hu et al., 1970; Lowenheim, 1963), are available for this purpose. Among the techniques used, evaporation is by far the most popular method for the deposition of metal systems on III–V compound semiconductors. In the case of a multicomponent contact material, either the sequential evaporation of various components to the required thickness or the evaporation of the evaporant (containing various components in proper proportion) to dryness must be carried out to ensure correct stoichiometry.

Despite frequent use of sputtering techniques for depositing dielectric films (Sharma, 1978), they have rarely been used for depositing the contact-metal system on III–V compound semiconductors because of low sputter rates, greater degree of surface damage, and difficulty in accurate monitoring of the thicknesses. The back-sputter cleaning of the semiconductor surface, often carried out prior to deposition, can also be a cause of surface damage. This kind of surface damage of III–V compound semiconductors, earlier suspected of being detrimental to ohmic contacts, may, however, have a beneficial influence on them. A few studies (Sullivan, 1976; Hu et al., 1970; Heime et al., 1974) indicate that back sputtering or sputtering induced surface damage may be helpful in improving the contact resistance in more than one way. In fact, many more investigations are needed to ascertain the usefulness of such surface damage for ohmic contacts because this application, combined with the use of ion etching (Spencer and Schmidt, 1971) and rf sputtering for pattern formation, may greatly help in improving III–V compound device technology. Takagi et al. (1975, 1976) have developed an ionized cluster beam technique for depositing highly adhesive metal and semiconductor films. This method has been used by Ishida et al. (1976) to deposit films of Au base alloys on III–V compound semiconductors, and they have reported that good ohmic contacts can be obtained at temperatures well below the one normally used (see Table II) and without any further postdeposition heat treatment.

Even though both electrolytic and electroless plating in a chemical solution are simple deposition techniques and are useful for selective area deposition, they have rarely been used for ohmic contact purposes. One of the main reasons for their limited use is that these techniques are very sensitive to

TABLE II

Metal Systems Used for Ohmic Contact to III–V Compound Semiconductors

Metal system	Semiconductor	Method	References
Ag–Ga	AlP (n)	AgGa preform alloyed at 150°C	Rideout, 1975
Ag–In	GaAs (n, p)	Sequential evaporation of In and Ag in the proportion 25% In–75% Ag and sintering at 500°C in H_2 atmosphere	Matino and Tokunaga, 1969
Ag–In–Ge	GaAs (n)	90% Ag–5% In–5% Ge evaporated onto a substrate kept at 180°C and alloyed at 600°C in forming gas atmosphere	Cox and Strack, 1967
		Addition of Ga improved contact	Sebestyen et al., 1974
Ag–In–Zn	GaAs (p)	80% Ag–10% In–10% Zn evaporated and alloyed at 600°C in forming gas atmosphere	Cox and Strack, 1967
		Cu, Ti, Pt deposited on ohmic contacts have detrimental effects on electrical characteristics	Chino and Wada, 1977
	GaP (p)	(80% Ag–10% In–10% Zn) alloy evaporated and alloyed at 650°C in Ar Atmosphere	Madou et al., 1977
Ag–Sn	GaAs (n)	Ag (2% Sn) evaporated and alloyed for 30 sec at 550–650°C in forming gas atmosphere	Knight and Paola, 1969
	GaAs (n)	Simultaneous evaporation of Ag and Sn in the ratio 1:2 and alloying in H_2 atmosphere	Edwards et al., 1972
Ag–Te–Ni	GaP (n)	Ag (1% Te) evaporated, followed by Ni film and alloyed at 600°C in H_2 atmosphere	Nakatsuka et al., 1971
Ag–Zn	GaAs (p)	90% Ag–10% Zn evaporated film alloyed at 450°C for 10 min in H_2 atmosphere	Ishihara et al., 1976
Al	InAs (p)	Evaporated Al contact	Borrello et al., 1971
	GaP (p)	Conventional evaporated Al comb-type contact	Stringfellow et al., 1975

TABLE II (Continued)

Metal system	Semiconductor	Method	References
Au	AlAs (n, p)	Preform alloyed at 700°C	Rideout, 1975
	GaN (p)	Evaporated Au contact	Jacob and Boris, 1977
	InAs (p)	Au wire contact using ultrasonic bonder	Takahashi et al., 1969
	InP (n, p)	Au or Pb plated Au wire alloyed by thermocompression	Oldham and Milnes, 1963
Au–Be	GaAs (p)	Au (1% Be) film deposited by ionized cluster beam technique at substrate temperature of 300°C	Ishida et al., 1976
	GaP (p)	Au (1% Be) film evaporated on GaP maintained at 250°C and then sintered in temperature range 420–570°C	Pfeifer, 1976; Gershenzon et al., 1966
		Au–Be eutectic + 5% Ni also used	Brantley et al., 1976
	GaP (p)	Au (1% Be) film deposited by ionized cluster beam technique at substrate temperature of 400°C	Ishida et al., 1976
Au–Cd	InP (p)	Au (2% Cd) contact sintered in forming gas in temperature range 400–450°C	Thiel et al., 1977
Au–Cr	InP (p)	Evaporated contact alloyed for 2 min at 420°C	Oe et al., 1977; Itaya et al., 1977
Au–Ge	GaAs (n)	Au (12% Ge) alloy evaporated and alloyed at a rapid heating rate at 450°C for 2.5 min in N_2 atmosphere	Yokoyama et al., 1975; Fukuta et al., 1976
	GaAs (n)	Au/Au–Ge/GaAs (n) structure prepared by evaporation of Au–Ge alloy followed by Au on GaAs substrate maintained at 150°C and alloyed at 450°C for 3 min in forming gas	Sharma et al., 1978
	GaP (n)	Au (3% Ge) alloyed contact	Gillessen et al., 1977
Au–Ge–Ni	GaAs (n)	Au (12% Ge) and Ni successively evaporated and alloyed at 480°C for 30 sec in H_2 ambient	Chino and Wada, 1977; Otsubo et al., 1977

TABLE II (*Continued*)

Metal system	Semiconductor	Method	References
		Au–Ge (88–12%) with about 5% Ni film evaporated and alloyed at 480°C for 3 min in H_2 ambient	Edwards et al., 1972
	InP (n)	Evaporated and alloyed at 500°C for 3 min in H_2 atmosphere	Guha and Hasegawa, 1977
Au–In–Ge	InP (n)	Evaporated and alloyed contact	White and Gibbons, 1972
Au–Ni	GaP (n)	Evaporated Au–Ni (20:1) alloyed at 550°C for 2 min in H_2 atmosphere	Fremunt et al., 1973
Au–Sb	AlAs (n)	Au (1% Sb) probe spark formed in vacuum	Whitaker, 1965
Au–Si	GaAs (n)	Alloying the evaporated Au–Si film at 425°C in H_2 atmosphere	Vandamme and Tijburg, 1976
	GaP (n)	Au (2% Si) evaporated and alloyed for 5 min at 600°C in inert atmosphere	Bergh and Strain, 1969; Schumaker and Rozgonyi, 1972
	GaP (n)	Evaporated Au (2% Si) alloy sintered at 500°C	Kasami et al., 1972
Au–Sn	GaAs (n)	Au (20% Sn) evaporated and alloyed for 2 min at 450°C in H_2 atmosphere followed by Ni electroless plating	Gulati et al., 1969; Rideout, 1975
	GaP (n)	Au (38% Sn) alloyed for 1–2 sec at 700°C in forming gas atmosphere	Lorenz and Pilkuhn, 1966
	GaP (n)	Au (2% Sn) alloyed contact	Haeri, 1973
	GaSb (n)	Pulse plating of Au–Sn and subsequent alloying at 350°C for 10 sec in H_2 atmosphere	Capasso et al., 1980
	InP (n)	Evaporated contact heat-treated at about 430°C for 2 min in forming gas	Oe et al., 1977; Itaya et al., 1977
Au–SnNi–Au	GaAs (n)	Alloying electroplated layers on Au–SnNi–Au at 300°C	Kelly and Wrixon, 1978
Au–Te	GaAs (n)	Au (10% Te) alloyed by laser beam	Pikhtin et al., 1970
		Au (2% Te) evaporated film alloyed at 500°C	Rideout, 1975
	GaP (n)	Au (10% Te) alloyed by laser beam	Pikhtin et al., 1970
	GaSb (n)	Au (10% Te) alloyed by laser beam	Pikhtin et al., 1970

TABLE II (Continued)

Metal system	Semiconductor	Method	References
Au–Zn	GaAs (p)	Au–Zn alloyed in inert atmosphere	Sharma and Mukerjee, 1975
	GaAs (p)	Au–Zn contact deposited by sequential sputtering of Zn and Au in Ar atmosphere and alloyed at 500°C in N_2	Gopen and Yu, 1971
	GaP (p)	Au (3% Zn) evaporated onto GaP substrate heated to 200°C and then alloyed at 505°C in forming gas	Yoneda, 1975
	GaP (p)	Au–Zn Alloy evaporated and alloyed at 650°C in H_2 or Ar atmosphere	Madou et al., 1977; Dierschke and Pearson, 1970
	InP (p)	Au (2% Zn) contact sintered in forming gas in temperature range 400–450°C	Thiel et al., 1977
	InP (p)	Electrodes of Au–Zn alloy heat-treated in vacuum for 5 min at 500°C	Yoshikawa et al., 1977
	InP (p)	Electroplated Au/Zn/Au sintered at 475°C	Schiavone and Pritchard, 1975; Wagner et al., 1976
In	AlAs (n)	Soldering of In dots	Whitaker, 1965
	AlP (n)	In dot ultrasonically soldered	Richman, 1968
	GaAs (n)	Melting In dot at 300°C in H_2 atmosphere	Wronski, 1969
	GaP (n)	Melting In at 400°C in H_2 stream	Madou et al., 1977
	GaSb (n)	Evaporated In film alloyed at 400°C for 5 min in H_2 atmosphere	Sharma and Suri, 1973
	GaSb (p)	In alloyed contacts	Nerou et al., 1976
	GaN (n)	Soldering In contact	Jacob and Boris, 1977; Shintani and Minagawa, 1976
	InAs (n)	Evaporated In contact alloyed in H_2 atmosphere	Sharma, unpublished work
	InP (p)	In preform alloyed in the temperature range 300–600°C	Rideout, 1975
		In does not result in an ohmic contact to moderately resistive p-type InP	Thiel et al., 1977
In–Al	GaAs (n)	Sequential evaporation of In and Al and micro alloying for 90 sec at 320°C in forming gas	Healy and Mattauch, 1976
	GaN (n)	In–Al preform used	Rideout, 1975

1. OHMIC CONTACTS TO III–V COMPOUND SEMICONDUCTORS

TABLE II (*Continued*)

Metal system	Semiconductor	Method	References
In–Au	GaAs (n)	Evaporated 90% In–10% Au film alloyed at 550°C for 30 sec in forming gas	Handu and Tyagi, 1972; Paola, 1970
	GaAs (n)	Evaporated In followed by Au alloyed in forming gas	Hakki and Knight, 1966
In–Ga	InP (n)	Soldering In–Ga	Williams et al., 1977
In–Ni	GaAs (n)	Evaporated In followed by electroless Ni plating and alloyed in forming gas	Hakki and Knight, 1966
In–Sn	GaP (n)	In–Sn dot alloyed	Purohit, 1967
	InP (n)	Evaporated 80% In–20% Sn and alloyed in vacuum	Williams et al., 1977
	InP (n)	Evaporated In–Sn alloyed contact	Pande and Roberts, 1976
In–Te	AlAs (n, p)	In–Te preform alloyed at 160°C	Rideout, 1975
In–Zn	GaAs (p)	In–Zn dot alloyed	Purohit, 1967
		Ohmic contact to semi-insulating GaAs	Suleimanov, 1977
	InP (p)	In (5% Zn) contact sintered in forming gas in temperature range 400–450°C	Thiel et al., 1977
	InP (p)	In–Zn contact alloyed at 500°C	Beppu et al., 1977
Ni–Sn	GaP (n)	Electroless Ni plating followed by evaporated Sn film alloyed at 325°C in H_2 atmosphere	Gershenzon et al., 1966
Pd–Ge	GaAs (n)	As deposited Pd–Ge–GaAs (n) sintered at 500°C for 2 hr	Sinha et al., 1975
Sn	GaP (n)	Sn dot alloyed	Sharma and Mukerjee, 1976
Sn–Ni	GaAs (n)	Evaporated Sn followed by electroless Ni plating and then alloyed in forming gas	Hakki and Knight, 1966
Sn–Sb	GaAs (n)	Sn (4% Sb) or Sn alloyed at 300–350°C for 15–30 sec in an inert atmosphere	Dale and Turner, 1963
Sn–Te	InAs (n)	Sn (1% Te) preform used for contact	Rideout, 1975
	InSb (n)	Sn (1% Te) preform used for contact	Rideout, 1975

surface cleanliness and very often give rise to undesirable interfaces. Sullivan (1976) has electroplated a number of metals (Ag, Au, Pt, Pd) on GaAs surfaces that have been lightly damaged by ion bombardment, and he has observed that the contacts are either ohmic or rectifying, depending on the bombardment species, the contact metal, and the sintering temperature; but the reliability and usefulness of such a technique can easily be questioned. The electroless plating technique has, however, been used frequently for depositing overlayers of Au, Ni, etc., on ohmic contacts as well as on Schottky contacts. Such overlayers are required for bonding thin wires with contact-metal systems without any alteration in the properties of the contacts. The wire bonding is usually carried out by either the thermocompression or the ultrasonic bonding technique (Fogiel, 1968).

Finally, the requirement of a buffer layer (e.g., an n^+ layer on an n-type semiconductor) between the contact-metal system and the semiconductor in order to ensure good ohmic contact can also be met by using epitaxial techniques (Milnes and Feucht, 1972; Sharma and Purohit, 1974), including solution regrowth (Otsubo et al., 1977; Andrews and Holonyak, 1972) and alloying (Sebestyen et al., 1975b; Basterfield et al., 1972; Dale, 1966) for producing such layers. However, it has been observed that it is difficult to produce by these techniques uniform layers that are less than a few tenths of a micron in thickness. Recently it has been shown that the molecular beam epitaxial technique (DiLorenzo et al., 1979; Barnes and Cho, 1978) can be used for growing such buffer layers and it may be possible, in the future, to do away with the subsequent alloying in some particular cases. Ion implantation (Mozzi et al., 1979; Tondon et al., 1979; Mizutani et al., 1977; Higgins et al., 1976; Hunsperger and Hirsch, 1975), on the other hand, promises to be a desirable alternative to epitaxial techniques for obtaining submicron layers without any undesirable interfaces. Sintering will, however, be required after ion implantation in these cases to obtain desired results. Lately this sintering is being carried out by using a pulsed laser or a pulsed electron beam. The published literature on the technique of laser annealing or sintering has been recently reviewed by Bell (1979). There is also some indication (Rideout, 1975; Sebestyen, 1977) that it may be possible to form an amorphous surface layer by ion implantation which, in turn, may act as a graded amorphous buffer layer (Sebestyen, 1977) suitable for ohmic contacts. Ion implantation, as such, may not be a suitable technique for depositing metal system layers because the rate of deposition is very low in this case. Based on the above-mentioned observations, it can be predicted that ion implantation (for producing a buffer layer) followed by evaporation (for depositing the metal system) may prove to be a useful method for making good reproducible ohmic contacts to semiconductors in the near future.

1. OHMIC CONTACTS TO III–V COMPOUND SEMICONDUCTORS   17

The methods used for fabricating ohmic contacts to III–V compound semiconductors can also be used for making contacts to epitaxial layers of mixed III–V compound crystals. The substrate temperature during deposition, alloying or sintering temperature and time, etc., for making ohmic contacts to mixed crystals may, however, be different from those required for III–V compound semiconductors. For the sake of completeness, the methods reported to have been used for making ohmic contacts to mixed crystals are listed in Table III.

TABLE III

METHODS USED FOR OTHMIC CONTACTS TO MIXED III–V COMPOUND SEMICONDUCTORS

Semiconductor	Type	Metal system	Method	References
(GaAl)P	n	Sn	Sn preform	Rideout, 1975
(GaAl)As	n	Au–Ge–Ni	Au–Ge–Ni alloy evaporated onto sample maintained at 250°C and later alloyed at a temperature between 490 and 540°C in forming gas	Shih and Blum, 1972
	n	Au–Sn	Evaporated Au–Sn contact alloyed at 500°C followed by evaporation of Au	Henshall, 1977
	p	Au–Zn	Au–Zn alloy evaporated keeping sample at 250°C and later alloying at 500°C in forming gas	Shih and Blum, 1972; Henshall, 1977
	p	Al	Al evaporated keeping sample at 250°C and later alloying at 500°C in forming gas	Shih and Blum, 1972
Ga(AsP)	n	Ni–In	Ni–In alloy evaporated contact sintered at 500°C in nonoxidizing ambient	Pancholy and Grannemann, 1977
	n	Au–Ge–Ni	Au–Ge–Ni alloy evaporated keeping sample at 250°C and later alloying at a temperature between 530 and 580°C in forming gas	Shih and Blum, 1972
	n	Au–Sn	Evaporated Au–Sn contact alloyed at 450°C	Rideout, 1975
	p	Au–Zn	Au–Zn evaporated keeping sample at 250°C and later alloyed at a temperature between 400 and 500°C in forming gas	Shih and Blum, 1972

TABLE III (*Continued*)

Semiconductor	Type	Metal system	Method	References
	p	Al	Al evaporated keeping sample at 250°C and later alloying at 500°C in forming gas	Shih and Blum, 1972
Ga(AsSb)	n	In–Te	In–Te contact alloyed at 450°C in H_2 atmosphere	Kozlov et al., 1971
	p	Au–In–Zn	Evaporated Au–In–Zn contact alloyed at 565°C in H_2 atmosphere	Mitsuhata, 1972
	p	Au–Cr	Contact formed by successive evaporation of Cr and Au films	Cho et al., 1977
(GaIn)As	n	In	In solder contacts	Conrad et al., 1967
(GaIn)Sb	n	In–Te	Soldering In dots doped with 1 wt% Te	Segawa et al., 1976
	n	In–Te	In–(1 wt%) Te evaporated and alloyed	Kawashima et al., 1976
In(AsP)	n	In	Soldered In contacts	Allen and Mehal, 1970
In(AsSb)	n	In–Te	In–Te preform	Rideout, 1975
In(GaAs)	n	In	Alloying In dots at 440°C in N_2 atmosphere	Sugino et al., 1976
	n	In	Alloying In film at 450°C in H_2 atmosphere	Takeda et al., 1978
	n	Sn	Sn preform	Rideout, 1975
	p	Ni–Au	Ni–Au plated contacts	Nuese et al., 1976

Although the advantages and disadvantages of various techniques that are used in the processes of making ohmic contacts are discussed in this section, it is clear from Tables II and III that evaporation of a metal system, followed by alloying in an inert atmosphere, is at present the most commonly used method for making ohmic contacts to III–V compound semiconductors. Recently, Q-switched ruby laser alloying has also been used for alloying (Margalit *et al.*, 1978) metal systems. The methods for determining the contact resistances and the structural characterization of the ohmic contacts will be discussed in the following sections.

IV. Methods for Evaluating Contact Resistance

It is clear from the earlier discussions that if a metal–semiconductor contact has no potential barrier at the interface, then it is an ideal ohmic contact having a linear current–voltage characteristic and its resistance is just that of the bulk semiconductor. However, an actual metal–semiconductor ohmic contact does have a fairly high potential barrier, but the greatly reduced depletion region width (caused by the presence of a highly doped thin layer between the metal and the bulk semiconductor) allows the electrons to tunnel easily through the barrier. This actual ohmic contact exhibits higher resistance than that expected from an ideal ohmic contact. According to Berger (1972a), this additional resistance may be considered to be a series resistor in the lead to the ideal contact and is referred to as "contact resistance."

If this contact resistance is much larger than the series bulk resistance, then it can easily be measured either by examining the current–voltage characteristic of a simple diode structure on a curve tracer or by using a lock-in amplifier to measure the small-signal ac current with a small-signal ac voltage applied to the diode at zero dc bias. When the contact resistance is comparable to the bulk resistance, these methods are clearly inadequate. Other methods, e.g., the twin-contact (Valdes, 1954), differential (Berger, 1972a), three-contact (Martinez, 1976; Berger, 1972b; Murrmann and Widmann, 1970), four-contact (Yokoyama et al., 1975), extrapolation (Otsubo et al., 1977; Yu, 1970; Lui and Yasui, 1970), four-dot array and planar back-contact (Cox and Strack, 1967) methods are available for this purpose. Some of these methods, used for the determination of contact resistances in the case of III–V compounds, are briefly discussed.

The twin-contact method (Berger, 1972a) consists essentially of measuring the resistance R_{AB} of a homogeneous semiconductor sample having two equal contacts A and B of area A_c on opposite faces (see Fig. 4). This type of contact, often referred to in the literature as the vertical type, gives rise to uniform current distribution. The contact resistance R_c, in this case, is given by (Berger, 1972a)

$$R_c = \tfrac{1}{2}(R_{AB} - \rho_B l/A_c), \tag{4}$$

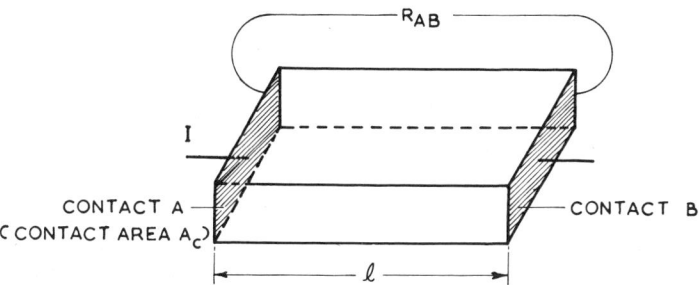

FIG. 4. Schematic representation of the twin contact method for determining contact resistance of a metal–semiconductor ohmic contact.

where ρ_B is the bulk resistivity of the sample and l is the distance between the contacts A and B. The specific contact resistance ρ_c (or contact resistivity) can be determined by using the relation $\rho_c = R_c A_c$. The contact resistance determined according to Eq. (4) is very sensitive to measurement errors since a small difference of large terms appears in the expression. In fact, it has been shown (Berger, 1972a; Joffe, 1946) that length measurement errors have the largest influence. Edwards *et al.* (1972) have used samples of various thicknesses with two planar-parallel contacts for obtaining the specific contact resistance of ohmic contacts to GaAs.

Another method is that referred to as the differential method (Berger, 1972a), in which direct comparison of the actual contact under investigation is made with the other two contacts having negligible contact resistance but having areas the same as that of the actual contact. Although this method, when compared with the twin-contact method, improves the accuracy by more than one order of magnitude, it has a number of practical difficulties (e.g., fabrication of test structure).

A much more accurate three-contact method (Berger, 1972b) has been used by Vyas (1976) for determining the specific contact resistance of ohmic contacts to *n*-type GaAs. This method utilizes a horizontal type of planar contacts in which the main current flows parallel to the surface, the magnitude of the current being kept small to avoid a thermal effect. The test structure, shown schematically in Fig. 5a, has three identical planar contacts 1, 2, and 3 that are arranged with different spacings l_1 and l_2. If a constant current I flows between two such contacts, then the voltage drop V between them can be expressed as (Vyas, 1976)

$$V = RI = (2R_c + R_s l/W)I, \tag{5}$$

where R is the total resistance between the two contacts, W and R_c are the width and contact resistance, respectively, of each contact, R_s is the sheet resistance (ohms per unit area) of the semiconductor beneath the contact metal, and l is the spacing between the two contacts.

In the actual determination of R_c, if a constant current I flows between contacts 1 and 2 and contacts 2 and 3, then the two resistances R_1 and R_2 can be obtained by measuring the corresponding voltage drops V_1 and V_2 across them (see Fig. 5b). Using Eq. (5), the contact resistance R_c can be expressed in terms of R_1 and R_2 as

$$R_c = \tfrac{1}{2}(R_2 l_1 - R_1 l_2)/(l_1 - l_2). \tag{6}$$

According to the transmission line model (TLM) of Berger (1972a,b), in this case the contact resistance R_c is related to the specific contact resistance ρ_c by the expression

$$\rho_c = (R_c^2 - R_T^2)A_c/\alpha d, \tag{7}$$

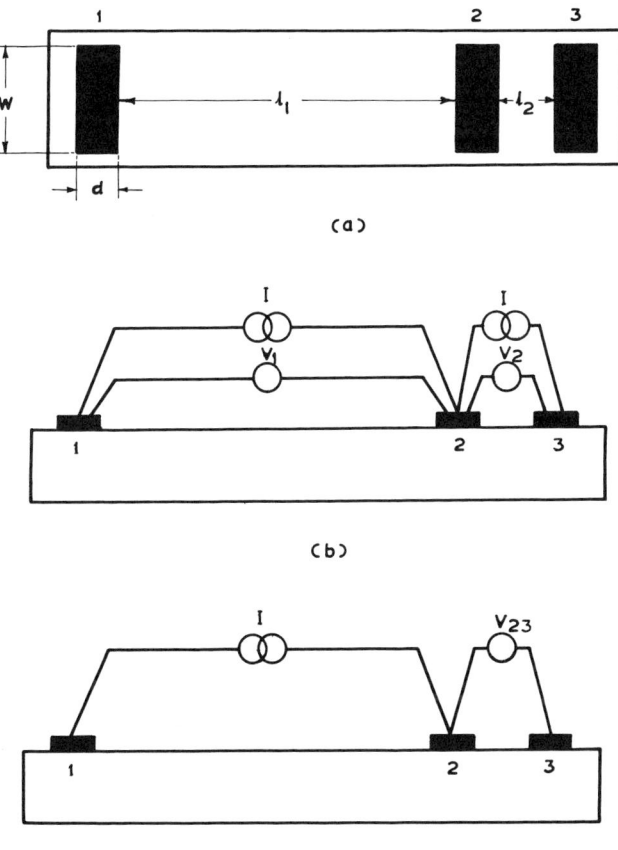

FIG. 5. Three-contact method for determining specific contact resistance. (a) Schematic arrangement of contacts, (b) measurement of R_1 and R_2, and (c) measurement of contact end resistance R_T.

where A_c is the contact area, d the contact length, α the attenuation constant such that $\cosh(\alpha d) = R_c/R_T$, and R_T the "contact end resistance." The contact end resistance $R_T \, (= V_{23}/I)$ can be determined by passing a constant current I between contacts 1 and 2 and measuring the voltage V_{23} between contacts 2 and 3 (see Fig. 5c).

A modified test structure having four contacts has been used by Yokoyama et al. (1975) for the determination of the specific contact resistance of Au–Ge alloy contacts to n-type GaAs. Based on a four-point probe method described by McNeil (1969), Shih and Blum (1972) have evaluated the specific contact resistances of ohmic contacts fabricated by various metal systems on n-type and p-type (AlGa)As, Ga(AsP), and GaP.

Another method for determining a reliable average value of the contact resistance R_c, first described by Shockley (1964) and referred to in the literature as the extrapolation method, has been developed and employed for the determination of the contact resistances of metal–Si contacts by Yu (1970) and of metal–GaAs contacts by Otsubo et al. (1977) for rectangular contacts and by Sifre (1976) for circular contacts in the case of metal–Si contacts. This method, which can be used for the determination of the contact resistance of a metal–n^+-semiconductor layer, is shown schematically in Fig. 6. It consists essentially of depositing a metal grid pattern having a number ($N > 4$) of identical equispaced metal contacts on the surface of the sample. The two extreme contacts are used for passing a constant current I through the sample, and voltage drops $V_1, V_2, \ldots, V_N$ are measured between contacts 1 and 2, 1 and 3, \ldots, 1 and N, respectively. The measured resistance R_N between contacts 1 and N can be expressed to a first approximation by

$$R_N = V_N/I = 2R_c + R_{SN} \simeq 2R_c + R_s X_N/W, \qquad (8)$$

where R_{SN} is the semiconductor layer contribution to the measured resistance, R_c the contact resistance of each contact, R_s the sheet resistance of the semiconductor layer, X_N the distance between contacts 1 and N, and W the width of the contact. It is clear from this expression that the extrapolation of the straight-line plot of measured resistance versus distance to zero separation (i.e., $X_N = 0$) gives rise to an intercept equal to $2R_c$. The

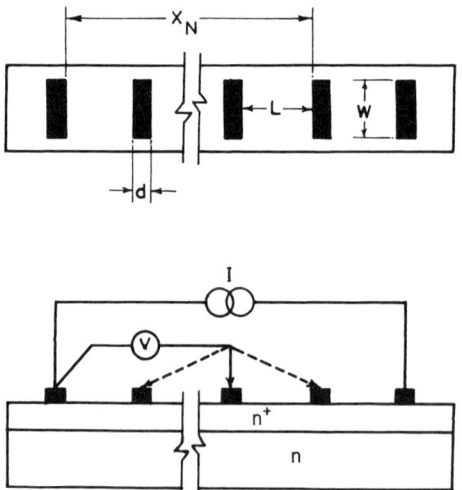

FIG. 6. Schematic representation of extrapolation method for determining the contact resistance of a metal–n^+-semiconductor ohmic contact.

main error is introduced in taking the correct value of X_N for different contact geometries. The error is reduced by taking the distance between two adjacent contacts much greater than the length of the contact (i.e., $L \gg d$) and using the corrected geometry-dependent expression for X_N (Martinez, 1976). The specific contact resistance ρ_c can be determined by using the relation $\rho_c = R_c(Wd)$. Sharma et al. (1978) have used this method for evaluating the specific contact resistance of Au–Ge alloy contacts to n-type GaAs.

A more accurate method, especially for devices having dot and planar-back type contacts, has been developed by Cox and Strack (1967). The test structure for this method consists of fabricating an array of circular contacts of different diameters on the top of a lightly doped, thin epitaxial layer and a comparatively large area back side contact to the heavily doped substrate. The back side contact may be common to all circular contacts. The measured resistance R between a particular circular contact and the large area back contact can be expressed as

$$R = \frac{\rho}{\pi d} \arctan \frac{4t}{d} + \frac{4\rho_c}{\pi d^2} + R_r, \qquad (9)$$

where ρ is the resistivity of the epitaxial material, d the contact diameter, t the epitaxial layer thickness, ρ_c the specific contact resistance, and R_r the residual resistance of the substrate and/or the contact resistance of the back side contact. In the case of a homogeneous wafer, t is the thickness of the wafer and ρ is its resistivity. The specific contact resistance ρ_c can be obtained in this case by using curve fitting techniques when the resistivity and thickness of the epitaxial layer are known.

From the above descriptions of various methods, it can be seen that methods are available for the evaluation of the specific contact resistance of metal–semiconductor contacts in various kinds of structures [e.g., in metal–n, metal–$(n^+ - n)$, metal–$(n - n^+)$, metal–$(n - p)$]. However, it is not always possible to achieve the desired accuracy because in addition to measurement errors it is difficult to incorporate the effects of current crowding and other geometrical constraints in the expressions used for the evaluation of specific contact resistances. Except for n-type GaAs (Edwards et al., 1972; Vyas, 1976), the dependence of specific contact resistance on doping concentration has not been reported in the literature. Even though there exists a wealth of experimental data in the literature on ohmic contacts to n-type GaAs, it is difficult to compare them because of the different fabrication methods and measuring techniques used. Edwards et al. (1972) have plotted their specific contact resistance values of ohmic contacts to n-type GaAs and also data of other workers for a range of doping concentrations.

They have observed that the specific contact resistance appears to be approximately proportional to the reciprocal of the doping concentration, as pointed out by Goldberg and Tsarenkov (1969). A similar relationship has also been observed for Si and Ge (Nibler, 1963). Vyas (1976) observed that for Au–Ge alloy ohmic contacts to n-type GaAs specific contact resistance varies as the reciprocal of the square root of the doping concentration, as reported by Martinez (1976) and Yu (1970) for Si. However, more systematic investigations are required to confirm whether or not the specific contact resistances for a given semiconductor are fairly independent of the contact metal system, fabrication methods, and measuring techniques.

V. Structural Characterization of Ohmic Contacts

In addition to the already discussed electrical characterization of ohmic contacts, their structural characterization is equally important, not only because the available knowledge of metal–semiconductor contact interfaces is very limited but also because the need for reproducible and reliable contacts is increasingly felt because of the considerable growth in the use of III–V compound semiconductor devices in recent years. In fact, it seems almost certain that the performance, reliability, product yield, and aging of many devices depend to a very great extent on the metallurgy of the contact systems. A number of methods developed for thin film and surface analysis (Maisel and Glang, 1970; Kane and Larrabee, 1974; Ibach, 1977; Benninghoven, 1973; Mayer and Turos, 1973) have been used for structural characterization of ohmic contacts. Some of the reported methods, used for investigating the surfaces and the metallurgy of contact regions, are listed in Table IV.

It can be seen from Table IV that more than one method is available for obtaining the types of information (e.g., surface defects, presence of surface impurities, composition and structure of the outermost layers, compositional and structural changes that accompany sintering or alloying, interdiffusion, distribution profiles of various constituents in the vicinity of the metal–semiconductor interface, structure of the regrowth layer at the interface, lateral distribution) generally needed for structural characterization of ohmic contacts. In fact, the complexity of the analysis, resulting from the use of multicomponent systems for ohmic contacts, often demands simultaneous use of more than one of the above-mentioned methods to obtain the necessary information. For example, scanning electron microscopy, used in conjunction with electronbeam induced x-ray analysis of beveled surfaces, provides an extremely useful combination for surface and in-depth analysis of ohmic contacts (Chino and Wada, 1977). Although very little information is available in the literature regarding the use of the above-mentioned

TABLE IV

Various Methods Used for Structural Characterization
of Ohmic Contacts to III-V Compounds

Method		Information
X-ray and electron diffraction analysis	(i)	Identification of film structures before and after alloying (Wittmer et al., 1977; Buene et al., 1976)
	(ii)	Identification of the regrowth layer structure after alloying and removal of metal films
	(iii)	Check crystallographic perfection of surface by LEED (Jona, 1965; Bayliss and Kirk, 1975)
	(iv)	Identification of intermetallic compounds within alloyed contacts (Brantley et al., 1976)
Optical microscopy	(i)	Examination of macroscopic and structural defects
	(ii)	Examination of cross-sectional view of metal–semiconductor interface (Gyulai et al., 1971)
Scanning electron microscopy (SEM)	(i)	Surface morphology of deposited metal films before and after alloying (Sharma et al., 1978; Buene et al., 1976; Gyulai et al., 1971)
	(ii)	Observation of regrowth layers after alloying and removal of metal films (Sharma et al., 1978; Yokoyama et al., 1975)
	(iii)	Examination of ohmic contact–semiconductor interface by exposing the cleaved and beveled surfaces (Hartnagel et al., 1976; Chino and Wada, 1977).
Transmission electron microscopy (TEM)		Changes in microstructures at interface caused by sintering or alloying (Magee and Peng, 1975; Brantley et al., 1976; Augustus and White, 1977)
Electron-probe microanalysis (EPMA and EDXA) (Birks, 1971)	(i)	Elemental analysis and lateral distribution (Wittmer et al., 1977; Yokoyama et al., 1975)
	(ii)	Elemental profiles in vicinity of interface by exposing cleaved or beveled surfaces (Hartnagel et al., 1976; Chino and Wada, 1977; Ohata and Ogawa, 1974)
Auger electron spectroscopy (AES) (Kane and Larrabee, 1974; Ertle and Kuppers, 1974)	(i)	Determining presence of foreign elements on surface (Shiota et al., 1977)
	(ii)	Chemical composition of upper 10–20 Å layer of surface
	(iii)	AES with controlled sputter etching for depth-composition profiles of contact system (Robinson, 1975)
Ion-beam backscattering analysis (He^+ ion) (Chu et al., 1973; Cookson and Pilling, 1973)	(i)	Low energy (1–2 keV) backscattering useful for surface analysis (e.g., trace impurities, adsorption, gettering)

TABLE IV (*Continued*)

Method	Information
	(ii) Low energy backscattering in conjunction with sputter etching used for depth profiles
	(iii) MeV He$^+$ ion backscattering in conjunction with angle lapped samples for deep profiles (Palmstrom *et al.*, 1977)
	(iv) MeV ion backscattering for mass sensitive depth microscopy (Wittmer *et al.*, 1977; Gyulai *et al.*, 1971)
	(v) Lateral variation of contact interdiffusion (Palmstrom *et al.*, 1977)
Proton-induced X-ray analysis (PIX) (Cairns *et al.*, 1973)	(i) Composition at outermost layers of sample
	(ii) Detection of trace amounts of elements (Gray *et al.*, 1973)
Secondary ion mass spectrometry (SIMS) (Giber, 1976)	(i) Distribution profiles of various constituents in vicinity of metal–semiconductor interface (Sharma *et al.*, 1978; Brantley *et al.*, 1975; Kim *et al.*, 1975; Vyas, 1976)
	(ii) Compositional characterization of monolayers near surface (this method covers whole mass range)
	(iii) Investigate slow diffusion in semiconductors (Vyas and Sharma, 1978)
Photo emission spectroscopy (Feuerbacher and Fitton, 1977)	Adsorption phenomena at surfaces
Internal friction technique (Nowick, 1953)	Investigate alloying temperatures of metal–semiconductor system (Nakanishi, 1973)

methods for analyzing ohmic contacts and, therefore, it is very difficult to envisage at present all the merits and demerits of these methods, nevertheless it is still worthwhile to include some of their general capabilities and limitations in this section.

Since depth profiles of the various constituents (i.e., the distribution of constituents as a function of depth) in the vicinity of a metal–semiconductor contact are considered to be most vital for the structural characterization of ohmic contacts, the first to be compared here are the methods used for depth profile determination. These methods can be subdivided into (a) surface analysis methods that rely on controlled sputter etching to give depth profiles, (b) methods involving x-ray generation, and (c) high energy backscattering methods, in which the depth distribution is based on the energy loss of the detected particles. Among the methods (Auger electron spectroscopy, secondary ion mass spectrometry, low energy backscattering) that utilize sputtering techniques to obtain depth profiles, secondary ion mass spectrometry can be considered to be the most suitable method for depth profile measurements over the whole mass range since direct qualitative as well as quantitative determination of secondary ions, very high detection sensitivity, and very fine depth and lateral resolutions are some of its salient features. Since this method has a sensitivity sufficient to measure trace concentrations, it can be used to obtain diffusion profiles of very slowly diffusing elements in semiconductors. However, this, as well as other methods that utilize sputtering, suffer from uncertainties in depth resolution because of the variation in sputtering rates and the nonuniform bottoms of sputtered erosion craters. Auger electron spectroscopy, in the form of the scanning Auger microprobe, can be used for measuring lateral distributions. Methods that involve x-ray generation (e.g., electron-probe microanalysis, ion induced x-ray analysis) are in general not suitable for determining depth profiles inasmuch as good depth resolutions cannot be achieved in these cases. Finally, the high energy (2 MeV) He^+ ion backscattering method, which is inherently capable of giving depth profiles without recourse to sample erosion by sputtering, has frequently been used (Wittmer et al., 1977; Buene et al., 1976; Gyulai et al., 1971) to obtain depth profiles and information about the interaction of constituent elements under heat treatment in the case of ohmic contacts to n-type GaAs. This method, in its conventional form, has a number of limitations (e.g., it provides information concerning the depth distribution of the elements, averaged over a relatively large beam spot area ($\simeq 1$ mm^2); it is difficult to use for deep profiles because of the uncertainties involved in resolving overlapping peaks in the spectrum and the energy straggling of the penetrating beam; it gives poor lateral resolution. The focused ion beam (about a 10-μm spot) backscattering method (Cookson and Pilling, 1973), which overcomes some of the above-mentioned limitations of conventional backscattering, has recently been used by Palmstrom et al. (1977) for studying the degradation of AuGe–GaAs(n) ohmic contacts. Based on this discussion, it can be suggested that useful information about the depth distributions of the constituents of a contact system can be obtained by using this method together with secondary ion mass spectrometric measurements.

Another important aspect of the structural characterization of contacts consists of identifying the presence of intermetallic compounds within alloying contacts. They are a potential source of stress in many semiconductor devices. A number of methods (e.g., transmission electron microscopy, x-ray and electron diffraction analysis, LEED) is available and these methods have been used for such investigations (Magee and Peng, 1975; Brantley et al., 1976; Augustus and White, 1977). In fact, an approximate value of the stress in a contact to GaP has been determined by Brantley et al. (1976) by using an x-ray diffraction lattice curvature measurement technique (Rozgonyi and Ciesielka, 1973).

Not only is the presence of surface impurities recognized as playing an important role in the behavior of metal-compound semiconductor contacts, but also surface cleanliness is generally considered to be essential for reliable, reproducible ohmic contacts (Gyulai et al., 1971; Paola, 1970; Hu et al., 1970; Pruniaux, 1971). An important contamination on the surface of a III–V compound semiconductor is the formation of a natural oxide. For example, it appears that natural oxide to a thickness of 20–30 Å is normally present on the surface of any GaAs sample kept under standard laboratory conditions (Adams and Pruniaux, 1973; Lukes, 1972; Schwartz, 1975). This problem of the formation of a natural oxide when the III–V compound surface is exposed to air is very acute for AlAs. Any of the surface analysis methods can be used to determine such contaminations. Shiota et al. (1977) have used Auger electron spectroscopy for determining the presence of foreign elements on the surface of GaAs samples and have suggested suitable surface preparation techniques to avoid them. As a matter of fact, a comparison of chemical profiles (obtained by Auger electron spectroscopy or by low energy ion-beam backscattering) of specimens, collected from various steps in the fabrication process, can be used to determine the source that is responsible for contaminating the surface or contact.

As an illustrative example, in order to understand the usefulness of various methods for structural characterization, the most extensively used Au–Ge contact system, with reference to n-type GaAs, is examined. Gold–germanium metal contacts are normally made either by the evaporation of Au–Ge eutectic alloy (12 wt % Ge) to dryness or by the coevaporation of Au and Ge or deposition of separate films of Au and Ge in the eutectic composition on the GaAs substrate. The alloying is then carried out by heating the sample in a forming gas atmosphere to a temperature between 450 to 500°C. The period of alloying depends on the alloying temperature and the heating and cooling cycles of the alloying system. However, two main problems (i.e., balling-up of the Au–Ge alloy and formation of Au–Ga alloy during the heat cycle of the alloying process) are frequently encountered that give rise to high specific contact resistance. Scanning electron microscopy (Gyulai et al., 1971; Sharma et al., 1978) can be used to observe the balling-up phenomenon. Usually a layer of Ni is evaporated on top of a Au–Ge film to suppress the balling-up of the Au–Ge alloy (Edwards et al., 1972; Heime et al., 1974). The balling-up phenomenon is not observed however if an oxide-free surface of GaAs (Shiota et al., 1977) and/or a rapid temperature rise alloying system (Yokoyama et al., 1975) are used. The formation of Au–Ga alloy (occurring around 300°C) is minimized (if not avoided) by using a rapid temperature rise alloying system in which the temperature of the sample is raised within $\sim$10 sec to 450°C (Yokoyama et al., 1975; Shiota et al., 1977). In fact, the combination of an oxide-free GaAs surface with a rapid temperature rise alloying system leads to excellent ohmic contacts to n-type GaAs with specific contact resistance as low as 1×10^{-6} Ω cm^2 (Fukuta, 1977). Low alloying temperature and short alloying times are preferred to keep the highly damaged regrown layer of GaAs as small as possible during the cooling cycle and to reduce the loss of volatile As during alloying. The As loss during heating and its influence on contact resistance have been studied by Szigethy et al. (1977), using in situ resistance and mass spectrometric measurement techniques. Their findings have suggested that this loss can be avoided for alloyed ohmic contacts by using an over-pressure of arsenic during alloying. After removal of alloys (e.g., Au–Ge, Au–Ga) by use of suitable etchants, examinations, by scanning electron microscope (Sharma et al., 1978), of the surfaces of samples that were subjected to comparatively long alloying times have provided an indication of the presence of a highly damaged GaAs surface layer which has to be minimized for good ohmic contact.

The alloying behavior of the Au–Ge/(n)GaAs system has been studied by determining depth-composition profiles by means of secondary ion mass spectrometry (Kim et al., 1975; Vyas, 1976; Sharma et al., 1978), He$^+$ ion scattering (Palmstrom et al., 1977; Gyulai et al., 1971), and Auger electron spectroscopy (Robinson, 1975). Considering the multiplicity of experimental approaches followed by various workers, it is difficult to suggest, based on their findings, an exact alloying mechanism for the Au–Ge system, but a qualitative picture can certainly be visualized. The metallurgical structure of the Au–Ge/(n)GaAs alloyed system can be visualized as consisting of a top region having a mixture of Au–Ge and Au–Ga alloys in varying compositions, followed by a region having a heavily doped regrown GaAs layer formed during the cooling cycle and/or a Au and Ge diffused portion of bulk GaAs formed during the alloying time. The SIMS depth profile measurements (Sharma et al., 1978; Vyas, 1976) clearly show that Au penetrates more and more into GaAs with increase in alloying time, and the distribution of Au in GaAs for short alloying times effectively follows a complementary error function law, leading to an effective diffusion coefficient of about 6×10^{-13} cm^2/sec at 450°C (Vyas and Sharma, 1978). It is interesting to note that the extrapolated effective diffusion coefficient at 835°C agrees well with the experimentally observed value of the slow diffusing component of Au into GaAs obtained by Larrabee and Osborne [in Kendall (1968)]. The Ge also follows more or less the same penetration depths as does Au (Vyas, 1976), but if a top layer of Au is evaporated on the Au–Ge/(n)GaAs structure before alloying (Sharma et al., 1978), then the penetration of Ge into GaAs is arrested with the increase in the alloying time at 450°C caused by a comparatively more dominant Ge interaction with the surface layer of Au. In fact, this "gettering" effect is one of the reasons for a sharp increase in the specific contact resistance for long alloying times (e.g., $t > 5$ min). Out diffusion of

Ga into and loss of volatile As through the Au-rich melt is also observed (Hartnagel et al., 1976; Sharma et al., 1978). The effective out-diffusion coefficients for Ga of between 10^{-13} and 10^{-12} cm^2/sec, estimated from the distribution profile measurements by SIMS (Vyas and Sharma, 1978), are found to be of the same order as the value estimated by Robinson (1975), using the AES method. In fact, a large amount of Ga out diffusion to the surface (Robinson, 1975; Weiss and Hartnagel, 1975) may be considered responsible for n^+ doping profiles measured below Au–Ge–Ni contacts (Harris et al., 1969) because Ge acts as a donor when an excess of As is present (Andrews and Holonyak, 1972). The out diffusion of Ga is further enhanced by alloying a Au/Au–Ge/(n)GaAs structure (Sharma et al., 1978) instead of a Ni/Au–Ge/(n)GaAs structure (Vyas, 1976) in a slow heating system because in this case Ga is also consumed by extra Au in the formation of a Au–Ga alloy during the process. In spite of the complexity of this multiple component system, it seems to be well established (Jaros and Hartnagel, 1975; Sebestyen et al., 1974; Sharma et al., 1978) empirically that Ge produces an n^+ GaAs layer between a metal system and n-type GaAs. Although the experimental results of Gyulai et al. (1971) and Magee and Peng (1975) indicate the formation of an amorphous or highly disordered layer beneath the metal system, this Ge-doped n^+ GaAs layer (Sharma et al., 1978) is supposed to lie beneath the highly disordered layer.

It is clear from the above investigations that an optimized condition with respect to the alloying cycle, which includes rise time, alloying time, and cooling time, has to be arrived at in order to minimize the specific contact resistance. In other words, out diffusion of Ga and As through the Au-rich melt, in diffusion of Au and Ge and the formation of compounds between these four constituents have to be adjusted so that the contact resistance is a minimum. In addition to the alloying cycle and alloying temperature, the thickness of the Au–Ge layer and of the over layer of either Au (Sharma et al., 1978) or Ni (Vyas, 1976; Ohata and Ogawa, 1974; Otsubo et al., 1977) also must be optimized in order to minimize the specific contact resistance.

Apart from the above-mentioned factors involved in the formation of the Au–Ge alloy ohmic contact to n-type GaAs, migration of contact materials during operation and other aging factors can degrade the device performance and shorten the mean time of the failure of the operational devices. Only a few studies of contact degradation that have been carried out in recent years (Chino and Wada, 1977; Irie et al., 1976; Chino et al., 1975; Ohata and Ogawa, 1974; Palmstrom et al., 1977). Palmstrom et al. (1977) have investigated the degradation of Au–Ge contacts that results from thermal and electromigration effects in GaAs transferred-electron devices by using a focused backscattering technique. Chino and Wada (1977), Chino et al. (1975), and Irie et al. (1976) have studied the gradual degradation of devices that is caused by metal (e.g., a heat sink in a Gunn diode or an IMPATT diode and a bonding pad for a lead wire in a MESFET or in a LED) in contact with an ohmic contact. With the help of electron probe microanalysis, Chino and Wada (1977) have analyzed the metallurgy of the contact region upon aging and have correlated the results with changes in electrical characteristics of the contacts. They have attributed the degradation primarily to the gradual consumption of the highly doped GaAs layer by GaAs reaction with the heat sink metal. Irie *et al.* (1976) have also observed, in MESFET life tests, the gradual degradation of source and drain ohmic contacts consisting of Pt/Au–Ge alloyed films and Au/Pt/Ti bonding pads, They have attributed this degradation to sheet resistance increase of the n-type layer beneath the contacts that is caused by alloy penetration and/or the formation of a high resistance layer beneath the contact because of crystal defect generation. Whatever may be the other reasons for ohmic contact degradation, one of the most important is the thermo- and electro-interdiffusion of various constituents at operating temperatures. In fact, the studies (Simic and Marinkovic, 1976, 1977) of interdiffusion of Au and Ga and of Au and In at room temperature do suggest interdiffusion as one of the probable reasons for the degradation of ohmic contacts. Wittmer *et al.* (1977) have also suggested that the existence of a high resistivity layer and its increase during aging may be caused by Ge out diffusion from the regrown GaAs layer during a long period of device operation. With the availability of detailed information (Poate *et al.*, 1978) regarding interdiffusion and/or reactions between metal–metal and metal–semiconductor thin films, it will be possible not only to have a clear understanding of the degradation problems associated with ohmic contacts but also to produce improvements in the development of metallization schemes for high reliability ohmic contacts. Since many III–V compound devices are used in systems for space applications, and are consequently subjected to a hostile environment, systematic studies of radiation-induced degradation of ohmic contacts also need considerable attention. One such reported investigation (Blundell *et al.*, 1977) deals with high energy proton or He induced degradation of ohmic contacts in GaAs transferred-electron devices.

From the above-mentioned observations, it is clear that the stability of metal–semiconductor interfaces (e.g., the Au–Ge/GaAs interface) is in doubt even near room temperature and thus requires extensive studies to evaluate the gradual degradation and to find remedial measures in terms of suitable buffer layers to prevent the same. Various methods (e.g., SIMS, AES, EPMA, SEM), used for investigating the metallurgical structure of the alloyed ohmic contacts, are available to assess the degree of degradation. The degradation by these methods can, however, be measured relative to a standard, i.e., the starting alloyed ohmic contact.

VI. Conclusions

Among the types of contacts discussed in this article, the one having a highly doped layer of low resistivity between metal system and semiconductor seems to be ideal for obtaining good ohmic contacts to III–V compound semiconductors. In this case, apart from a considerable reduction in the contact resistance, the influence of injection, if any, by the ohmic contact is also diminished because the lifetime and diffusion length of the minority carrier decrease appreciably with an increase in doping density. However, since our knowledge of the fundamental processes at the surfaces and interfaces is rather limited and systematic information about the metallurgical structure of various ohmic contacts to III–V compound semiconductors is lacking, it is very difficult to pinpoint at present the most probable model encountered in actual practice. It may, however, be mentioned that with the availability of molecular beam epitaxial and ion implantation techniques it may be possible to have more information about the ohmic contacts in the near future.

Inspite of the fact that a variety of analytical methods are now available for the structural characterization of ohmic contacts, the multielement systems and numerous other factors involved in the process of making ohmic contacts make it rather difficult to analyze exactly the role of various elements in reducing contact resistance. Depending on the type of ohmic contact (e.g., horizontal or vertical) encountered in an actual device, a number of methods (discussed in this chapter) is available for the evaluation of specific contact resistance. Although these sophisticated analytical methods are necessary for initial structural characterization and evaluation of contact resistance, for in-process characterization and evaluation nondestructive methods they are preferred and must be developed to fit in the future planar device technology of III–V compounds.

Finally, it can be concluded that the progress in this field of ohmic contacts to III–V compound semiconductors is still in its infancy, and a considerable amount of research work is needed to standardize and optimize the reported ohmic contacts and to discover new ohmic contacts to fit into the growing device technology of III–V compounds.

References

Adams, A. C., and Pruniaux, B. R. (1973). *J. Electrochem. Soc.* **120**, 408.
Allen, H. A., and Mehal, E. W. (1970). *J. Electrochem. Soc.* **117**, 1081.
Andrews, A. M., and Holonyak, N. Jr. (1972). *Solid-State Electron.* **15**, 601.
Augustus, P. D., and White, P. M. (1977). *Thin Solid Films* **42**, 111.
Bardeen, J. (1947). *Phys. Rev.* **71**, 717.
Barnes, P. A., and Cho, A. Y. (1978). *Appl. Phys. Lett.* **33**, 651.

Basterfield, J., Josh, M. J., and Burgess, M. R. (1972). *Acta Electronica* **15**, 5.
Bayliss, C. R., and Kirk, D. C. (1975). *Thin Solid Films* **29**, 97.
Bell, A. E. (1979). *RCA Review* **40**, 295.
Bennett, A. J., and Duke, C. B. (1967). *Phys. Rev.* **162**, 578.
Benninghoven, A. (1973). *Appl. Phys.* **1**, 3.
Beppu, T., Iwamoto, M., Naito, M., and Kasami, A. (1977). *IEEE Trans. Electron Devices* **24**, 951.
Berger, H. H. (1972a). *J. Electrochem. Soc.* **119**, 507.
Berger, H. H. (1972b). *Solid-State Electron.* **15**, 145.
Bergh, A. A., and Strain, R. J. (1969). In "Ohmic Contacts to Semiconductors" (B. Schwartz, ed.), p. 115. Electrochem. Soc., New York.
Bernstein, L. (1962). *J. Electrochem. Soc.* **109**, 270.
Bessot, J. J. (1976). *Thin Solid Films* **32**, 19.
Bethe, H. A. (1942). *Mass. Inst. Technol., Radiat. Lab., Rep.* No. 43-12.
Birks, L. A. (1971). "Electron Probe Microanalysis," 2nd ed. Wiley (Interscience), New York.
Blundell, R., Morgan, D. V., and Howes, M. J. (1977). *Electron. Lett.* **13**, 483.
Borrello, S. R., Pruett, G. R., and Sawyer, J. D. (1971). *Proc. Photoconduct. Conf., 3rd, Stanford, Calif.*, 1969 p. 385.
Brantley, W. A., Schwartz, B., Keramidas, V. G., Kammlott, G. W., and Sinha, A. K. (1975). *J. Electrochem. Soc.* **122**, 434.
Brantley, W. A., Keramidas, V. G., Schwartz, B., Reed, M. H., and Petroff, P. M. (1976). *J. Electrochem. Soc.* **123**, 1582.
Brodsky, M. H., and Dohler, G. H. (1975). *Crit. Rev. Solid State Sci.* **5**, 591
Buene, L., Finstad, T., Rimstad, K., Lonsjo, O., and Olsen, T. (1976). *Thin Solid Films* **34**, 149.
Cairns, J. A., Marwick, A. D., and Mitchell, I. V. (1973). *Thin Solid Films* **19**, 91.
Campbell, D. S. (1976). *Thin Solid Films* **32**, 3.
Capasso, F., Panish, M. B., Sumski, S., and Foy, P. W. (1980). *Appl. Phys. Lett.* **36**, 165.
Chang, C. Y., Fang, Y. K., and Sze, S. M. (1971). *Solid-State Electron.* **18**, 541.
Chapman, B. N., and Anderson, J. C. (1974). "Science and Technology of Surface Coatings." Academic Press, New York.
Cheung, D. T., Chiang, S. Y., and Pearson, G. L. (1975). *Solid-State Electron.* **18**, 263.
Chino, K., and Wada, Y. (1977). *Jpn. J. Appl. Phys.* **16**, 1823.
Chino, K., Wada, Y., and Fukuda, K. (1975). *J. Jpn. Soc. Appl. Phys.* **44**, Suppl., 149.
Cho, A. Y., Casey, H. C. Jr., and Foy, P. W. (1977). *Appl. Phys. Lett.* **30**, 397.
Chopra, K. L. (1969) "Thin Film Phenomena." McGraw-Hill, New York.
Chu, W. K., Mayer, J. W., Nicolet, M.-A., Buck, T. M., Amsel, G., and Eisen, F. (1973) *Thin Solid Films* **17**, 1.
Conrad, R. W., Hoyt, P. L., and Martin, D. D. (1967). *J. Electrochem. Soc.* **114**, 164.
Cookson, J. A., and Pilling, F. D. (1973). *Thin Solid Films* **19**, 381.
Cowley, A. M., and Sze, S. M. (1965). *J. Appl. Phys.* **36**, 3212.
Cox, R. H., and Hasty, T. E. (1969). In "Ohmic Contacts to Semiconductors" (B. Schwartz, ed.), p. 88. Electrochem. Soc., New York.
Cox, R. H., and Strack, H. (1967). *Solid-State Electron.* **10**, 1213.
Crowell, C. R. (1974). *J. Vac. Sci. Technol.* **11**, 951.
Crowell, C. R., and Rideout, V. L. (1969a). *Solid-State Electron.* **12**, 89.
Crowell, C. R., and Rideout, V. L. (1969b). *Appl. Phys. Lett.* **14**, 85.
Crowell, C. R., and Sze, S. M. (1966). *Solid-State Electron.* **9**, 1035.
Dale, J. R. (1966). *Phys. Status Solidi* **16**, 351.
Dale, J. R., and Turner, R. G. (1963). *Solid-State Electron.* **6**, 388.
Dierschke, E. G., and Pearson, G. L. (1970). *J. Appl. Phys.* **41**, 329.

DiLorenzo, J. V., Niehaus, W. C., and Cho, A. Y. (1979). *J. Appl. Phys.* **50**, 951.
Duke, C. B. (1970). *J. Vac. Sci. Technol.* **7**, 22.
Edwards, W. D., Hartman, W. A., and Torrens, A. B. (1972). *Solid-State Electron.* **15**, 387.
Ertl, G., and Kuppers, J. (1974). "Low Energy Electron and Surface Chemistry." Verlag Chemie, Weinheim.
Farrow, R. C. (1977). *J. Phys. D* **10**, L135.
Feuerbacher, B., and Fitton, B. (1977). *In* "Electron Spectroscopy for Surface Analysis" (H. Ibach, ed.), p. 151. Springer-Verlag, Berlin and New York.
Fogiel, M. (1968). "Microelectronics." Res. Educ. Assoc., New York.
Fremunt, R., Kotran, J., and Janouskova, O. (1973). *J. Phys. D* **6**, L96.
Fukuta, M. (1977). Cited in Shiota *et al.* (1977).
Fukuta, M., Suyama, K., Suzuki, H., and Ishikawa, H. (1976). *IEEE Trans. Electron Devices* **23**, 390.
Garcia-Moliner, F., and Flores, F. (1976). *J. Phys. C* **9**, 1609.
Gershenzon, M., Logan, R. A., and Nelson, D. F. (1966). *Phys. Rev.* **149**, 580.
Giber, J. (1976). *Thin Solid Films* **32**, 295.
Gillessen, K., Marshall, A. J., Schuller, K.-H., and Gramann, W. (1977). *IEEE Trans. Electron Devices* **24**, 944.
Goldberg, Yu. A., and Tsarenkov, B. V. (1969). *Fiz. Tekh. Poluprovodn.* **3**, 1718 [Engl. transl.: *Sov. Phys.—Semicond.* **3**, 1447 (1970)].
Gopen, H. J., and Yu, A. Y. C. (1971). *Solid-State Electron.* **14**, 515.
Gray, T. J., Lear, R., Dexter, R. J., Schwettman, F. N., and Wimer, K. C. (1973). *Thin Solid Films* **19**, 103.
Guha, S., and Hasegawa, F. (1977). *Solid-State Electron* **20**, 27.
Guha, S., Arora, B. M., and Salvi, V. P. (1977). *Solid-State Electron.* **20**, 431.
Gulati, R., Purohit, R. K., and Chandra, I. (1969). *J. Inst. Telecommun. Eng., New Delhi* **15**, 815.
Gupta, S. C., Sharma, B. L., and Sreedhar, A. K. (1971). *Solid-State Electron.* **14**, 427.
Gyulai, J., Mayer, J. W., Rodriguez, V., Yu, A. Y. C., and Gopen, H. J. (1971). *J. Appl. Phys.* **42**, 3578.
Hackman, R., and Harrop, P. (1972). *IEEE Trans. Electron Devices* **19**, 1231.
Haeri, S. Y. (1973). *Electron. Lett.* **9**, 279.
Hakki, B. W., and Knight, S. (1966). *IEEE Trans. Electron Devices* **13**, 94.
Handu, V. K., and Tyagi, M. S. (1972). *J. Inst. Telecommun. Eng., New Delhi* **18**, 527.
Harris, J. S., Nannichi, Y., Pearson, G. L., and Day, G. F. (1969). *J. Appl. Phys.* **40**, 4575.
Hartnagel, H., Tomozawa, K., Herron, L. H., and Weiss, B. L. (1976). *Thin Solid Films* **36**, 393.
Healy, M. F., and Mattauch, R. J. (1976). *IEEE Trans. Electron Devices* **23**, 374.
Heime, K., Konig, U., Kohn, E., and Wortmann, A. (1974). *Solid-State Electron.* **17**, 835.
Heine, V. (1965). *Phys. Rev.* **138**, A1689.
Henshall, G. D. (1977). *Solid-State Electron.* **20**, 595.
Higgins, J. A., Welch, B. M., Eisen, F. H., and Robinson, G. D. (1976). *Electron. Lett.* **12**, 17.
Hillegas, W. J., Jr., and Schnable, G. L. (1963). *Electrochem. Technol.* **1**, 228.
Holland, L. (1961). "Vacuum Deposition of Thin Films." Chapman & Hall, London.
Hu, A. Y. C. (1970). *Solid-State Electron.* **13**, 239.
Hu, A. Y. C., Gopen, H. J., and Watts, R. K. (1970). Tech. Rep. No. AFAL-TR-70-196, Air Force At. Lab., Air Force Syst. Command, Wright-Patterson Air Force Base, Ohio.
Hunsperger, R. G., and Hirsch, N. (1975). *Solid-State Electron.* **18**, 349.
Ibach, H., ed. (1977). "Electron Spectroscopy for Surface Analysis." Springer-Verlag, Berlin and New York.
Irie, T., Nagasako, I., Kohgu, H., and Sekido, K. (1976). *IEEE Trans. Microwave Theory Tech.* **24**, 321.

Irvin, J. C., and Vanderwal, N. C. (1969). *In* "Microwave Semiconductor Devices and Their Circuit Applications" (H. A. Watson, ed.), p. 340. McGraw-Hill, New York.
Ishida, I., Wako, S., and Ushio, S. (1976). *Thin Solid Films* **39**, 227.
Ishihara, O., Nishitana, K., Sawano, H., and Mitsui, S. (1976). *Jpn. J. Appl. Phys.* **15**, 1411.
Itaya, Y., Suematsu, Y., and Iga, K. (1977). *Jpn. J. Appl. Phys.* **16**, 1057.
Jacob, G., and Boris, D. (1977). *Appl. Phys. Lett.* **30**, 412.
Jaros, M., and Hartnagel, H. L. (1975). *Solid-State Electron.* **18**, 1029.
Joffe, A. V. (1946). *J. Phys. USSR* **10**, 49.
Johnscher, A. K., and Hill, R. M. (1975). *Phys. Thin Films* **8**, 196.
Jona, F. (1965). *IBM J. Res. Dev.* **9**, 375.
Jung, G. (1975). *Electron Technol.* **8**, 63.
Kane, P. F., and Larrabee, G. B., eds. (1974). "Characterization of Solid Surfaces." Plenum, New York.
Kasami, A., Naito, M., and Toyama, M. (1972). *IEEE Trans. Electron Devices* **19**, 1090.
Kawashima, M., Ohta, H., and Katacka, S. (1976). *Electron. Lett.* **12**, 71.
Kelly, W. M., and Wrixon, G. T. (1978). *Electron. Lett.* **14**, 80.
Kendall, D. L. (1968). *In* "Semiconductors and Semimetals" (R. K. Willardson and A. C. Beer, eds.), Vol. 4, p. 234. Academic Press, New York.
Kim, H. B., Sweeney, G. G., and Heng, T. M. S. (1975). *Proc. Int. Symp. Gallium Arsenide Relat. Comp., 5th, Deauville, 1974*; *Inst. Phys. Conf. Ser.* **24**, 307.
Knight, S., and Paola, C. (1969). *In* "Ohmic Contacts to Semiconductors" (B. Schwartz, ed.), p. 102. Electrochem. Soc., New York.
Korwin-Pawlowski, M. L., and Heasell, E. L. (1975). *Solid-State Electron.* **18**, 849.
Kozlov, Yu. M., Pichakhchi, G. I., Sidorov, V. G., and Ukhanov, Y. I. (1971). *Sov. Phys.—Semicond.* **4**, 1571 [*Fiz. Tekh. Poluprovodn.* **4**, 1824 (1970)].
Kurtin, S., McGill, T. C., and Mead, C. A. (1969). *Phys. Rev. Lett.* **22**, 1433.
Lawless, K. R. (1967). *Phys. Thin Films* **4**, 191.
Linden, K. J. (1976). *Solid-State Electron.* **19**, 843.
Lorenz, M. R., and Pilkuhn, M. (1966). *J. Appl. Phys.* **37**, 4094.
Lowenheim, F. A., ed. (1963). "Modern Electroplating." Wiley, New York.
Lui, K., and Yasui, R. K. (1970). *Conf. Rec., IEEE Photovoltaic Spec. Conf., 8th, Seattle, Wash.*, p. 62.
Lukes, F. (1972). *Surf. Sci.* **30**, 91.
Madou, M. J., Cardon, F., and Gomes, W. P. (1977). *J. Electrochem. Soc.* **124**, 1623.
Magee, T. J., and Peng, J. (1975). *Phys. Status Solidi A* **32**, 695.
Maisel, L. I., and Glang, R., eds. (1970). "Handbook of Thin Film Technology." McGraw-Hill, New York.
Many, A., Goldstein, Y., and Grover, R. B. (1971). "Semiconductor Surfaces." North-Holland Publ., Amsterdam.
Margalit, S., Fekete, D., Pepper, D. M., Chien-Ping Lee, and Yariv, A. (1978). *Appl. Phys. Lett.* **33**, 346.
Martinez, A. (1976). Thesis, Univ. Paul Sabatier de Toulouse, Toulouse, France.
Matino, H., and Tokunaga, M. (1969). *J. Electrochem. Soc.* **116**, 709.
Mayer, J. W., and Turos, A. (1973). *Thin Solid Films* **19**, 1.
McNeil, G. (1969). *In* "Ohmic Contacts to Semiconductors" (B. Schwartz, ed.), p. 305. Electrochem. Soc., New York.
Mead, C. A. (1969). *In* "Ohmic Contacts to Semiconductors" (B. Schwartz, ed.), p. 3. Electrochem. Soc., New York.
Michaelson, H. B. (1950). *J. Appl. Phys.* **21**, 536.
Milnes, A. G., and Feucht, D. L. (1972). "Heterojunctions and Metal-Semiconductor Junctions." Academic Press, New York.

Mitsuhata, T. (1972). *J. Jpn. Soc. Appl. Phys.* **41**, Suppl., 110.
Mizutani, T., Handa, T., Yamazaki, H., and Fuzimato, M. (1977). *Solid-State Electron.* **20**, 443.
Moroney, W. J., and Anand, Y. (1971). *Proc. Int. Symp. Gallium Arsenide, 3rd, London, 1970* p. 259.
Mozzi, R. L., Fabin, W., and Piekashi, F. J. (1979). *Appl. Phys. Lett.* **35**, 337.
Murrmann, H., and Widmann, K. (1970). *IEEE Trans. Electron Devices* **17**, 1022.
Nakanishi, T. (1973). *Jpn. J. Appl. Phys.* **12**, 1818.
Nakatsuka, H., Domenico, A. J., and Pearson, G. L. (1971). *J. Jpn. Soc. Appl. Phys.* **40**, *Suppl.*, 58.
Nannichi, Y., and Pearson, G. L. (1969). *Solid-State Electron.* **12**, 341.
Nerou, J. P., Filion, A., and Girard, P.-E. (1976). *J. Phys. C* **9**, 479.
Nibler, F. (1963). *J. Appl. Phys.* **34**, 1572.
Nishitani, K., Ishihara, O., Sawano, H., Ishii, T., Mitsui, S., and Miki, H. (1977). *Jpn. J. Appl. Phys.* **16**, *Suppl. 1*, 93.
Nowick, A. S. (1953). *Prog. Met. Phys.* **4**, 1.
Nuese, C. J., Olsen, G. H. Henberg, M. E., Gannon, J. J., and Zamerowski, T. J. (1976). *Appl. Phys. Lett.* **29**, 807.
Oe, K., Ando, S., and Sugiyama, K. (1977). *Jpn. J. Appl. Phys.* **16**, 1693.
Ogawa, M., Shinoda, D., Kawamura, N., Nozaki, T., and Asanabe, S. (1971). *Proc. Int. Symp. Gallium Arsenide, 3rd, London, 1970* p. 268.
Ohata, K., and Ogawa, M. (1974). *Proc. Annu. IEEE Reliab. Phys. Symp., 12th, Las Vegas, Nev.*
Oldham, W. G., and Milnes, A. G. (1963). *Solid-State Electron.* **6**, 121.
Otsubo, M., Kuambe, H., and Miki, H. (1977). *Solid-State Electron.* **20**, 617.
Padovani, F. A. (1971). *In* "Semiconductors and Semimetals" (R. K. Willardson and A. C. Beer, eds.), Vol. 7A, p. 7. Academic Press, New York.
Palmstrom, C. J., Morgan, D. V., and Howes, M. J. (1977). *Electron. Lett.* **13**, 504.
Pancholy, R. K., and Grannemann, W. W. (1977). *J. Electrochem. Soc.* **124**, 430.
Pande, K. P., and Roberts, G. G. (1976). *J. Phys. C* **9**, 2899.
Panish, M. B. (1967). *J. Electrochem. Soc.* **114**, 516.
Paola, C. R. (1970). *Solid-State Electron.* **8**, 1189.
Pellegrini, B., and Salardi, G. (1975). *Solid-State Electron.* **18**, 791.
Peterson, K. E., and Adler, D. (1976). *IEEE Trans. Electron Devices* **23**, 471.
Pfeifer, J. (1976). *Solid-State Electron.* **19**, 927.
Phillips, J. C. (1974). *J. Vac. Sci. Technol.* **11**, 947.
Pikhtin, A. N., Popov, V. A., and Yas'kov, D. A. (1970). *Sov. Phys.—Semicond.* **3**, 1383 [*Fiz. Tekh. Poluprovodn.* **3**, 1646 (1969)].
Poate, J. M., Tu, K. M., and Mayer, J. W., eds. (1978). "Thin Films—Interdiffusion and Reactions." Wiley (Interscience), New York.
Popovic, R. S. (1978). *Solid-State Electron.* **21**, 1133.
Powell, C. F., Oxley, J. H., and Blocher, J. M., Jr., eds. (1966). "Vapour Deposition." Wiley, New York.
Pruniaux, B. R. (1971). *J. Appl. Phys.* **42**, 3575.
Purohit, R. K. (1967). *Phys. Status Solidi* **24**, K57.
Raymond, R. M., and Hayes, R. E. (1977). *J. Appl. Phys.* **48**, 1359.
Riben, A. R., and Feucht, D. L. (1966). *Int. J. Electron.* **20**, 583.
Richman, D. (1968). *J. Electrochem. Soc.* **115**, 945.
Rideout, V. L. (1975). *Solid-State Electron.* **18**, 541.
Roberts, G. G., and Pande, K. P. (1977). *J. Phys. D* **10**, 1323.
Robinson, G. Y. (1975). *Solid-State Electron.* **18**, 331.
Rozgonyi, G. A., and Ciesielka, T. J. (1973). *Rev. Sci. Instrum.* **44**, 1053.

Sato, Y., Uchida, M., Shimada, K., Ida, M., and Imai, T. (1970). *Rev. Electr. Commun. Lab.* **18**, 638.
Schiavone, L. M., and Pritchard, A. A. (1975). *J. Appl. Phys.* **46**, 452.
Schultz, W. (1954). *Z. Phys.* **138**, 598.
Schumaker, N. E., and Rozgonyi, G. A. (1972). *J. Electrochem. Soc.* **119**, 1233.
Schwartz, B., ed. (1969). "Ohmic Contacts to Semiconductors." Electrochem. Soc., New York.
Schwartz, B. (1975). *Crit. Rev. Solid State Sci.* **6**, 609.
Schwartz, B., and Sarace, J. C. (1966). *Solid-State Electron.* **9**, 859.
Sebestyen, T. (1977). *In* "Amorphous Semiconductors '76" (I. Kosa Somogyi, ed.), p. 321. Akadémiai Kiado, Budapest.
Sebestyen, T. (1978). Preprint Communication.
Sebestyen, T., Hartnagel, H., and Herron, L. H. (1974). *Electron. Lett.* **10**, 372.
Sebestyen, T., Hartnagel, H., and Herron, L. H. (1975a). *Proc. Int. Symp. Gallium Arsenide Relat. Compd., 5th, Deauville, 1974; Inst. Phys. Conf. Ser.* **24**, 77.
Sebestyen, T., Hartnagel, H. L., and Herron, L. H. (1975b). *IEEE Trans. Electron Devices* **22**, 1073.
Sebestyen, T., Menyhard, M., and Szigethy, D. (1976). *Electron. Lett.* **12**, 96.
Segawa, K., Miki, H., Otsubo, M., Shirahata, K., and Fujibayashi, K. (1976). *Electron Lett.* **12**, 124.
Shannon, J. M. (1976). *Solid-State Electron.* **19**, 537.
Sharma, B. L. (1978). *Solid-State Technol.* **21**, 48; 122.
Sharma, B. L., and Mukerjee, S. N. (1975). *Phys. Status Solidi A* **29**, K141.
Sharma, B. L., and Mukerjee, S. N. (1976), *J. Inst. Electron. Telecommun. Eng., New Delhi* **21**, 479.
Sharma, B. L., and Purohit, R. K. (1974). "Semiconductor Heterojunctions." Pergamon, Oxford.
Sharma, B. L., and Suri, S. K. (1973). *Phys. Status Solidi A* **16**, K47.
Sharma, B. L., Bharti, P. L., Mukerjee, S. N., and Mohan, S. (1978). *Indian J. Pure Appl. Phys.* **16**, 727.
Shih, K. K., and Blum, J. M. (1972). *Solid-State Electron.* **15**, 1177.
Shintani, A., and Minagawa, S. (1976). *J. Electrochem. Soc.* **123**, 1725.
Shiota, I., Motoya, K., Ohmi, T., Miyamoto, M., and Nishizawa, J. (1977). *J. Electrochem. Soc.* **124**, 155.
Shockley, W. (1964), Final Tech. Rep. No. AL-TDR-64-207. Air Force At. Lab., Air Force Syst. Command, Wright-Patterson Air Force Base, Ohio.
Sifre, G. (1976). Cited in Martinez (1976).
Simic, V., and Marinkovic, Z. (1976). *Thin Solid Films* **34**, 179.
Simic, V., and Marinkovic, Z. (1977). *Thin Solid Films* **41**, 57.
Sinha, A. K., Smith, T. E., and Levinstein, H. J. (1975). *IEEE Trans. Electron Devices* **22**, 218.
Spencer, E. G., and Schmidt, P. H. (1971). *J. Vac. Sci. Technol.* **8**, 552.
Spenke, E. (1958). "Electronic Semiconductors." McGraw-Hill, New York.
Stringfellow, G. B., Weiner, M. E., and Burmeister, R. A. (1975). *J. Electron. Mater.* **4**, 363.
Sugino, T., Inoue, M., Shirafuji, J., and Inuishi, Y. (1976). *Jpn. J. Appl. Phys.* **15**, 991.
Suleimanov, S. G. (1977). *Sov. Phys.—Semicond.* **11**, 844 [*Fiz. Tekh. Poluprovodn.* **11**, 1433 (1977)].
Sullivan, A. B. J. (1976). *Electron. Lett.* **12**, 133
Sze, S. M. (1969). "Physics of Semiconductor Devices." Wiley (Interscience), New York.
Szigethy, D., Sebestyen, T., Mojzes, I., and Gergely, G. (1977). *Proc. Int. Vac. Congr., 7th, Int. Conf. Solid Surf., Vienna* **3**, 1959.
Takagi, T., Yanada, I., and Sasaki. A. (1975). *J. Vac. Sci. Technol.* **12**, 1128.

Takagi, T., Yanada, I., and Sasaki, A. (1976). *Thin Solid Films* **39**, 207.
Takahashi, K., Baker, W. D., and Milnes, A. G. (1969). *Int. J. Electron.* **27**, 383.
Takeda, Y., Sasaki, A., Imamura, Y., and Takagi, T. (1978). *J. Electrochem. Soc.* **125**, 130.
Thiel, F. A., Bacon, D. D., Buehler, E., and Bachmann, K. J. (1977). *J. Electrochem. Soc.* **124**, 317.
Tondon, J. L., Nicolet, M. A., Tseng, W. F., Eisen, F. H., Campisano, S. U., Foti, G., and Rimini, E. (1979). *Appl. Phys. Lett.* **34**, 597.
Tyagi, M. S. (1977). *Jpn. J. Appl. Phys.* **16**, Suppl. 1, 333.
Valdes, L. B. (1954). *Proc. IRE* **42**, 420.
Vandamme, L. K. J., and Tijburg, R. P. (1976). *J. Appl. Phys.* **47**, 2056.
Vyas, P. D. (1976). Thesis, Univ. Paul Sabatier de Toulouse, Toulouse, France.
Vyas, P. D., and Sharma, B. L. (1978). *Thin Solid Films* **51**, L21.
Wagner, S., Shay, J. L., Bhar, T. N., Schiavone, L. M., Bachmann, K. J., and Buehler, E. (1976). *Appl. Phys. Lett.* **29**, 431.
Weast, R. C., ed. (1972). "Handbook of Chemistry and Physics," 52nd ed., p. E-69. Chem. Rubber Publ. Co., Cleveland, Ohio.
Weibmantel, C., Fielder, O., Hecht, G., and Reibe, G. (1972). *Thin Solid Films* **13**, 359.
Weiss, B. L., and Hartnagel, H. L. (1975). *Electron. Lett.* **11**, 263.
Wey, H. Y. (1976). *Phys. Rev. B* **13**, 3495.
Whitaker, J. (1965). *Solid-State Electron.* **8**, 649.
White, P. M., and Gibbons, G. (1972). *Electron. Lett.* **8**, 166.
Williams, R. H., Varma, R. R., and McKinley, A. (1977). *J. Phys. C* **10**, 4545.
Wittmer, M., Pretorious, R., Mayer, J. W., and Nicolet, M.-A. (1977). *Solid-State Electron.* **20**, 433.
Womac, J. F., and Rediker, R. H. (1972). *J. Appl. Phys.* **43**, 4129.
Wronski, C. R. (1969). *RCA Rev.* **30**, 314.
Yokoyama, N., Ohkawa, S., and Ishikawa, H. (1975). *Jpn. J. Appl. Phys.* **14**, 1071.
Yoneda, K., Takesada, H., and Yamamuro, M. (1975). *J. Jpn. Soc. Appl. Phys.* **44**, Suppl., 51.
Yoshikawa, A., Ishizaki, O., Kasai, H., and Nishimaki, M. (1977). *Jpn. J. Appl. Phys.* **16**, 2267.
Yu, A. Y. C. (1970). *Solid-State Electron.* **13**, 239.

CHAPTER 2

The Theory of Semiconducting Junctions

Allen Nussbaum

	List of Symbols	39
I.	Introduction	41
II.	A Description of the PN Junction prior to Injection	41
	1. Basic Concepts	41
	2. The Equilibrium Junction	48
III.	The PN Junction in the Low-Injection Regime	59
	3. Low-Level Injection in a PN Junction	59
	4. The Space-Charge Region Under Low-Level Injection	68
	5. The Effect of Generation–Recombination Processes on Device Characteristics	72
	6. Boundary Conditions	90
IV.	The PN Junction under High Injection	104
	7. Approximate and Semiempirical High-Injection Theories	104
	8. The Numerical Analysis of the van Roosbroeck Model	111
V.	Heterojunctions and Contacts	135
	9. Ideal Heterojunctions	135
	10. Unified Theories of Contacts and Junctions	162
VI.	Summary	189
	References	190

List of Symbols[†]

A	current coefficient for heterojunction	E_C	conduction-band edge
A^*	modified Richardson constant	E_V	valence band edge
		E_I	intrinsic level
C	capacitance, integration constant	E_F	Fermi level
		E_G	gap width
		$E_C^{(Sp)}, E_F^{(Sp)}, E_V^{(Sp)}$	energies with respect to Spenke's choice of reference level
D	electric displacement, diffusion constant		
D^*	ambipolar diffusivity	E_{pot}	bottom of potential well in a solid
E	electric field intensity		

[†] *Note about subscripts*: Many of the quantities listed can accept the subscripts e or h for electrons or holes, respectively, N or P for N region or P region, respectively, or one of each. For example, J_{eP} denotes the electron current density in the P region. These possibilities will not be specifically enumerated. *Note about primes*: Primed quantities, as used in Section 8, are defined in Table I.

E_T	trap level	m	mass of free electron (with subscript, effective mass of electron or hole); excess majority carrier concentration
E_{vac}	vacuum level		
G	density of states		
I	electric current		
J	current density		
J_S	reverse saturation current density	n	density of free electrons (with subscript 0, equilibrium value)
J_R	recombination current density		
		n_i	intrinsic concentration
J_E	drift current density	n_0^*	n_0 for $E_T = E_F$
J_D	diffusion current density	p	density of free holes (with subscript 0, equilibrium value), momentum
K	chemical potential		
L	diffusion length		
L_D	Debye length		
N_S	effective density of states	p_0^*	value of p_0 for $E_T = E_F$
N_C, N_V	effective density of states in conduction and valence bands, respectively	q	electric charge
		q_v	charge per unit volume
		r	relative distance, recombination rate
P	pressure		
Q	heat energy	t	time
R	net recombination rate, $\varepsilon_P n_{iP}/\varepsilon_N n_{iN}$	u	relative chemical potential; equilibrium value of u
		u_0	
R_m	radius in phase space	$u(0)$	value of u at junction
S	entropy	v	volume, velocity
T	temperature	w	width of space-charge region
U	internal energy		
V	electrostatic potential	x	coordinate in a one-dimensional junction
V_A	applied bias		
V_J	junction potential	α	slope of diode characteristic
V_{gal}	Galvani potential		
V_{vta}	Volta potential	δ	potential drop in bulk regions of junction
V_{st}	stopping potential		
V_0	equilibrium barrier potential	ε	slope of electrochemical potentials, dielectric constant
V_D	diffusion potential		
W	work function, work, bulk region width	ζ	$(E_F - E_{pot})/kT$
		η	$(E - E_{pot})/kT$
		μ	mobility
a	relative net impurity concentration, lattice constant	μ^*	ambipolar mobility
		$\bar{\mu}$	electrochemical potential
		ν	frequency
c	ratio of electron to hole mobility	ρ	electrical resistivity, $N_A \varepsilon_P / N_D \varepsilon_N$
		σ	electrical conductivity
d	$\exp[-e(V_A - V_D)/kT]$	τ	lifetime; τ_{e0} and τ_{h0} are lifetimes for heavy doping
d'	$\exp[(eV_D - eV_A - \varepsilon)/kT]$		
e	magnitude of electronic charge; base of natural logarithms	χ	electron affinity
		ψ	relative electrostatic potential
$f(E)$	Fermi function	Γ	hole injection coefficient
g	generation rate	Δn	density of injected electrons
h	Planck's constant	Δp	density of injected holes
k	Boltzmann's constant	$\Delta E_C, \Delta E_V, \Delta E_I$	energy level discontinuities

I. Introduction

The first theories of rectification in solid-state devices based on energy-band theory were attempts to explain the behavior of metal–semiconductor junctions. Independent proposals involving quantum mechanical tunneling were advanced in 1932 by A. H. Wilson in England, L. W. Nordheim in Germany, and J. Frenkel and A. Yoffe in the U.S.S.R. Tunneling theory was accepted until 1938, when it was pointed out by Davydov (1938) that it predicts rectification in the wrong direction. It is worthwhile reproducing here the abstract to this paper, which represents the beginning of modern diode theory:

> The conditions are considered under which a current passes through a contact between two electronic semiconductors, possessing conductivities of different type (free electrons and "holes"). It is shown that, according to the direction of the current, there is an increase or decrease in the concentration of free charges near the contact. The result is rectification. General diffusion equations are solved for one special case. The formulae thus obtained give satisfactory agreement with the known data for the Cu_2O rectifier, if it is assumed that the blocking layer possess normal electronic conductivity. Rectification must be accompanied by a thermal effect, associated with the dissociation and recombination of free electrons and "holes". It is shown that the theories of rectification previously proposed (tunnel effect, cold emission) cannot explain the observed phenomena.

II. A Description of the *PN* Junction prior to Injection

1. Basic Concepts

Junction theory is based on concepts from several disciplines. We shall briefly review those ideas which we need, and show how they combine to produce a fundamental set of differential equations.

a. *Energy-Band Theory*

The energy-band description of a solid suitable for use with *PN*-junction theory may be obtained from the Schrödinger equation, assuming a square-well model (Wannier, 1960). The bands appropriate to a semiconductor are shown in Fig. 1, with the most important energy levels labeled. An arbitrary position below the bottom of the valence band is taken as the reference level (corresponding to zero energy), and the intrinsic level E_I, which we shall define exactly below, is distinct from the center of the forbidden band. The width of the forbidden band is

$$E_G = E_C - E_V,$$

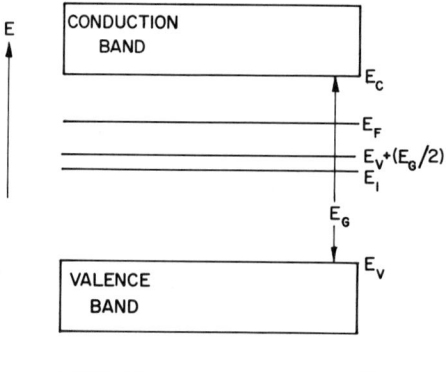

FIG. 1. Energy-band diagram for a semiconductor.

where E_C and E_V are measured with respect to the reference level, as are E_F and E_I. At $T = 0°K$, this material is considered to be an insulator, for the valence band (and all bands below it) are fully occupied, whereas the conduction band (and all those above it) are completely empty. Another possibility is that the uppermost occupied band is only partially full, in which case the material is a conductor and the Fermi level E_F separates the occupied and unoccupied levels (at $T = 0°K$) in the single band with which we are concerned.

b. Equilibrium Statistics

At some reasonable temperature (say $T = 300°K$), the conditions described above are altered. For the insulator, for example, electrons are thermally excited into the conduction band, leaving behind free holes in the valence band. We shall assume the equilibrium carrier densities are low enough so that the probability of occupancy of a given level is governed by Maxwell–Boltzmann statistics and that the density of available levels may be obtained from a solution of the three-dimensional Schrödinger equation for a free electron confined to a potential box. The resulting expression for the equilibrium density n_0 of free electrons in the conduction band is (Nussbaum, 1962)

$$n_0 = N_C \exp[(E_F - E_C)/kT] \qquad (1)$$

and the equilibrium density p_0 of holes in the valence band is

$$p_0 = N_V \exp[(E_V - E_F)/kT], \qquad (2)$$

where E_C and E_V are defined via Fig. 1, k is Boltzmann's constant, and N_C and N_V are the effective density of states in the conduction and valence band, respectively, given by

$$N_C = 2(2\pi m_e kT)^{3/2}/h^3, \quad (3)$$
$$N_V = 2(2\pi m_h kT)^{3/2}/h^3, \quad (4)$$

where m_e and m_h are the corresponding effective mass of an electron and a hole, while h is Planck's constant. The Fermi level shown in Fig. 1 lies at a position appropriate to N-type material. We note from (1) and (2) that

$$n_0 p_0 = N_C N_V \exp(-E_G/kT). \quad (4a)$$

Defining the intrinsic level E_I by

$$E_I = \tfrac{1}{2}(E_C + E_V) + \tfrac{1}{2}kT \log_e(N_V/N_C) \quad (5)$$

and the intrinsic concentration n_i by

$$n_i = \sqrt{N_C N_V} \exp(-E_G/2kT) \quad (6)$$

permits Eqs. (1) and (2) to be written as

$$n_0 = n_i \exp[(E_F - E_I)/kT], \quad (7)$$
$$p_0 = n_i \exp[(E_I - E_F)/kT]. \quad (8)$$

The use of the identity $E_C - E_V = E_G$ with (5) yields the useful relations

$$\begin{aligned}E_C - E_I &= \tfrac{1}{2}E_G - \tfrac{1}{2}kT \log(N_V/N_C),\\ E_I - E_V &= \tfrac{1}{2}E_G + \tfrac{1}{2}kT \log(N_V/N_C).\end{aligned} \quad (8a)$$

As indicated in Fig. 1, the intrinsic level E_I is close to the center of the forbidden band; it lies precisely at the center when $m_e = m_h$, as may be seen from (3)–(5). Since $N_V/N_C = (m_h/m_e)^{3/2}$ and this latter quantity is greater than unity for most materials, E_I generally lies a little higher than the center of the gap. Equations (7) and (8) imply that the density of free carriers in each band is the product of two quantities, the intrinsic concentration n_i, characteristic of pure material, and a Boltzmann factor which represents the shift of the Fermi level above or below the approximate center of the gap due to the addition of impurities.

We shall express the ratio N_C/N_V in terms of mobility ratios, assuming a scattering regime such that the carrier relaxation time τ is independent of energy, i.e., $\mu \sim m^{-1}$ in accordance with the relation (Nussbaum, 1962) $\mu \sim e\tau/m$. This situation is approximately true for certain doping levels that produce an appropriate combination of scattering by lattice vibrations and ionized impurities (Nussbaum, 1962). Then from (3) and (4) we have

$$N_V/N_C = (m_h/m_e)^{3/2} = (\mu_e/\mu_h)^{3/2}. \quad (8b)$$

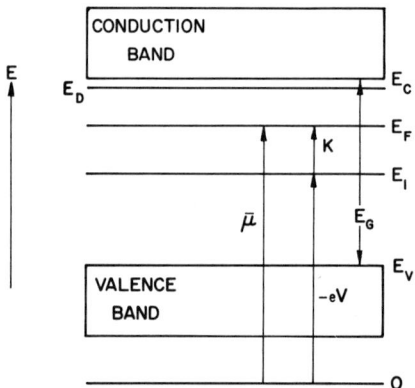

FIG. 2. Thermostatic potentials on an energy-band diagram.

Then Eqs. (8a) assume the form

$$E_C - E_I = \tfrac{1}{2}E_G - \tfrac{3}{4}kT \log(\mu_e/\mu_h),$$
$$E_I - E_V = \tfrac{1}{2}E_G + \tfrac{3}{4}kT \log(\mu_e/\mu_h) \qquad (8c)$$

using the fact that the mobility has been taken as inversely proportional to the effective mass for each type of carrier. Finally, (7) and (8) together with (1) and (2) show that

$$n_i = N_C \exp[(E_I - E_C)/kT] = N_V \exp[(E_V - E_I)/kT]. \qquad (8d)$$

Although replacing effective masses with mobilities as indicated above is admittedly an approximation, we justify this procedure by arguing first of all that we usually use it in connection with mobility ratios rather than individual values, and further, that these ratios for the most part appear as arguments of logarithmic terms which are small. In Fig. 2 we see a donor level E_D fairly close to the bottom of the conduction band. For a P-type material, we would replace this with an acceptor level E_A in an equivalent position above the valence band.

c. Electromagnetic Theory

The continuity equation, which expresses charge conservation in a current-carrying medium, has the form

$$\text{div } \mathbf{J} + \partial q_v/\partial t = 0, \qquad (9)$$

where $\mathbf{J}$, the current density, is related to the electric field $\mathbf{E}$ through Ohm's law

$$\mathbf{J} = \sigma \mathbf{E} \qquad (10)$$

in its simplest form; we shall generalize this for semiconductors in Section 3.

The symbol q_v is the charge per unit volume, or charge density, and σ is the conductivity. Another modification for semiconductors is that we must have distinct continuity equations for electrons and holes. Furthermore, the possibility of charge generation or recombination must be considered, so that (9) becomes, in one dimension,

$$(1/e)(dJ_h/dx) + \partial p/\partial t + R_h = 0 \tag{11}$$

and

$$-(1/e)(dJ_e/dx) + \partial n/\partial t + R_e = 0, \tag{12}$$

where R_h and R_e are the net recombination rates of holes and electrons, respectively, J_h and J_e are the individual current densities, and the symbols p and n without subscript denote nonequilibrium values. (The quantities R_e and R_h will be precisely defined in Section 5.)

Electric field intensity **E** and electrostatic potential V are related by the definition

$$\mathbf{E} = -\operatorname{grad} V, \tag{13}$$

and when this is combined with the first Maxwell equation

$$\operatorname{div} \mathbf{E} = q_v/\varepsilon$$

we obtain Poisson's equation

$$\nabla^2 V = -q_v/\varepsilon, \tag{14}$$

where ε is the permittivity. (Note that we are using the symbol E both for field strength and for energy; the context should be sufficient to remove any ambiguities.)

d. Thermostatics

It has been shown by Callen (1960) that equilibrium thermodynamics may be placed on a postulatory foundation by assuming that the internal energy U of a homogeneous substance can be expressed as a function of the entropy S, volume v, number N of particles, and total electric charge q through the relation

$$U = U(S, v, N, q).$$

Differentiating gives

$$dU = \left.\frac{\partial U}{\partial S}\right|_{v,N,q} dS + \left.\frac{\partial U}{\partial v}\right|_{S,N,q} dv + \left.\frac{\partial U}{\partial N}\right|_{S,v,q} dN + \left.\frac{\partial U}{\partial q}\right|_{S,v,N} dq. \tag{15}$$

We may identify the first two partial derivatives on the right-hand side by temporarily holding N and q constant. Comparing the resulting form of (15) with the first law of thermodynamics, namely

$$dU = dQ - dW,$$

where dU is the change in energy for a process which involves the adsorption of an amount of heat dQ while doing the work dW, shows that

$$\left.\frac{\partial U}{\partial S}\right|_{v,N,q} = T = \text{absolute temperature},$$

$$\left.\frac{\partial U}{\partial v}\right|_{S,N,q} = -P = \text{pressure}.$$

The other two derivatives are given the following names and symbols:

$$\left.\frac{\partial U}{\partial N}\right|_{S,v,q} = K = \text{chemical potential},$$

$$\left.\frac{\partial U}{\partial q}\right|_{S,v,N} = V = \text{electrostatic potential}.$$

Hence, (15) becomes

$$dU = T\,ds - P\,dv + K\,dN + V\,dq. \tag{16}$$

We may regard the four terms on the right as the thermal, mechanical, chemical, and electrical contributions, respectively, to the internal energy change. For a system with N electrons, dq becomes

$$dq = -e\,dN, \tag{17}$$

where $e = 1.6 \times 10^{-19}$ coul is the magnitude of the charge on the electron. Then (16) becomes

$$dU = T\,dS - P\,dv + \bar{\mu}\,dN, \tag{18}$$

where the electrochemical potential $\bar{\mu}$ is defined as (Domenicali, 1954)

$$\bar{\mu} = K - eV. \tag{19}$$

This relation may be interpreted as implying that the total energy of an electron is the algebraic sum of the chemical and the electrostatic contributions.

To indicate how these quantities fit onto an energy-band diagram, we first identify the chemical potential K with the position of the Fermi level, as expressed by the quantity $(E_F - E_I)$ in Eqs. (7) and (8). That is, it seems very reasonable that K should measure the amount by which the position of E_F is shifted above (or below) its position E_I, which is appropriate to an undoped material. In fact, we write (7) and (8) as

$$n_0 = n_i \exp(K/kT) \tag{20}$$

and
$$p_0 = n_i \exp(-K/kT), \tag{21}$$
where (as in Fig. 2)
$$K = E_F - E_I \tag{21a}$$
and E_I is given by (5) and (8a). We note also that (20) and (21) lead to
$$p_0 n_0 = n_i^2 \tag{22}$$
corresponding to (4a) combined with (6). Equations (20) and (21) then express the very reasonable idea that the free-carrier concentrations in a doped material are obtained from the common concentration n_i of a pure material and a Boltzmann factor representing the shift of the Fermi level due to the impurities.

Since the reference level for electrostatic potential V is a matter of choice and since we are using E_I as the reference for the chemical potential K, it is logical to have the intrinsic level serve the function of establishing the relative values of V for the entire band structure. That is, we take the height of E_I as a measure of the electrostatic potential in energy units by identifying this distance as $-eV$ (Fig. 2). We then see from (19) that we have postulated an equivalence between the Fermi level E_F of equilibrium statistics and the electrochemical potential of $\bar{\mu}$ of thermostatics. An analytic justification of this comes out of the conventional derivation of the Fermi–Dirac distribution, using undetermined multipliers to maximize the expression $S = k \log W$, where W is the thermodynamic probability (Nussbaum, 1962). We note that the arrows in Fig. 2 involve a specific sense, so that $-eV$ denotes a positive energy, and (19) is equivalent to the relation
$$E_F = \bar{\mu} = K + (-eV), \tag{23}$$
where, to be consistent with Eq. (21a), we must have that
$$-eV = E_I. \tag{23a}$$
Further, for a P-type material, E_F would move below E_I (in contrast to the situation of Fig. 2), so that K in Eq. (2) actually goes negative, leading to a value of p_0 much larger than that of n_i. By the same token, K is positive in an N-type material, and n_0 increases from n_i by an amount specified by the Boltzmann factor $\exp(K/kT)$ of Eq. (20).

Both Eqs. (20) and (21) imply that the Boltzmann factor is a reasonable approximation to the Fermi–Dirac distribution, which has the form
$$f(E) = 1/\{\exp[(E - E_F)/kT] + 1\}, \tag{24}$$
where $f(E)$ is the probability that an energy level E will be occupied. The quantity E_F is that level which has a 50% probability of occupancy, for when

$E = E_F$, then $f(E_F) = \frac{1}{2}$. The Fermi energy is often identified (Moore, 1962) with the chemical potential K, but this is correct only for processes in which changes of electrostatic energy can be ignored (e.g., chemical reactions). The reason for this comes by comparing (16) for constant q with (18). Under these circumstances

$$K = \bar{\mu} = E_F.$$

This result, however, is equivalent to regarding the electron as a pure particle; its charge is ignored. In general, we must take into account both the particle and the electrical attributes of the electron, and we shall see, in fact, that these two attributes show up in connection with transport properties.

2. THE EQUILIBRIUM JUNCTION

An isolated system is said to be in thermal equilibrium when the temperature is uniform; similarly, we specify mechanical and electrochemical equilibrium via the second and third terms of Eq. (18), so that when S and v are fixed we have the relation

$$\text{grad } \bar{\mu} = 0. \tag{25}$$

This is equivalent to the requirement that the Fermi level be constant throughout the junction. Let us combine this fact with the prediction that when a P-type semiconductor is in good electrical contact with an N-type semiconductor, forming a PN junction, the holes from the P-type material will tend to diffuse into the N-type region and vice versa. The cause of this redistribution is both electrical (electrons mutually repel one another, and so do holes) and chemical (either type of carrier will diffuse away from regions of high concentration). The readjustment is a self-canceling process, however, since holes going into the N region not only repel any additional positive charges but they leave behind negatively ionized acceptors which also impede hole motion. Combining this with a similar phenomenon for electrons, we realize that when equilibrium in the PN junction has been established, there exists a double or dipole layer of charge enclosing the junction, which— because it inhibits any further motion—is called a barrier layer. It is also known as the space-charge layer, since this region is no longer neutral. The loss of electrons by the N-type side of the junction is equivalent to an apparent reduction in impurity concentration. That is, the magnitude of the chemical potential K has to decrease, in accordance with Eq. (20). This means that E_F should move down closer to E_1, but Eq. (25) requires E_F to be fixed. Hence, the intrinsic level must bend upwards, and taking into account the analogous effect on the P-type side of the junction, we obtain Fig. 3, which shows the situation for a junction which is symmetrically doped and

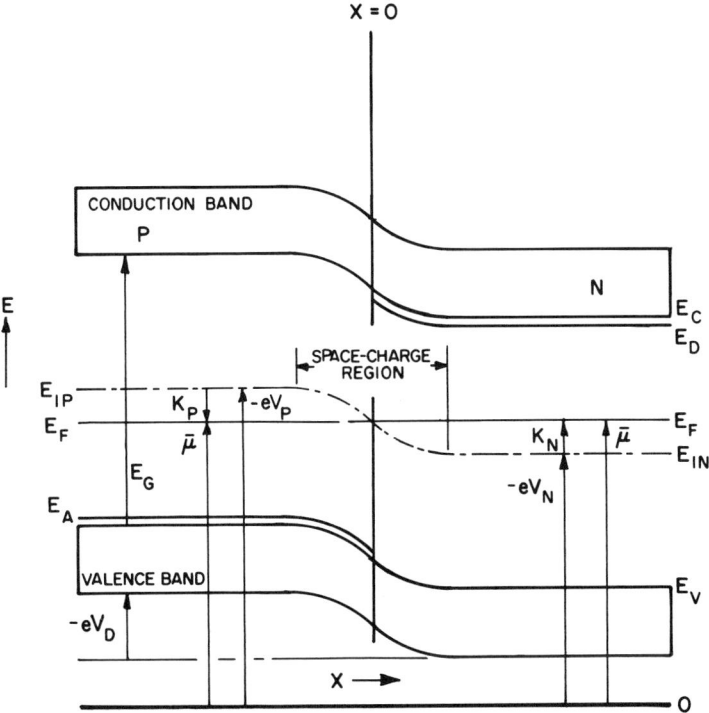

FIG. 3. Equilibrium energy-band diagram for an abrupt, symmetrical *PN* junction.

where the transition from donors to acceptors is abrupt. The diagram shows $\bar{\mu}$, K, and $-eV$ (with appropriate subscripts) as defined via Fig. 2 for the regions where the bands are unaltered. We have shown a smooth transition in the bands as the junction is crossed, with the levels on the *P*-type side lying a distance $-eV_D$ above those on the reference side. This quantity, which we emphasize is positive, is called the diffusion, barrier, or built-in potential and is defined as

$$-eV_D = -eV_P - (-eV_N). \tag{26}$$

The specific shape of the bands in the space-charge region can be determined by solving Eq. (14). To convert it to dimensionless form, we first differentiate (19), using the fact that $\bar{\mu}$ is constant, and obtain

$$d^2K/dx^2 = e\,d^2V/dx^2. \tag{26a}$$

The charge q_v per unit volume in Eq. (14) is

$$q_v = e(N_D - N_A + p_0 - n_0), \tag{27}$$

where N_D and N_A are the donor and acceptor concentrations, respectively (assumed to be fully ionized at room temperature). By Eqs. (20) and (21)

$$q_V = 2n_i e[a - \sinh(K/kT)], \qquad (28)$$

where

$$a = (N_D - N_A)/2n_i \qquad (29)$$

specifies the relative net impurity concentration. Next, define a relative chemical potential u by

$$u = K/kT. \qquad (30)$$

The use of (26a), (28), and (29) in the one-dimensional form of Eq. (14) gives the Poisson–Boltzmann equation (Passau and van Styvendael, 1960)

$$d^2u/dx^2 = (1/L_D^2)(\sinh u - a), \qquad (31)$$

where the intrinsic Debye length L_D is

$$L_D^2 = \varepsilon kT/2n_i e^2. \qquad (32)$$

It is convenient to define a dimensionless length

$$r = x/L_D \qquad (33)$$

so that (31) simplifies to

$$d^2u/dr^2 = \sinh u - a. \qquad (34)$$

We shall solve Eq. (34) for the abrupt germanium junction specified by

$$N_D = 10^{15}/\text{cm}^3, \quad N_A = 0 \qquad \text{for} \quad x \geq 0,$$
$$N_D = 0, \quad N_A = 10^{18}/\text{cm}^3 \qquad \text{for} \quad x \leq 0.$$

The constants are

$$n_i = 2.4 \times 10^{13}/\text{cm}^3,$$
$$E_G = 0.70 \quad \text{eV},$$
$$N_C/N_V = (m_e/m_h)^{3/2} = (\mu_h/\mu_e)^{3/2} = (1800/3600)^{3/2},$$

where μ_e and μ_h are the electron and the hole mobility, respectively. By (8a), we find that

$$E_I - E_V = 0.363 \quad \text{eV},$$

or in dimensionless units

$$(E_I - E_v)/kT = 14.29.$$

This quantity locates E_{IN} with respect to the top of the valence band. To find the height of E_{IP} above this level, we again take advantage of the constant value of $\bar{\mu}$ to write

$$(K - eV)_N = (K - eV)_P$$

2. THE THEORY OF SEMICONDUCTING JUNCTIONS

or, using (20) and (26),

$$-eV_P - (-eV_N) = -eV_D = kT \log \frac{n_{ON}}{n_i} - kT \log \frac{n_{OP}}{n_i}$$

$$= kT \log \frac{n_{ON} p_{OP}}{n_i^2} \qquad (34)$$

$$= kT \log \frac{(10^{15})(10^{18})}{(2.4 \times 10^{13})^2} = 0.365 \quad \text{eV}.$$

The positive value conforms with the sense of $-eV$ as indicated by the arrow of Fig. 3.

To position the bands with respect to the Fermi level, we use the fact that the chemical potentials for materials in equilibrium are constant. That is, as Fig. 3 suggests, we have the conditions

$$d^2u/dr^2 = 0 \quad \text{for} \quad r = \pm\infty,$$

where the limits on r are justified by the fact that the dimensions of the diode are generally much greater than L_D. Then Eq. (34) leads to the result

$$\sinh u = a, \qquad (35)$$

but this is actually two equations since a changes abruptly at $r = 0$. By (29),

$$a_N = N_D/2n_i = 20.830,$$
$$a_P = -N_A/2n_i = -2.083 \times 10^4. \qquad (36)$$

Designating the equilibrium value of the dimensionless chemical potential u in the N region by u_{ON}, the transcendental equation (35) has a solution

$$u_{ON} = 3.730 \qquad (37)$$

and similarly

$$u_{OP} = -10.637 \qquad (38)$$

from which the equilibrium chemical potentials are

$$K_N = 0.095 \quad \text{eV}, \quad K_P = -0.270 \quad \text{eV}. \qquad (39)$$

We note that the algebraic sum of the two quantities in (39) should be

$$-eV_D = K_N + |K_P| = 0.365,$$

agreeing with (34a).

The values of K_N and K_P locate E_{IN} and E_{IP}, respectively in Fig. 3; this is a simple approach to the determination of Fermi levels. Actually, Eq. (35) is equivalent to the use of charge conservation—that is, to the statement that $q_v = 0$—but the special form used here is valid only for the case of fully ionized impurities in a nondegenerate material. The more general situation requires the use of Fermi statistics (Nussbaum, 1962); at temperatures well above 300°K, other complications enter (Jain and van Overstraeten, 1974).

We continue with our calculation by integrating Eq. (34) to obtain

$$\tfrac{1}{2}(du/dr)^2 = \cosh u - au + C, \tag{40}$$

where C is a constant of integration, or

$$du/dr = [2(\cosh u - au + C)]^{1/2}, \tag{41}$$

and to determine C, we use the boundary condition that du/dr vanish on either side of the space-charge region, again from Fig. 3. That is,

$$du/dr = 0 \quad \text{for} \quad r = \pm\infty \tag{42}$$

so that

$$C = au_0 - \cosh u_0, \tag{43}$$

where u_0 is the equilibrium value of u. Since (31) is actually two equations, one involving a_N and the other a_P, (43) should be written as

$$\begin{aligned} C_N &= a_N u_{0N} - \cosh u_{0N}, \\ C_P &= a_P u_{0P} - \cosh u_{0P}. \end{aligned} \tag{44}$$

The second integration step, by Eq. (41), is

$$r = \int_{u(0)}^{u_0} \frac{du}{[2(\cosh u - au + C)]^{1/2}}, \tag{45}$$

where $u(0)$ is the value of u at the junction ($r = 0$), and where u_0, C, and a have different values for the N and the P regions. The only constant in (45) that is not yet known is $u(0)$; to determine it, we use the condition that the normal component of **D** must be continuous, which is equivalent to the relation

$$du/dr|_N = du/dr|_P \tag{46}$$

or by (41),

$$\cosh u(0) - a_N u(0) + C_N = \cosh u(0) - a_P u(0) + C_P$$

from which

$$u(0) = (C_N - C_P)/(a_N - a_P). \tag{47}$$

From (36)–(38) and (44), we find that

$$C_N = 56.8, \qquad C_P = 2.01 \times 10^5,$$

and by (47),

$$u(0) = -9.64.$$

We may now evaluate the two integrals in (45), one in each region, and obtain the plot of $u = K/kT$ versus $r = x/L_D$ shown in Fig. 4a. In order to

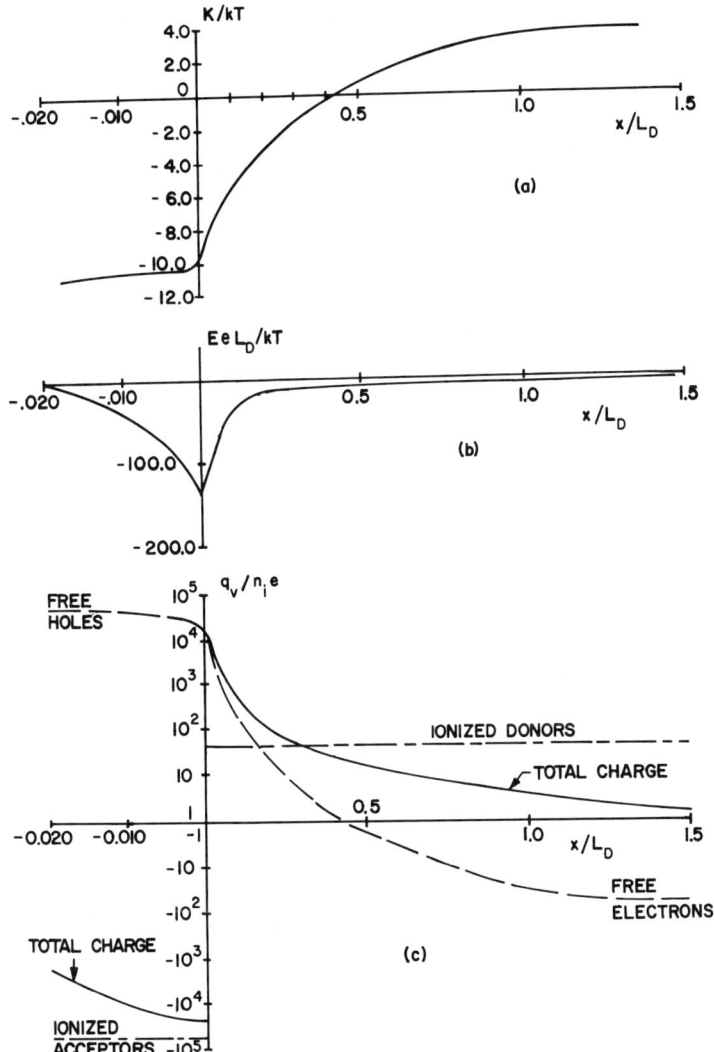

FIG. 4. Equilibrium potential, field, and charge as a function of distance for an abrupt, asymmetrical PN junction ($N_A = 10^{18}/\text{cm}^3$ in P region, $N_D = 10^{15}/\text{cm}^3$ in N region).

indicate the behavior of K in both regions, we have to make an abrupt change in scale as r goes from positive to negative values.

The same program used to calculate K can also determine the derivative du/dr, from which we find that

$$-du/dr = (eL_D/kT)\,dV/dx = eL_DE/kT. \tag{49}$$

This quantity is shown in Fig. 4b, and in a similar way we obtain the normalized charge density from the derivative of the field (Fig. 4c).

Since the chemical potential K determines the position of E_F with respect to E_I, we can work back from Fig. 4a to plot the shape of the intrinsic level as it varies across the space-charge layer. All the other levels (E_C, E_D, E_A, and E_V) will behave in the same way as E_I, leading to the energy-band diagram of Fig. 5. The shape of E_I versus r will be recognized as the reflection of u versus r with respect to the horizontal axis (including the scale change). An interesting feature of Fig. 5 is the presence of the inversion layer; this is the section in the N-type region adjacent to the junction for which the material is metallurgically N-type but electrically is P-type. That is, the chemical potential K_{in} is actually negative, although the doping is N-type.

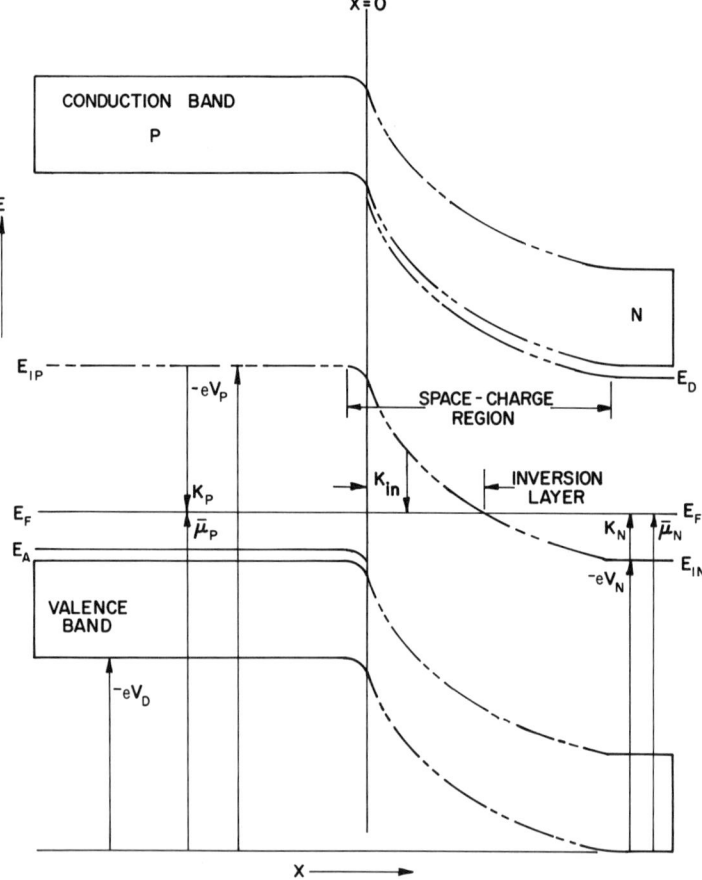

FIG. 5. Energy-band diagram for the abrupt, asymmetrical PN junction.

The physical basis for this phenomenon is that the more heavily doped P-type side of the junction can not sustain a large potential difference or a field because its conductivity is orders of magnitude lower than that of the N-type material. Hence, all but a small fraction of the transition layer lies to the right of the junction. As Fig. 4 indicates, the point where $K = 0$ and the majority carriers change from holes to electrons is at approximately $0.5L_D$ into the N region. For completeness, we show the corresponding situation for the symmetric junction in Fig. 6.

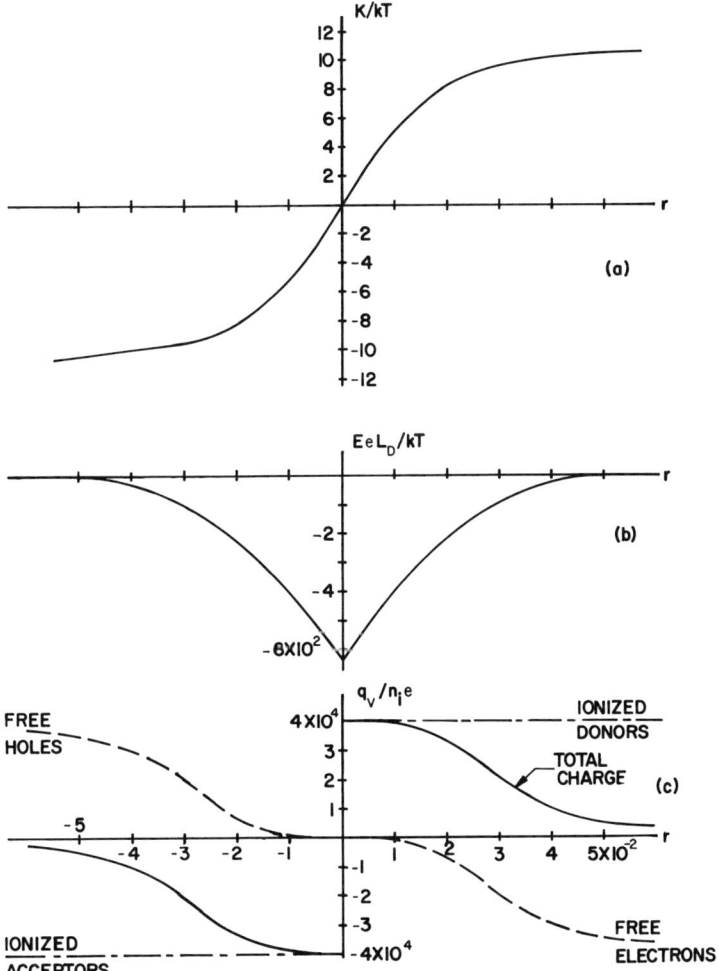

FIG. 6. Equilibrium potential, field, and charge as a function of distance for an abrupt, symmetrical PN junction ($N_A = 10^{18}/\text{cm}^3$ in P region, $N_D = 10^{18}/\text{cm}^3$ in N region).

A number of suggestions has been made for simplifying the integral in (45) over a limited range, so that an approximate, analytic evaluation is possible [see, e.g., Baranov et al. (1968) or Hauser and Littlejohn (1968)], but the numerical approach is so simple (and, of course, accurate) that these special methods are unnecessary.

For convenience in some applications, however, it is often assumed that the barrier field is strong enough to sweep all the free electrons and holes from the space-charge region, leaving behind a uniformly ionized region of acceptors to the left of the junction and donors to the right. This assumption is the famous depletion approximation of Schottky (1942) and is justified to some extent by Fig. 4c, which shows that the average of the total charge on the N-type side of the junction might approximate the constant donor contribution. The Poisson equation, Eq. (14), then reduces to

$$d^2V/dx^2 = (e/\varepsilon)(N_A - N_D) \qquad (50)$$

and this must be solved for the charge distribution shown in Fig. 7. Integrating twice for each region and applying the boundary conditions

$$dV_N/dx = 0, \qquad V_N = V_{ON} \quad \text{at} \quad x = x_N \qquad (51)$$

and

$$dV_P/dx = 0, \qquad V_P = V_{OP} \quad \text{at} \quad x = -x_P \qquad (52)$$

we obtain

$$V_N(x) = -(eN_D/2\varepsilon)(x - x_N)^2 + V_{ON} \qquad (53)$$

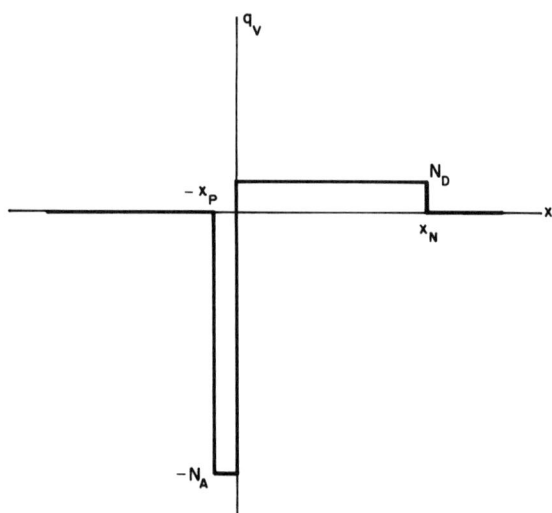

FIG. 7. Charge distribution corresponding to the depletion approximation.

and
$$V_P(x) = (eN_A/2\varepsilon)(x + x_P)^2 + V_{OP}. \tag{54}$$

Using the condition that the normal component of **D** is continuous at $x = 0$, it is easy to show from the derivatives of V_N and V_P that
$$N_A x_P = -N_D x_N, \tag{55}$$

which indicates that the areas of the two parts of Fig. 7 are equal. Further, we may obtain the charge q_s per unit area forming the dipole sheet as
$$q_{sN} = eN_D x_N, \quad q_{sP} = -eN_A x_P. \tag{56}$$

Let us also require V to be continuous at the junction. Then by (53) and (54),
$$V_{ON} - V_{OP} = (e/2\varepsilon)(N_D x_N^2 + N_A x_P^2),$$

or by (56),
$$-eV_D = -eV_{OP} - (-eV_{ON}) = (q_s^2/2\varepsilon)[(1/N_D) + (1/N_A)]. \tag{57}$$

If we define the width w of the space-charge region as
$$w = x_N + |-x_P| \tag{58}$$

then
$$w = (q_s/e)[(1/N_D) + (1/N_A)] \tag{59}$$

or
$$w^2 = [2\varepsilon(-eV_D)/e^2][(1/N_D) + (1/N_A)]. \tag{60}$$

This double layer is also equivalent to a capacitor; the associated depletion capacitance C_{depl} per unit area is
$$C_{\text{depl}} = dq_s/dV_D$$

or by (57),
$$C_{\text{depl}} = \left(\frac{e^2 \varepsilon}{2(-eV_D)(1/N_D + 1/N_A)}\right)^{1/2}. \tag{61}$$

We shall compare the predictions of the depletion approximation with other kinds of approximations at several places in subsequent sections.

It is possible to solve Poisson's equation for doping profiles other than the abrupt transitions corresponding to Figs. 4 and 6. A simple example is the symmetric, linearly graded junction of Fig. 8. We assume that we have arranged the variation of the impurity concentration with distance in such a way that the total charge q_v changes linearly in the space-charge region rather than being abrupt as in Figs. 4c and 6c. Since the two curves in Fig. 8

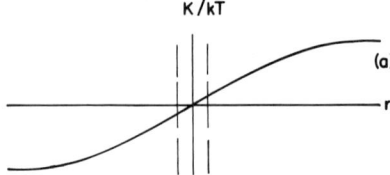

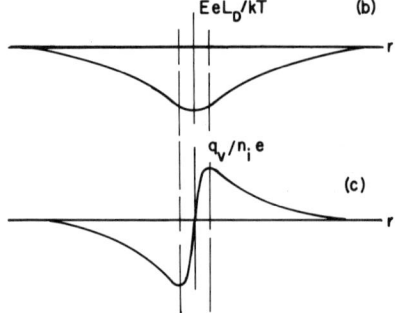

FIG. 8. Equilibrium potential, field, and total charge for a symmetric, linearly graded $\overline{PN}$ junction.

are obtained as derivatives of the one above, the shape of q_V in Fig. 8c implies that E in Fig. 8b will have inflection points as indicated by the vertical lines. We note that it is very difficult to see any difference in the behavior of K between the dashed lines of Fig. 8a and the corresponding region in Fig. 6a, but the behavior of the electric field and the total charge are significantly different.

Although the abrupt and linearly graded junctions are useful models for the study of junction theory, more complicated distributions occur in the fabrication of diffused junctions. Calculations of charge, field, and potential for Gaussian and error-function impurity distributions have been performed by Jain and Al-Rifai (1967). They found, however, that the Poisson–Boltzmann equation could not be solved for the region which was to the right of the junction and within a distance of $0.4L_D$ of the junction, even with double precision. Hence, they used an approximate solution in this range. The built-in voltage for junctions of this type over a broader range of parameters was calculated by Wilson (1969), and the Poisson–Boltzmann equation for an ion-implanted junction, which involves quadratic grading, has been solved by Klopfenstein (1975). It is clear that there are some practical problems associated with certain types of doping profiles.

Another form of approximation that is sometimes used in connection with highly asymmetric structures involves the assumption that the relative chemical potential u deviates slightly from its equilibrium value at a distance L_N from the physical junction. That is, we let

$$u = u_{0N} + \delta.$$

In the identity

$$\sinh(u_{0N} + \delta) = \sinh \delta \cosh u_{0N} + \cosh \delta \sinh u_{0N}$$

when δ is small, we have

$$\sinh(u_{0N} + \delta) = \delta \cosh u_{0N} + \sinh u_{0N}$$
$$= \delta \sqrt{1 + a_N^2} + a_N.$$

Then (31) becomes

$$d^2\delta/dx^2 = (1/L_D^2)(\delta\sqrt{1 + a_N^2}) = \delta/L_N^2,$$

where the quantity

$$L_N = L_D/(1 + a_N^2)^{1/4}$$

is called the extrinsic Debye length in the N region. For the asymmetric junction of Fig. 5, a_N is large enough to permit the approximation

$$L_N = L_D/\sqrt{a_N} = \sqrt{\varepsilon kT/e^2 N_D}. \tag{62}$$

A similar reation holds for the corresponding quantity L_P in the P region. Using L_N and L_P in the associated regions rather than L_D throughout the junction would eliminate the necessity for a scale change in Fig. 4.

III. The *PN* Junction in the Low-Injection Regime

3. Low-Level Injection in a *PN* Junction

Although the Shockley low-level theory for a *PN* junction (Shockley, 1949, 1950) is very well covered in many places, a critical review of the basic assumptions is worthwhile, since this approach needs improvement. We shall therefore introduce these assumptions in connection with a general set of equations and explain the consequence of each one of them.

The application of a difference in potential V_A to the terminals of a *PN* junction establishes nonequilibrium conditions, requiring a modification of the concepts presented in the preceding section. For one thing, the Fermi level E_F has no meaning in those regions where equilibrium conditions do not exist. Following Domenicali (1954), we rewrite (20) and (21) as

$$K = kT \log(n_0/n_i), \tag{63}$$
$$K = kT \log(n_i/p_0), \tag{64}$$

where n_0 and p_0 explicitly denote equilibrium concentrations. For nonequilibrium conditions, (63) and (64) suggest that we define distinct chemical potentials K_e and K_h for electrons and holes, respectively, by

$$K_e = kT \log(n/n_i), \tag{65}$$
$$K_h = kT \log(n_i/p), \tag{66}$$

where n and p are nonequilibrium concentrations. If Δn denotes the change in the electron concentration due to the applied potential difference, then

$$n = n_0 + \Delta n, \qquad (67)$$

and similarly

$$p = p_0 + \Delta p, \qquad (68)$$

where Δn and Δp are usually (but not always) positive.

We may also generalize (19) by defining separate electrochemical potentials as

$$\bar{\mu}_e = K_e - eV, \qquad (69)$$
$$\bar{\mu}_h = K_h - eV. \qquad (70)$$

These are the quantities which Shockley (1950, 1958) calls the quasi-Fermi levels or imrefs; the terminology given above seems preferable since it emphasizes that K_e and K_h and hence $\bar{\mu}_e$ and $\bar{\mu}_h$ are specified in terms of ratios of concentrations, rather than being indeterminate deviations from equilibrium energies. The definitions (65) and (66) are deliberately formulated, however, so that K_e and K_h reduce to the common value K when equilibrium is restored. It should also be noted that these definitions apply explicitly to nondegenerate materials. The use of (65) and (66) with (8d) and (23a) yields

$$n = N_C \exp[(\bar{\mu}_e - E_C)/kT], \qquad (70a)$$
$$p = N_V \exp[(E_V - \bar{\mu}_h)/kT], \qquad (70b)$$

These are the nonequilibrium versions of Eqs. (1) and (2).

To obtain the basic boundary conditions used in diode theory, we subtract (70) from (69). The result is

$$\bar{\mu}_e - \bar{\mu}_h = kT \log(np/n_i^2) \qquad (71)$$

or

$$np = n_i^2 \exp[(\bar{\mu}_e - \bar{\mu}_h)/kT]. \qquad (72)$$

In equilibrium, with $n = n_0$ and $p = p_0$, then $K_e = K_h$ and $\bar{\mu}_e = \bar{\mu}_h = \bar{\mu}$. Hence, (72) reduces to

$$n_0 p_0 = n_i^2 \qquad (73)$$

and this is identical to the product of (20) and (21).

The introduction of nonequilibrium electrochemical potentials enables us to postulate a generalized Ohm's law for each type of carrier in a semiconductor. As an example, we write the electron current density J_e as

$$J_e = (\sigma_e/e)(d\bar{\mu}_e/dx), \qquad (74)$$

2. THE THEORY OF SEMICONDUCTING JUNCTIONS

where σ_e, the electron contribution to the conductivity, is given in terms of the electron mobility by (Nussbaum, 1966)

$$\sigma_e = ne\mu_e. \tag{75}$$

Using (65) and (75), (74) becomes

$$\begin{aligned}J_e &= n\mu_e(dK/dx - e\,dV/dx) \\ &= \mu_e kT\,dn/dx - ne\mu_e\,dV/dx = \mu_e kT\,dn/dx + ne\mu_e E.\end{aligned} \tag{76}$$

In a similar way, the hole current density is

$$\begin{aligned}J_h &= (\sigma_h/e)\,d\bar{\mu}_h/dx = -\mu_h kT\,dp/dx - pe\mu_h\,dV/dx \\ &= -\mu_h kT\,dp/dx + pe\mu_h E,\end{aligned} \tag{77}$$

However, when a hole concentration gradient dp/dx exists, we would expect these charge carriers to diffuse towards a region of lower concentration and the resulting diffusion current density J_{hD} should be proportional to dp/dx. This is expressed in terms of Fick's law as

$$J_{hD} = -eD_h\,dp/dx, \tag{78}$$

where D_h is the diffusion constant for holes, the factor e introduces the charge carried by each hole, and the negative sign implies that the diffusion current is along the direction of the decreasing gradient. If we compare (78) with the first term on the right of (77), we see that these two expressions for diffusion current are identical if

$$eD_h = \mu_h kT, \tag{79}$$

which is the Einstein relation. It is sometimes said (Nussbaum, 1966) that this expression is a consequence of Boltzmann statistics, but the argument just given demonstrates that the connection is through the definitions (65) and (66). That is, we have chosen our nonequilibrium relations to have a form which reduces to the correct nondegenerate expressions when equilibrium is restored. It should be noted that Eq. (79), which we shall use in connection with the numerical theory of PN-junction diodes, is valid only for nondegenerate materials and in the limit of low frequencies. The more general form, which we do not need here, has been derived and discussed by van Vliet and van der Ziel (1977).

The other term on the right of (77) represents the drift current density J_{hE}, which is the motion of charge carriers under the influence of an electron field, or

$$J_{hE} = -\sigma_h\,dV/dx = pe\mu_h E = \sigma_h E. \tag{80}$$

We recognize (80) as the customary form of Ohm's law, Eq. (10), and the generalizations to include drift and diffusion terms are

$$\mathbf{J}_h = (\sigma_h/e)\,\text{grad}\,\bar{\mu}_h \tag{81}$$

and

$$\mathbf{J}_e = (\sigma_e/e)\,\mathrm{grad}\,\bar{\mu}_e. \tag{82}$$

That is, both the particle and the charge aspects of the hole or the electron must be taken into account.

One significant consequence of Shockley's treatment can be explained on the basis of these extended forms of Ohm's law. Consider, for example, a homogeneous N-type semiconductor subject to an applied potential V_A (Fig. 9). Ignoring effects at the contacts, we may assume that the material behaves somewhat like a metal; that is, application of a field causes only a drift current to flow from the negative to the positive terminal. This corresponds to the condition that

$$n = n_0 \tag{83}$$

for then (76) reduces to

$$J_e = n_0 e \mu_e E. \tag{84}$$

By (82), we see that a finite value of J_e leads to the condition

$$d\bar{\mu}_e/dx \neq 0 \tag{85}$$

so that $\bar{\mu}_e$ will have the nonzero slope shown in the energy-band diagram of Fig. 9. The potential $-eV$ has the same property, accounting for the tilting of the bands in the figure. Although $\bar{\mu}$ lies at a distance from the intrinsic level equal to $K_n = K_p = K$, where K is the equilibrium value, we note that E_F is not defined for the nonequilibrium case; it simply cannot be identified with any tilted energy level. However, we do have the interesting situation that the presence or absence of $-eV_A$ does not affect the equilibrium concentrations n_0 and p_0.

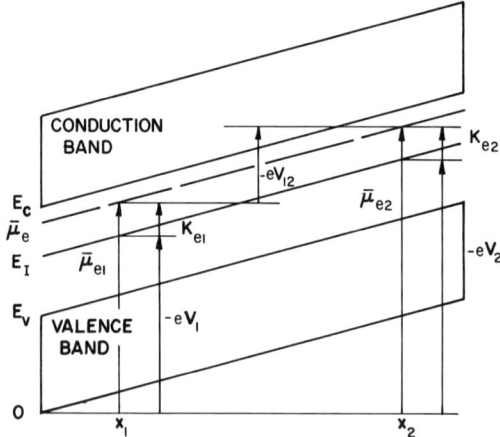

FIG. 9. Energy-band diagram for N-type semiconductor with an applied bias.

2. THE THEORY OF SEMICONDUCTING JUNCTIONS

Let us consider two arbitrary points in Fig. 9 with coordinates x_1 and x_2. If we compute the algebraic difference in the value of $\bar{\mu}$ at these two points, we obtain

$$\begin{aligned}\bar{\mu}_1 - \bar{\mu}_2 &= (K - eV)_1 - (K - eV)_2 \\ &= -eV_1 - (-eV_2) \\ &= -eV_{12},\end{aligned} \quad (86)$$

where $-eV_{12}$ is the rise in E_I from x_1 to x_2. Since E_I is a measure of $-eV$, this quantity is clearly the difference in electrostatic potential between these two points. Hence, (86) indicates that the change in the value of $\bar{\mu}$ between two points is proportional to the corresponding applied potential difference. This correlates with Eq. (82); a nonzero slope for the electrochemical potential implies a charge transfer.

When a potential difference V_A is applied to a PN junction in such a way that it reduces the diffusion barrier $-eV_D$ of Fig. 5 (a positive polarity for V_A at the P region corresponds to what is known as forward bias), the associated energy $-eV_A$ will cause significant changes in the energy-band diagram. Let us fix the position of the reference level at the N-region contact and allow the levels in the transition region and the P region to adjust themselves in accordance with requirements to be discussed. First of all, the barrier decreases from $-eV_D$ to $-eV_D - (-eV_A)$, as indicated in Fig. 10,

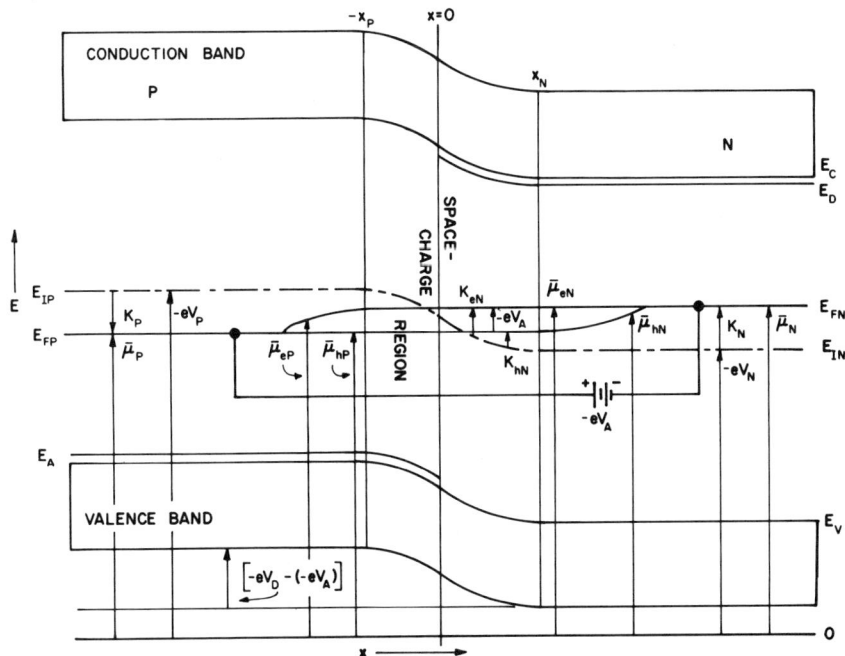

FIG. 10. Energy-band diagram for a PN junction under injection.

and all the equilibrium levels in the vicinity of the P-region contact are lowered by an amount $-eV_A$. The equilibrium Fermi levels E_{FP} and E_{FN} at each of the two contacts are no longer identical, even though the two regions may essentially be in equilibrium. (There will actually be deviations from equilibrium, as indicated in connection with Fig. 9, but we will assume them to be small.)

The lowering of the barrier at the junction enhances the diffusion of electrons from the P region to the N region and of holes in the opposite direction, establishing the phenomenon of injection; we intuitively regard electrons as particles falling down a barrier and holes as bubbles which rise at the barrier. Directing our attention to hole injection for the moment, we may write the hole concentration in the N region as $p = p_0 + \Delta p$, by Eq. (68), where Δp is the increase in p_0 due to injection. Then, by (66), the chemical potential K_{hN} for holes in the N region just to the right of the junction (the plane $x = 0$) becomes

$$K_{hN} = kT \log[n_i/(p_0 + \Delta p)]. \tag{87}$$

The presence of an extra amount Δp of positive charge should attract Δn electrons into the N region from the right-hand contact. It is customary to assume that

$$\Delta n = \Delta p, \tag{88}$$

and this relation expresses the hypothesis of space-charge neutrality. However, we realize that Δn is much smaller than the majority-carrier concentration n_0, so that to a very high degree of accuracy we have

$$K_{eN} = K_N = kT \log(n_0/n_i). \tag{89}$$

This quantity will also be constant in the space-charge region; as we shall show, this is a narrow layer compared to the regions in which injection is occurring. Similar considerations hold in the P region, with the carriers interchanging roles. It then follows that the electrochemical potential $\bar{\mu}_{eN}$ for electrons in the N region remains constant in that entire region, because $n_N \sim n_{0N}$, and keeps this value also in the very narrow space-charge layer. Similar arguments apply to the corresponding electrochemical potential $\bar{\mu}_{hP}$ for holes in the P region and space-charge layer as well.

We shall make use of these arguments in connection with the solution of the continuity equations, Eqs. (11) and (12). We define nonequilibrium, steady-state conditions by the requirement that n and p settle down to new, constant values after the application of the bias V_A so that $\partial p/\partial t = 0$ and (11) becomes

$$(dJ_h/dx) + eR_h = 0. \tag{90}$$

2. THE THEORY OF SEMICONDUCTING JUNCTIONS

We now make the assumption (whose validity will be examined in great detail later) that the recombination rate is proportional to the excess number of holes, or

$$R_h = \Delta p/\tau_h = (p - p_0)/\tau_h, \tag{91}$$

where $1/\tau_h$ is the proportionality constant and τ_h is known as the hole lifetime. Then (90) becomes

$$dJ_h/dx + e(p - p_0)/\tau_h = 0. \tag{92}$$

We make the further approximation that the drift current is negligible compared to the diffusion contribution so that (77) becomes

$$J_h = -eD_h \, dp/dx. \tag{93}$$

Applying (92) to the injection process in the N region, we have

$$d^2 p_{0N}/dx^2 - (p_N - p_{0N})/L_h^2 = 0, \tag{94}$$

where the notation p_{0N} explicitly indicates the equilibrium concentration of holes in the N region, and

$$L_h = \sqrt{D_h \tau_h} \tag{95}$$

is called the diffusion length of holes. The solution to (94) is

$$p_N - p_{0N} = A \exp(x/L_h) + B \exp(-x/L_h). \tag{96}$$

One of the two boundary conditions needed to evaluate the arbitrary constants A and B comes from Fig. 10 and the assumed behavior of $\bar{\mu}_e$ and $\bar{\mu}_h$. We realize that $\bar{\mu}_h$ in the entire distance $-\infty \leq x \leq x_N$ (the P region plus the space-charge region) is constant and depressed below $\bar{\mu}_n$ by an amount equal to $-eV_A$. That is, the electrochemical potentials are constant and parallel across the space-charge region, with a separation specified by the bias, or

$$\bar{\mu}_e - \bar{\mu}_h = -eV_A. \tag{97}$$

Inserting this in (72) gives

$$pn = n_i^2 \exp(-eV_A/kT). \tag{98}$$

For the N region, we have already seen that n_N is approximately equal to n_{0N}, and since $n_i^2 = n_{0N} p_{0N}$, (98) reduces to

$$p_N = p_{0N} \exp(-eV_A/kT). \tag{99}$$

We shall apply this boundary condition at $x = 0$ rather than at $x = x_N$, assuming that the space-charge region is very thin. For the second condition, we assume an infinite recombination-rate contact. That is, let

$$p_N = p_{0N} \tag{100}$$

at $x = \infty$. Then (99) and (100) determine A and B in (96), giving

$$p_N - p_{0N} = p_{0N}[\exp(-eV_A/kT) - 1]\exp(-x/L_h). \tag{101}$$

We thus see that

$$\begin{aligned}K_{hN} &= kT\log(n_i/p_{0N}) - kT\log\{[\exp(-eV_A/kT) - 1]\exp(-x/L_h) + 1\}\\ &= K_N - kT\log\{[\exp(-eV_A/kT) - 1]\exp(-x/L_h) + 1\}.\end{aligned} \tag{102}$$

To interpret this expression, it is convenient to postulate low-level injection, which means that the positive quantity $-eV_A/kT$ is small enough to use the Taylor expansion

$$\exp(-eV_A/kT) - 1 \sim -eV_A/kT$$

and (102) becomes

$$K_{hN} = K_N - kT\log[(-eV_A/kT)\exp(-x/L_h) + 1]. \tag{103}$$

For $x \gg L_h$, (103) reduces to

$$K_{hN} = K_N.$$

That is, the chemical potential for holes in the N region returns to the common equilibrium value at a distance of a few diffusion lengths from the junction. In the region $x \sim L_h$, the variation of K_{hN} with distance is approximately linear and when x has a range specified by

$$L_D < x < L_h, \tag{104}$$

where L_D was defined in Eq. (32), then K_{hN} varies in the way specified by Eq. (103) and indicated in a general way in Fig. 10. The horizontal scale in this figure is very misleading, for the Debye length in either germanium or silicon is several orders of magnitude smaller than the diffusion lengths.

To find the injected minority-carrier current, we evaluate (93) at $x = 0$, obtaining

$$\begin{aligned}J_{hN} &= -eD_h(dp_N/dx)_{x=0}\\ &= (eD_h p_{0N}/L_h)[\exp(-eV_A/kT) - 1].\end{aligned} \tag{105}$$

A similar calculation for the minority current injected into the P region gives

$$J_{eP} = (eD_e n_{0P}/L_e)[\exp(-eV_A/kT) - 1]. \tag{106}$$

The total current J across the junction is the sum of these two expressions, so that

$$J = J_S[\exp(-eV_A/kT) - 1], \tag{107}$$

where the reverse saturation current J_S is

$$J_S = eD_h p_{0N}/L_h + eD_e n_{0P}/L_e. \tag{108}$$

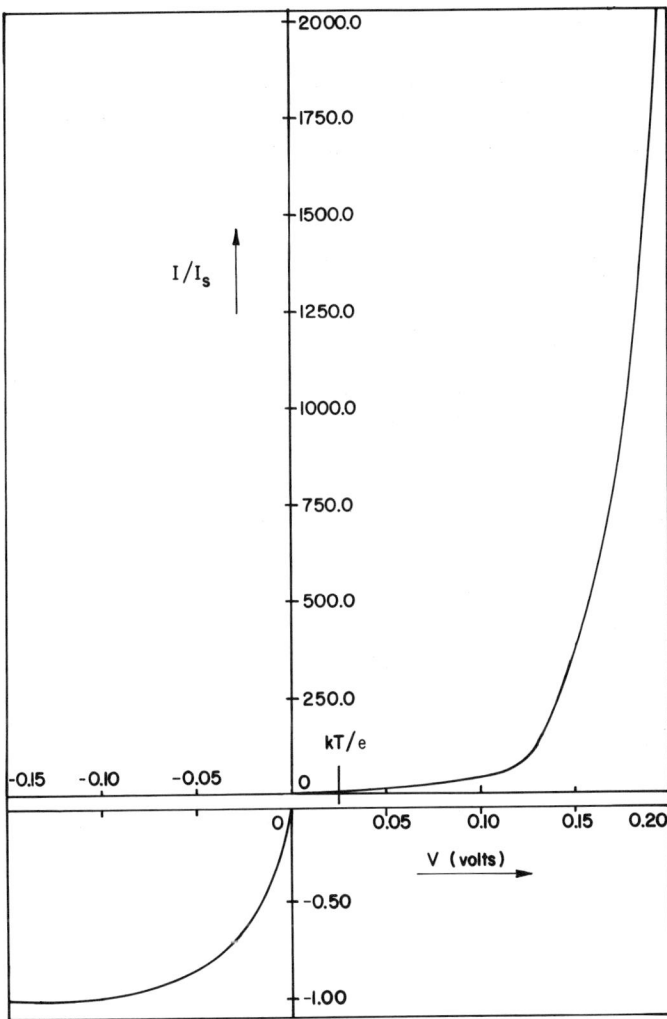

FIG. 11. Universal characteristic curve for a PN junction.

Equation (107) is the well-known characteristic equation (Shockley, 1949) of the junction, plotted in Fig. 11. The negative sign in the exponent, which disagrees with virtually all the literature, is necessary to insure that Eqs. (76) and (77) are consistent with common usage. That is, the negative sign in (70) leads to the correct sign in (77). Since the exponent $-eV_A/kT$ is cumbersome, this situation has been handled in one of the following two ways:

(a) A negative potential ψ is defined as $\psi = -V_A$ so that (107) becomes

$$J = J_s[\exp(e\psi/kT) - 1]$$

and ψ is identified as the forward voltage. This approach, used for example by Moll (1964), seems illogical, since it regards a given quantity and its negative as two distinct entities with separate symbols.

(b) Another approach is to simply drop the negative sign on $-eV_A$ when it is used in equations like (107). This involves remembering that one convention is used in discussing energy-band diagrams and another in considering characteristic curves.

We may now summarize the assumptions of the Shockley low-level theory, as follows:

(a) Drift currents are negligible; only diffusion currents need be considered.

(b) Majority-carrier concentrations remain constant and so do the corresponding electrochemical potentials.

(c) Recombination rates are proportional to the amount of minority-carrier injection.

(d) The space-charge region is of negligible thickness.

4. THE SPACE-CHARGE REGION UNDER LOW-LEVEL INJECTION

The review of standard low-level injection theory given in the preceding section pointed out that the space-charge region is taken to have a negligible thickness. We shall see later that—although this approximation may be reasonably correct—this very thin region has a great deal to do with the electrical properties of the device. At the same time, we realize that there is another, implicit assumption in Fig. 10 connected to the problem of what is happening in the transition region. What we have done is to join the equilibrium positions of E_C, E_I, and E_V on each side of the junction by smooth curves passing through the space-charge region. We have no right

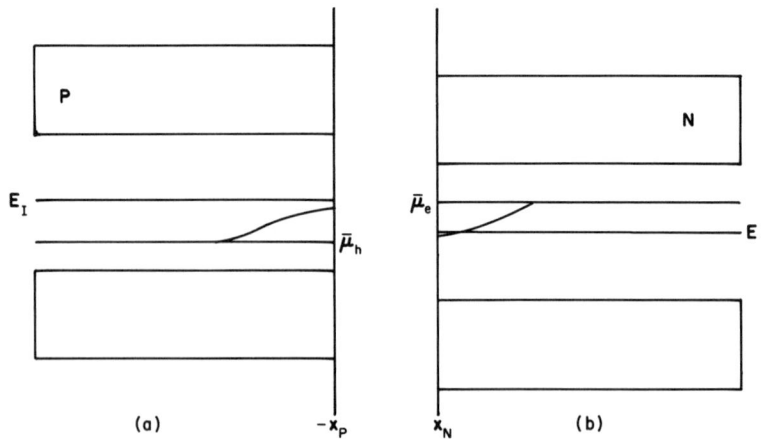

FIG. 12. Energy bands for a *PN* junction as determined by the Shockley low-level theory.

2. THE THEORY OF SEMICONDUCTING JUNCTIONS

to do this, other than an intuitive idea of how the bands might vary in the region $-x_P \leq x \leq x_N$, and this layer should really be regarded as *terra incognita*, as indicated by Fig. 12.

Returning to Section 2, we recall that we were able to determine the equilibrium band shapes in the transition region from a numerical solution of the Poisson–Boltzmann equation, Eq. (31). To obtain the corresponding equation when injection is occurring, let us define a relative potential ψ as

$$\psi = eV/kT \tag{109}$$

and (14) will simplify to

$$\frac{d^2\psi}{d^2r} = [(n-p)/2n_i] - a. \tag{110}$$

This equation involves the use of Eqs. (29) and (32). By definition, the carrier concentrations are

$$n = n_i \exp(K_e/kT) = n_i \exp[(\bar{\mu}_e + eV)/kT] \tag{111}$$

and

$$p = n_i \exp(-K_h/kT) = n_i \exp[-(\bar{\mu}_h + eV)/kT] \tag{112}$$

so that (110) becomes

$$\frac{d^2\psi}{dr^2} = \frac{\exp[(\bar{\mu}_e/kT) + \psi] - \exp\{-[(\bar{\mu}_h/kT) + \psi]\}}{2} - a. \tag{113}$$

This is an unusual problem; we have an equation with three unknown quantities. To obtain a solution we use a procedure of Fulkerson (1968). Let us choose μ_e at the right-hand contact as the reference level via the condition

$$\bar{\mu}_e = 0 \tag{114}$$

and then express (97) as

$$-\psi_A = -\bar{\mu}_h/kT \tag{115}$$

so that (113) simplifies to

$$d^2\psi/dr^2 = \tfrac{1}{2}[\exp(\psi) - \exp(-\psi_A - \psi)] - a, \tag{116}$$

which we recognize as a slight generalization of (34). The first integral is

$$d\psi/dr = [\exp(\psi) + \exp(-\psi_A - \psi) - 2a\psi + 2C]^{1/2}. \tag{117}$$

Evaluating the constant C at the equilibrium value ψ_0 of ψ gives

$$C = a\psi_0 - \tfrac{1}{2}[\exp(\psi_0) + \exp(-\psi_A - \psi_0)]$$

and (117) then becomes

$$d\psi/dr = \{[\exp(\psi) - \exp(\psi_0)] + \exp(-\psi_A)[\exp(-\psi) - \exp(-\psi_0)] \\ - 2a(\psi - \psi_0)\}^{1/2}. \tag{118}$$

To integrate once more—using the numerical methods that were necessary in connection with Eq. (45)—we must determine the value ψ_J of ψ at the junction and we must also find a_N, a_P, C_N, and C_P. The details of such a calculation are given in the next section; the results are shown by the solid curves of Fig. 13, corresponding to large forward or reverse bias and to equilibrium. For comparison, the dotted curves represent the parabolas obtained from the depletion approximation, which were adjusted to match the solid curves at the right-hand edge of the depletion layer. We note that the shape of the forward-bias curves ($-\psi_A = 8.0$), either solid or dotted, is very similar to the equilibrium behavior of E_I, so that the speculation inherent in Fig. 10 is reasonable under the assumption used. An energy-band diagram incorporating the forward-bias solution is shown in Fig. 14.

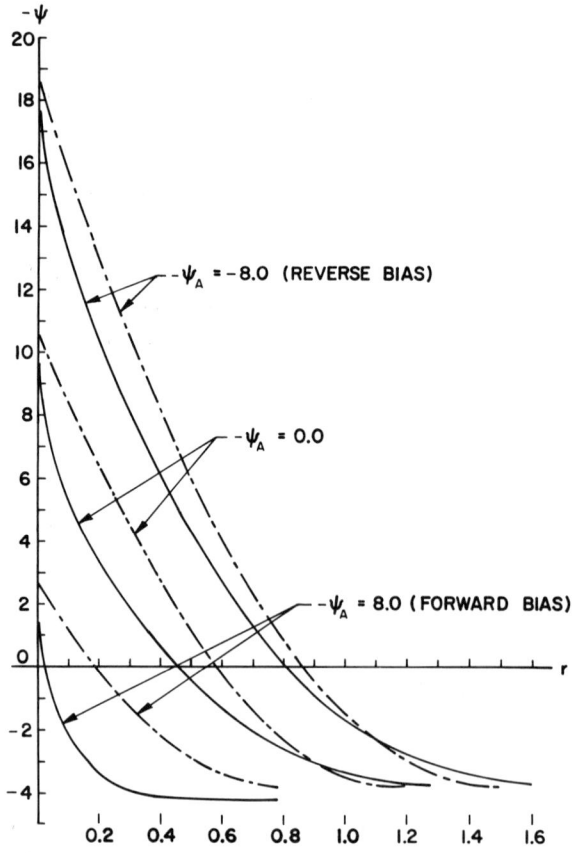

FIG. 13. Band shapes in the transition region of an asymmetric, abrupt PN junction.

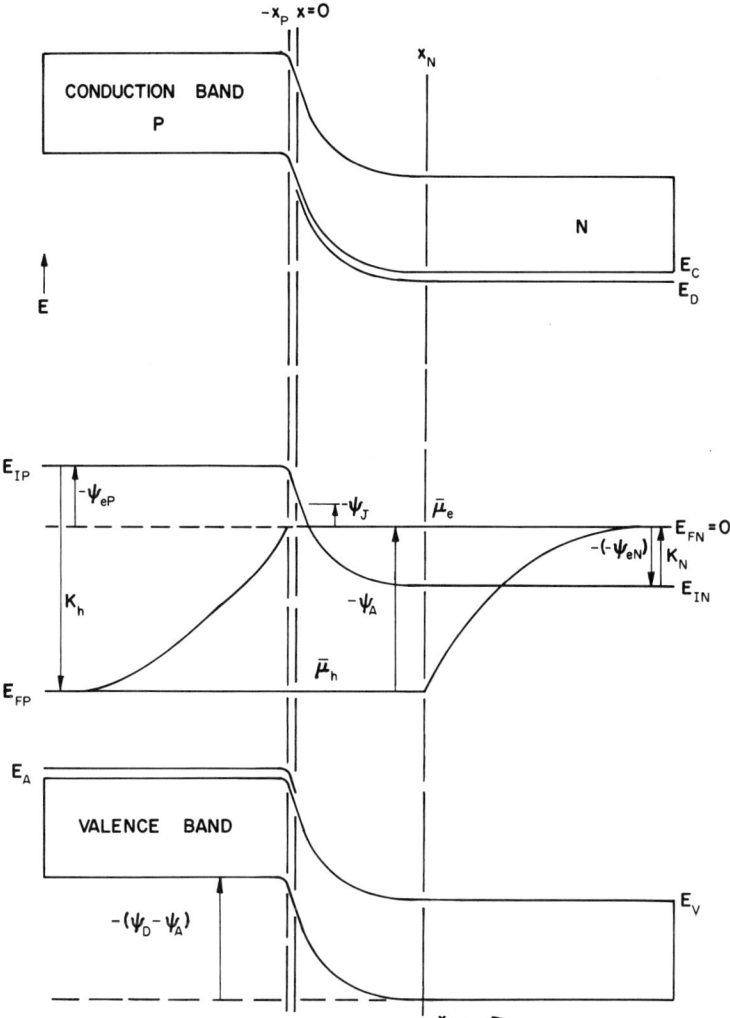

FIG. 14. Energy-band diagram for asymmetric, abrupt junction under forward bias.

It is also possible to apply the depletion approximation to the determination of the way in which the capacitance C_{depl} of Eq. (61) varies with applied potential. For a forward bias, for example, the diffusion barrier has been reduced from $(-eV_D)$ to $[-eV_D - (-eV_A)]$, and the width of the barrier layer decreases in accordance with Eq. (60) (this change in width shows up on Fig. 13). Hence, we need modify (61) so that it becomes

$$C_{\text{depl}}(V_A) = \left(\frac{e\varepsilon}{2(V_A - V_D)[(1/N_D) + (1/N_A)]} \right)^{1/2}. \tag{119}$$

It was mentioned in connection with the equilibrium junction that approximations can be made for particular kinds of doping profiles and similar ideas have been advanced in connection with the junction under bias. For example, van de Wiele and Demoulin (1970) have neglected minority carriers only on each side of an abrupt, asymmetric junction. They write the charge density in the P region as

$$q_{VP} = e(p + N_D - N_A) \tag{120}$$

with a corresponding expression on the N side. This is not as drastic as the depletion approximation and, when combined with Taylor expansions of the form

$$\exp \psi = 1 + \psi + \tfrac{1}{2}\psi^2$$

enables them to obtain explicit expressions for the potential, the field, and the depletion capacitance. As expected on the basis of Fig. 13, their intermediate approximation deviates very little from either the complete depletion approximation or the exact numerical calculation.

It would appear, then, that the depletion approximation should be a perfectly satisfactory basis for the theory of asymmetric junctions. However, as we shall show, this is true only for those quantities deduced from a knowledge of the band structure. In the case of electrical properties, such as the current–voltage characteristic, we shall find a strong disagreement between theory and observed behavior. Before indicating how this comes about, however, we must consider in more detail the nature of the generation–recombination mechanism.

5. The Effect of Generation–Recombination Processes on Device Characteristics

Silicon and germanium are semiconductors for which generation and recombination of electrons and holes take place via intermediate levels located (usually) deep in the forbidden gap; that is, these levels—known as recombination centers or traps—are fairly close to the intrinsic level E_I. The levels may be ascribed to impurities other than the normal substitutional dopants from Columns III and V—gold, for example, goes in interstitially—or they may come from dislocations and other imperfections. Loosely speaking, a deep level is considered to be an electron trap when the probability of a captured electron from the conduction band returning to this band is greater than the probability that it will drop down to the valence band and recombine with a hole (with a corresponding description for hole traps). For some applications, such as the theory of transient photoconductive processes (Nussbaum *et al.*, 1977), it is necessary to distinguish carefully between traps and recombination centers, but for *PN*-junction theory, we can refer to these levels, regardless of function, simply as traps.

Let us define the generation rate g_e for electrons as the number of electrons per unit volume per second produced by some thermal or optical activation process, let r_e be the number per unit volume per second which disappear by recombining with holes, and let

$$R_e = r_e - g_e \qquad (121)$$

be the net recombination rate. We then define the lifetime τ_e of an electron as

$$\tau_e = \Delta n/R_e = (n - n_0)/R_e. \qquad (122)$$

This quantity is often confused with the capture lifetime T_e, which is the length of time between generation and recombination that an electron is free. It may be expressed as

$$T_e = n/r_e, \qquad (123)$$

whereas

$$\tau_e = (n - n_0)/(r_e - g_e). \qquad (124)$$

Comparing (123) and (124) indicates that τ_e might have the same physical meaning as T_e for low injection and a small rate of thermal generation.

We shall now derive the Hall–Shockley–Read (HSR) (Hall, 1952; Shockley and Read, 1952) relation for R_e (and for R_h, the corresponding net recombination rate for holes). Although this result has been presented in a number of ways (van der Ziel, 1976), the treatment which follows represents a very simple approach and examines the underlying assumptions.

Figure 15 shows the four possible transition processes between the trap levels, of energy E_T and density N_T, and the two bands. We can compute, as an example, the recombination rate for electrons dropping into the traps from the conduction band by assuming that this rate r_{CT} is proportional to the density of free electrons in the conduction band and the density p_T of trapped holes, or

$$r_{CT} = K_{CT} n p_T = K_{CT} n [N_T(1 - f_T)], \qquad (125)$$

where K_{CT} is a proportionality constant and f_T is the probability of occupancy of a trap. Note that f_T is not the Fermi function $f(E_T)$ of Eq. (24), since Fig. 15 may represent a nonequilibrium situation. In the same way, the generation rate g_{TC} for the converse process is

$$g_{TC} = K_{TC} n_T (N_C - n) = K_{TC} (N_T f_T)(N_C - n), \qquad (126)$$

where $(N_C - n)$ represents the available states in the conduction band. For most materials, $n \ll N_C$, so that

$$g_{TC} = K_{TC} N_T f_T N_C. \qquad (127)$$

If we now require the system to be in equilibrium, then all concentrations remain constant, and the two processes expressed through Eqs. (125) and

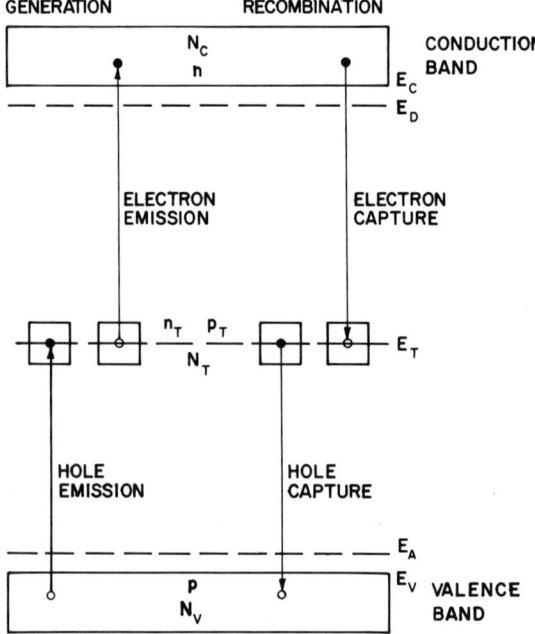

FIG. 15. Generation and recombination processes via a single trap level.

(127) must have equal rates, so that

$$r_{CT0} = g_{TC0} \tag{128}$$

or

$$K_{CT} n_0 N_T (1 - f_{T0}) = K_{TC} N_T f_{T0} N_C, \tag{129}$$

where f_{T0} is simply $f(E_T)$, the Fermi function for $E = E_T$, giving

$$(1 - f_{T0})/f_{T0} = (1/f_{T0}) - 1 = \exp[(E_T - E_F)/kT]. \tag{130}$$

Using (1) as well, we have

$$K_{TC} = K_{CT} \exp[(E_T - E_C)/kT] \tag{131}$$

and, by (121),

$$R_e = r_{CT} - g_{TC} = K_{CT} N_T [n(1 - f_T) - n_0^* f_T], \tag{132}$$

where

$$n_0^* = N_C \exp[(E_T - E_C)/kT]. \tag{133}$$

is a convenient abbreviation. Although we can interpret n_0^* as the value of n_0 when E_F falls on E_T, this is highly unlikely for the usual kind of doping concentrations and trap positions. In the same way, we find the net rate for holes as

$$R_h = r_{TV} - g_{VT} = K_{TV}N_T[pf_T - p_0^*(1 - f_T)], \quad (134)$$

where p_0^* is the value of p_0 for $E_F = E_T$. If we now return to the continuity equations, impose steady-state conditions, and subtract (12) from (11), we obtain

$$(1/e)\,dJ/dx + (R_h - R_e) = 0, \quad (135)$$

where the total current density is

$$J = J_h + J_e. \quad (136)$$

For a one-dimensional situation, J will be independent of x. Equation (135) shows that

$$R_h = R_e, \quad (137)$$

and we designate the common value of R_e and R_h by R. More generally, however, we have to worry about div **J** and the terms $\partial p/\partial t$ and $\partial n/\partial t$. An inherent part of the HSR theory is to assume that (137) is always valid.

If this is the case, then we rewrite (132) as

$$R = [n(1 - f_T) - n_0^* f_T]/\tau_{e0}) \quad (138)$$

where

$$\tau_{e0} = 1/N_T K_{CT} \quad (139)$$

and (134) as

$$R = [pf_T - p_0^*(1 - f_T)]/\tau_{h0}, \quad (140)$$

where

$$\tau_{h0} = 1/N_T K_{TV}. \quad (141)$$

Then multiply (138) by $\tau_{e0}(p + p_0^*)$, (140) by $\tau_{h0}(n + n_0^*)$, and add to obtain

$$R = (pn - p_0^* n_0^*)/[\tau_{h0}(n + n_0^*) + \tau_{e0}(p + p_0^*)]. \quad (142)$$

This is the form of R which must be used in the continuity equations in place of Shockley's original rates $\Delta n/\tau_e$ or $\Delta p/\tau_h$. This is not unexpected, for $n_0^* p_0^*$ is simply n_i^2 (since the equilibrium product is independent of the position of E_F for nondegenerate material). Hence the numerator of (142) represents the term $(pn - p_0 n_0)$, and the denominator is a variable lifetime depending on p, n, and the trap position, rather than a constant τ_e or τ_h.

We may identify the constants τ_{e0} and τ_{he} by considering, for example, a heavily doped P-type region under low injection. Then the numerator in (142) becomes

$$pn - p_0 n_0 = (p_0 + \Delta p)(n_0 + \Delta n) - p_0 n_0 \sim p_0 \Delta n$$

since $\Delta n \sim \Delta p$, $\Delta n \Delta p$ is negligible, and $n_0 \ll p_0$. In the same way, the denominator reduces to $p_0 \tau_{e0}$, because all other concentrations are small compared to p_0. Hence

$$R = \Delta n/\tau_{e0}.$$

But by (122),

$$R = \Delta n/\tau_e,$$

so that τ_{e0} represents the limiting value of the minority-carrier lifetime τ_e when the material is very strongly P-type (and similarly for τ_{h0}).

To apply HSR recombination theory to the problem of device characterization, let us start with some simplifying assumptions (which we shall later remove). First, we write (142) as

$$R = \frac{n_i^2\{\exp[(\bar{\mu}_e - \bar{\mu}_h)/kT] - 1\}}{\tau_{h0}\{n_i \exp[(\bar{\mu}_e + eV)/kT] + n_i \exp[(E_T - E_I)/kT]\} + \tau_{e0}\{n_i \exp[-(\bar{\mu}_h + eV)] + n_i \exp[(E_I - E_T)/kT]\}}, \quad (143)$$

where, by (133) and (7),

$$n_0^* = N_C \exp[(E_T - E_C)/kT] = n_i \exp[(E_T - E_I)/kT], \quad (144)$$

with a similar relation

$$p_0^* = N_V \exp[(E_V - E_T)/kT] = n_i \exp[(E_I - E_T)/kT] \quad (145)$$

for holes. Without too much error, we can place E_T right on E_I, so that $n_0^* = p_0^* = n_i$, and we shall also let $\tau_{e0} = \tau_{h0} = \tau_0$. We again use the assumption about $\bar{\mu}_e$ and $\bar{\mu}_h$ contained in Eq. (97), the definition of Eq. (109), and the reference convention of Eqs. (114) and (115) to obtain

$$R = n_i[\exp(-\psi_A) - 1]/\tau_0\{[\exp(\psi) + 1] + [\exp(-\psi_A - \psi) + 1]\}. \quad (146)$$

It is easy to see that the denominator of this expression will have a minimum—and R itself will have a sharp maximum—when

$$\psi = -\psi_A/2. \quad (147)$$

Since we are using the level E_{FN} of Fig. 14 as a reference, Eq. (146) indicates that R has this maximum at the approximate center of the transition region; this is the point where the electrostatic potential is roughly half the applied potential difference. Hence, the recombination process may be significant not only over distances of about one diffusion length from the junction but right at the junction itself.

This concept lies at the basis of the Sah *et al.* (1957) (SNS) theory regarding the effect of generation–recombination current on the characteristics of

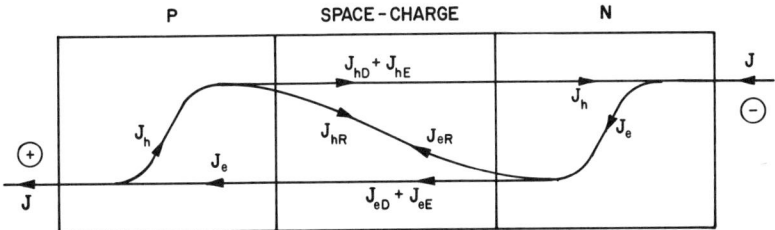

FIG. 16. Generation–recombination currents in the space-charge region of a PN junction.

silicon diodes. Figure 16 shows schematically (rather than graphically) the recombination in the space-charge region for a diode under forward bias. The injected hole current J_h in the N region, for example, represents the difference between what enters the space-charge region and the generation–recombination component J_{hR}. We realize that J_{hR} (which is equal to J_{eR}) may be a significant fraction of the total hole current J_h. Further, the current density J which appears in the external circuit or in the regions of the diode near the contacts has three contributing mechanisms—drift, diffusion, and recombination—so that it should be expressed as

$$J = J_{hD} + J_{hE} + J_{hR} + J_{eD} + J_{eE} + J_{eR}. \tag{148}$$

For a given applied potential, the value of J is then larger than what the low-level theory predicts and the characteristic curves consequently show deviations from the standard form of (107).

If the generation–recombination current is dominant in the transition region, (148) simplifies to

$$J = J_{hR} + J_{eR} = 2J_{hR} = J_R \tag{149}$$

since $J_{hR} = J_{eR}$. Hence, the continuity equation for injected holes becomes

$$dJ_R/dx = 2eR \tag{150}$$

or

$$J_R = 2e \int_0^w R\, dx, \tag{151}$$

where w, the width of the space-charge region, has to be specified in a somewhat arbitrary (but reasonable) fashion.

We can estimate J_R by assuming that R is of significance only at the center of the space-charge region, where it has its maximum value R_{max}. Using (147) in (146) gives

$$R_{max} = n_i[\exp(-\psi_A) - 1]/2\tau_0[\exp(-\psi_A/2) + 1] \tag{152}$$

or for reasonably large forward bias

$$R_{max} = (n_i/2\tau_0)\exp(-\psi_A/2). \tag{153}$$

Substituting this into (151) gives

$$J_R = (n_i ew/\tau_0) \exp(-\psi_A/2). \tag{154}$$

This expression is seen to be a special case of the equation

$$J_R = [n_i^2 ew/(\tau_{h0} n_0^* + \tau_{e0} p_0^*)][\exp(-\psi_A/2) - 1] \tag{155}$$

as given by Hauser (1965); the algebra without the simplifying assumptions we have used is tedious.

For fairly low-injection levels, Eq. (142) indicates an exponential dependence which differs from (154). In a heavily doped P region, for example, we neglect n, n_0^*, and p_0^* in comparison to p_0 (an approximation to p), so that (142) reduces to

$$R = n_i^2 [\exp(-\psi_A) - 1]/\tau_{e0} p_{OP}$$

and by (151)

$$J_R = (2ewn_i^2/\tau_{e0} p_{OP})[\exp(\psi_A) - 1]. \tag{156}$$

By (107) and (108), the diffusion current J_D for a junction, which we shall take to be symmetric ($D_e = D_h$, $L_e = L_h$, $p_{ON} = n_{OP}$) for the purpose of doing a rough calculation, is

$$J_D = \frac{2eD_e n_{OP}}{L_e}[\exp(-\psi_A) - 1] = \frac{2eL_e n_{OP}}{\tau_e} \exp(-\psi_A). \tag{157}$$

Then the ratio of the initial generation–recombination current to the low-level diffusion current is, by (154) and (157),

$$J_R/J_D = (n_i/n_{OP})(w/2L_e) \exp(\psi_A/2). \tag{158}$$

Using

$$n_i = 1.5 \times 10^{10}/\text{cm}^3,$$
$$n_{OP} = n_i^2/n_{ON} = 2.25 \times 10^{20}/10^{15} = 2.25 \times 10^5/\text{cm}^3,$$
$$w = 10^{-6} \text{ m},$$
$$L_e = 60 \times 10^{-6} \text{ m}$$

shows that J_R is at least 10 times greater than J_D when the forward bias $-\psi_A$ is 9.0 or smaller. Of course, for a very small forward bias, we can expand the exponential term in (152) to show that

$$\exp(-\psi_A) - 1 \sim -\psi_A \tag{159}$$

so that when we start at zero bias and build up to large forward values, the initial portion of the characteristic curve should be ohmic. As the bias gets larger, this approximation is not valid, and the characteristic becomes like

2. THE THEORY OF SEMICONDUCTING JUNCTIONS

that of (154). Thus, the generation–recombination current is the main contributor to the overall dc behavior of the diode for values of ψ_A in the lower part of the operating range, and it is predicted to have the form

$$J_R = J_{RS} \exp(-\psi_A/\alpha), \tag{160}$$

where $\alpha = 1.0$ at very low bias but goes over to $\alpha = 2.0$ as the bias rises. Then, when J_D becomes more significant than J_R, Eq. (157) specifies the behavior, so that we can say that α returns to its initial value of 1.0. The change of α from 1.0 to 2.0 in the middle of the generation–recombination regime can be ascribed to trap saturation. As traps become less effective in producing a recombination contribution, the rise in bias for a given change in J_h is correspondingly greater. However, the predicted transition of α from 1.0 to 2.0 and back to 1.0 is not, in fact, observed. To see why, we must identify and eliminate the approximations inherent in the SNS approach.

We start by writing (150) as

$$dJ_R/dx = (dJ_R/d\psi)\, d\psi/dx = 2eR$$

or

$$J_R = 2e \int \frac{R\, d\psi}{d\psi/dx}. \tag{161}$$

It is then assumed that $d\psi/dx$ is approximately constant and can be removed from the integral. This is justified by the behavior of K as shown, for example, in Fig. 4a. Then (161) combined with (143) and the assumption that $E_T = E_I$ gives

$$J_R = \frac{2en_i}{\tau_{h0}\, d\psi/dx} \int \frac{[\exp(-\psi_A) - 1]\, d\psi}{(\tau_{e0}/\tau_{h0})[\exp(-\psi_A - \psi) + 1] + [\exp(\psi) + 1]}. \tag{162}$$

Under moderate bias, the numerator can be simplified because $\exp(-\psi_A) \gg 1$; the two terms in parentheses in the denominator can likewise be simplified, for—as the original form (142) of R indicates—this follows from the fact that both n and p are large compared to n_i in the space-charge region when injection takes place. Hence, (162) reduces to

$$J_R = \frac{2en_i \exp(-\psi_A)}{\tau_{h0}\, d\psi/dx} \int \frac{\exp(\psi)\, d\psi}{(\tau_{e0}/\tau_{h0})\exp(-\psi_A) + \exp(2\psi)}. \tag{163}$$

The integral in (163) can be evaluated between the limits of 0 to ∞ under the assumption that the integrand is significant only when R is close to $R_{\max}$. Then

$$\int_0^\infty \frac{dz}{z^2 + a^2} = \frac{1}{a} \arctan(z/a) \Big|_0^\infty = \frac{\pi}{2a},$$

where

$$z = \exp\psi, \quad a = \sqrt{\tau_{e0}/\tau_{h0}}\exp(-\psi_A/2),$$

and

$$J_R = \{n_i e\pi \exp(-\psi_A/2)/(d\psi/dx)\sqrt{\tau_{e0}\tau_{h0}}\}. \tag{164}$$

This procedure thus leads to an expression which is of the same form as (154). The SNS calculation improved upon (164) by recognizing that the depletion layer width w varies with applied bias V_A. This feature was incorporated in two ways; first, by assuming that ψ depends on ψ_A in accordance with the approximation

$$d\psi/dx = [-\psi_D - (-\psi_A)]/w \tag{165}$$

and second, by assuming that w will decrease under forward bias as specified by (60). This variation is expressed as

$$w = \left[\frac{2\varepsilon kT}{e^2}(\psi_A - \psi_D)\left(\frac{1}{N_A} + \frac{1}{N_D}\right)\right]^{1/2}. \tag{166}$$

The result of incorporating (143), (165), and (166) into (161) works out to be (Sah et al., 1957; Chou, 1971)

$$J_R = [n_i e/2(\tau_{e0}\tau_{h0})^{1/2}]w_0 f(-\psi_A), \tag{167}$$

where w_0 is the value of w for zero bias and the function $f(-\psi_A)$ has the form

$$f(b) = B\int_{z_1}^{z_2}\frac{dz}{z^2 + 2bz + 1} \tag{168}$$

with

$$\begin{aligned}B &= 2\sinh\psi_A/(\psi_D - \psi_A),\\ b &= \exp(-\psi_A/2)\cosh[E_T - E_I/kT + \tfrac{1}{2}\log(\tau_{h0}/\tau_{e0})],\\ \left.\begin{matrix}z_1\\ z_2\end{matrix}\right\} &= (\tau_{h0}/\tau_{e0})^{1/2}\exp[\pm(\psi_D + \psi_A)/2].\end{aligned} \tag{169}$$

Equation (164) has formed the basis for a large number of studies of device theory in relation to measured properties. As typical examples, we may cite the papers by Choo (1968, 1971), Chou (1971), Brancus and Dolocan (1972), Chevychekelov (1960), and Agakhangan (1966).

The approximations we have considered indicate that α in (160) should make a transition from its initial value of 1.0 to a value of 2.0 in the upper part of the generation–recombination regime and then go back to 1.0, by

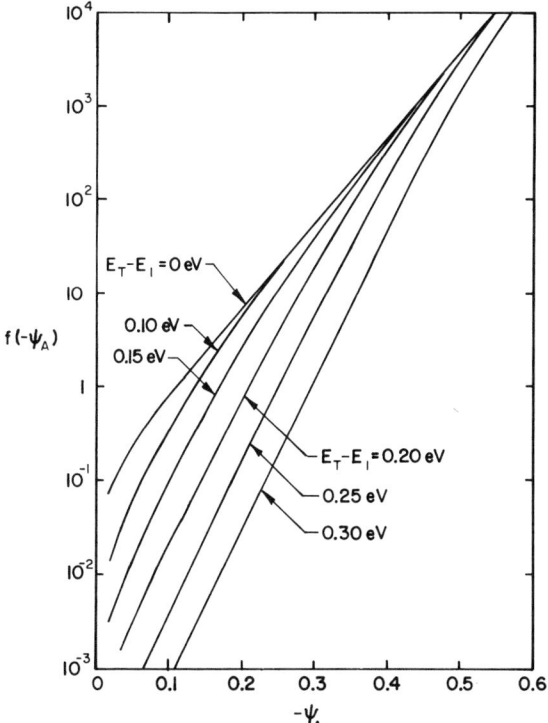

FIG. 17. The Sah–Noyce–Shockley function.

(157). (For high injection, which we discuss later, it is found that α makes a second transition to 2.0.) It is pointed out by SNS that these discontinuous sections should combine to give an effective α between 1.0 and 2.0 in the lower part of the bias range. However, a numerical evaluation of the SNS function $f(-\psi_A)$ in (167) by Chou (1971) shows the behavior illustrated in Fig. 17. He considered a silicon junction with a built-in potential difference of 0.7 eV and he used six different trap positions, ranging from $E_T - E_I = 0.0$ eV to $E_T - E_I = 0.30$ eV. His results show that J_R has a characteristic with $\alpha = 1.0$ at low-injection levels and makes a transition to 2.0 at a bias roughly equal to $2(E_T - E_I)$.

To continue with our procedure of eliminating all possible assumptions, let us apply a suggestion of Fulkerson (1968). He pointed out that it is possible to express $d\psi/dx$ analytically, without recourse to the depletion approximation, through the use of Eq. (118). We, therefore, substitute this into (161), along with the complete form of R from (143) and the assumption in (97) that the electrochemical potentials are constant, to obtain the rather formidable expression

$$J_R = \frac{eL_D n_i}{\tau_{h0}}$$
$$\times \int_{\psi_J}^{\psi_0} \frac{[\exp(-\psi_A) - 1]\,d\psi}{\{(\tau_{e0}/\tau_{h0})[\exp(-\psi_A - \psi) + \exp(-u_T)] + [\exp(\psi) + \exp(u_T)]\}},$$
$$\times [\exp(\psi) + \exp(-\psi_A - \psi) - 2a\psi + 2C]^{1/2}$$
(170)

where

$$u_T = (E_T - E_I)/kT.$$

The lower limit ψ_J is the value of ψ at the abrupt junction ($r = 0$) and the upper limit ψ_0 is the maximum or equilibrium value at either edge of the space-charge region. To compute ψ_J and the values of C and ψ_0 at the edges, we find ψ_{0N} and ψ_{0P}, the maximum displacements of the intrinsic level from the reference level at the two edges from (35), giving

$$a_N = \sinh \psi_{0N} \quad \text{and} \quad a_P = \sinh \psi_{0P}. \quad (171)$$

The relations equivalent to (44) are

$$C_N = a_N \psi_{0N} - \tfrac{1}{2}\{\exp(\psi_{0N}) + \exp[-(\psi_A + \psi_{0N})]\} \quad (172)$$

and

$$C_P = a_P \psi_{0P} - \tfrac{1}{2}\{\exp(\psi_{0P}) + \exp[-(\psi_A + \psi_{0P})]\}, \quad (173)$$

and in the same fashion as (47), the value ψ_J for ψ at the junction is

$$\psi_J = (C_N - C_P)/(a_N - a_P). \quad (174)$$

We now apply (170) to a typical abrupt silicon diode, which—according to Grove (1967)—will have an impurity distribution given by

$$N_A = 1.5 \times 10^{18}/\text{cm}^3 \ (P \text{ region}),$$
$$N_D = 1.5 \times 10^{14}/\text{cm}^3 \ (N \text{ region}),$$

and for which

$$n_i = 1.5 \times 10^{10}/\text{cm}^3.$$

The limiting lifetimes were taken as (Ross and Madigan, 1961)

$$\tau_{e0} = 1.1 \times 10^{-6} \text{ sec}, \quad \tau_{h0} = 0.4 \times 10^{-6} \text{ sec}.$$

Using applied potentials of ± 8.0 units, the dimensionless potential $-\psi_J$ at the junction has values as follows:

$-\psi_A$	0	-8.0	8.0
$-\psi_J$	17.4181	9.4189	25.4173

The use of four decimal places shows that, to a very good approximation, Eqs. (172)–(174) simplify to

$$\psi_J = \psi_{0J} \pm \psi_A, \tag{175}$$

where ψ_{0J} is the equilibrium value of ψ_J. Equation (175) is perfectly acceptable for determining energy-band shapes, but in the numerical integration of equations involving the variable in the exponent, it is not good enough and was not used in the calculation involving Eq. (170), with results as shown in Fig. 18. Three different trap positions were used, and for that position which Chou (1971a) feels to be the most probable from experimental evidence, three different ratios of limiting lifetime were used. The corresponding values of α over the straight-line regions of these lines are as follows:

Curve	$(E_T - E_I)/kT$	τ_{e0}/τ_{h0}	α
A	0.0	2.75	1.65
B	4.0	2.75	1.50
C	8.0	2.75	1.00
A'	4.0	0.275	1.50
C'	4.0	27.5	1.50

The sequence of curves A, B, and C shows that the exponent is a very sensitive function of trap position, whereas the set A', B, C' indicates that it is independent of the lifetime ratio. A similar calculation for reverse bias is shown in

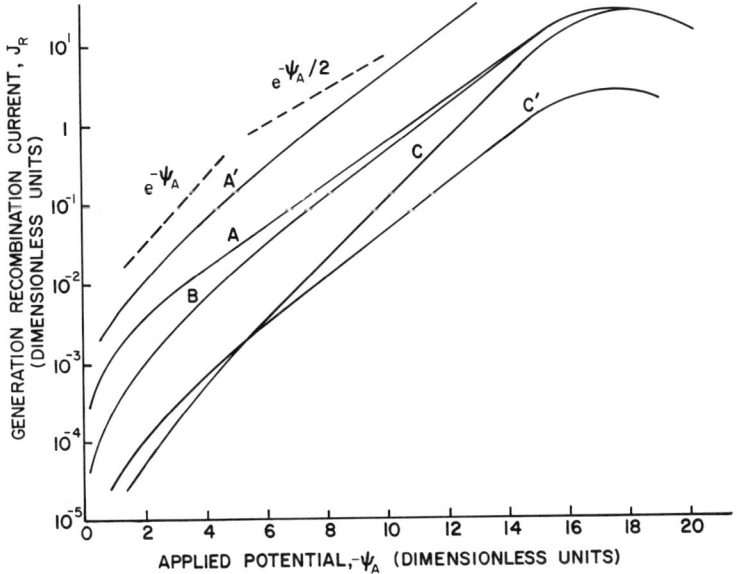

FIG. 18. Generation–recombination current versus forward bias for silicon junctions.

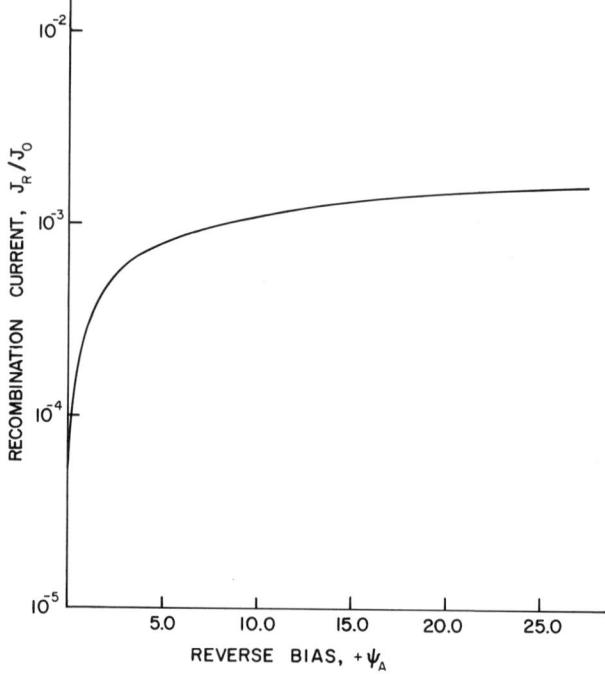

Fig. 19. Generation–recombination current versus reverse-bias for a silicon junction (Case B of Fig. 18).

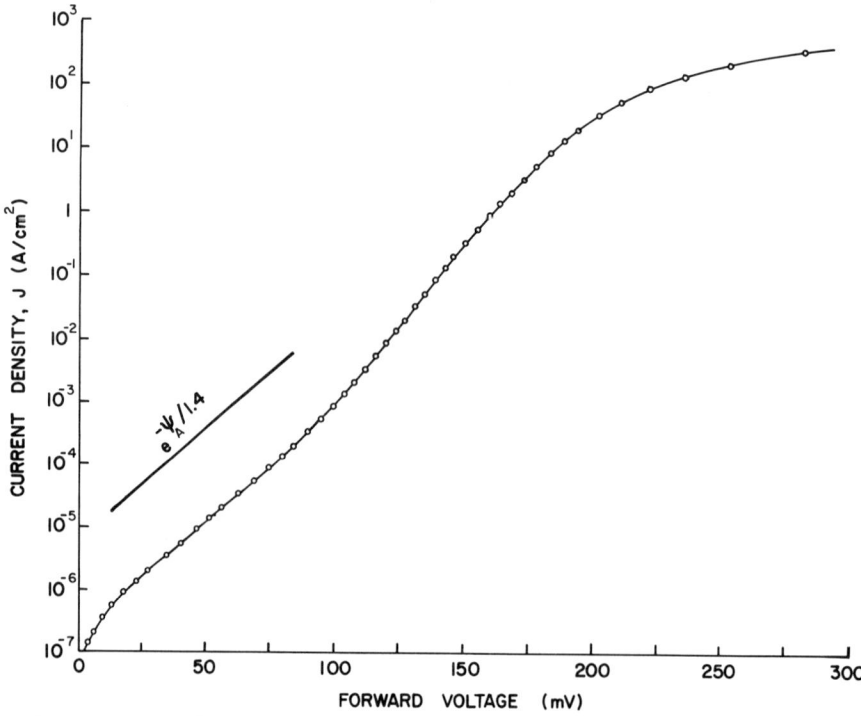

Fig. 20. Measured current–voltage characteristic for an InSb diode at 77°K. (After Barber, 1968.)

Fig. 19, using the constants of curve B, above. The nonsaturating behavior due to the generation–recombination processes is evident.

The results of Figs. 18 and 19 represent the culmination of a series of improved approximations for J_R, starting with (155), followed by (164), (167), and finally (170). To compare predictions with experiment, we show in Fig. 20 the measurements made by Barber (1969) on an InSb diode. We note that there is a three-decade straight-line section in the generation–recombination region for which $\alpha = 1.4$, so that the Sah–Noyce–Shockley theory does not give correct values of α for the region in which the generation–recombination contribution is the dominant one.

Barber (1968) did a similar study of the silicon diode whose impurity profile is shown in Fig. 21; the characteristic curve of Fig. 22 shows that $\alpha = 1.5$ throughout the entire generation–recombination region and then goes over to $\alpha = 1.0$ in the region where diffusion dominates. Hence, the trap position of curve B, Fig. 18, is matched for this junction.

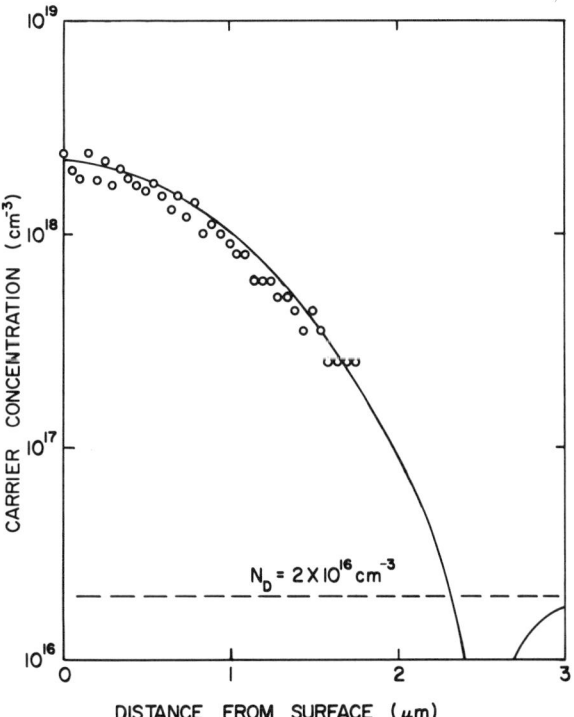

FIG. 21. Impurity profile of silicon junction. (After Barber, 1968.)

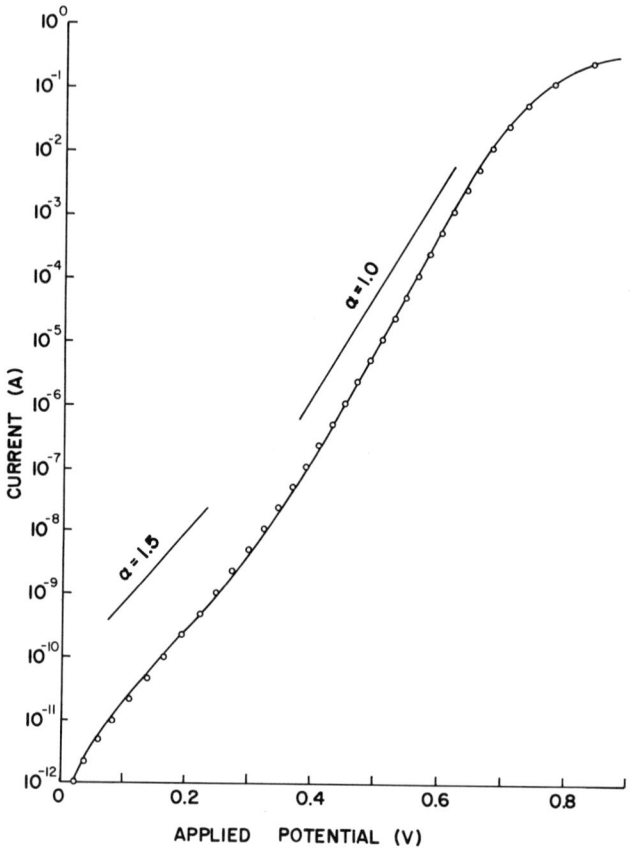

FIG. 22. Static characteristics of the silicon junction whose doping profile is shown in Fig. 21.

A direct method for measuring α has been devised by Howes and Read (1972), who report a value of 1.54 over three decades for an abrupt silicon diode with $N_A = 10^{18}$ acceptors/cm^3 and $N_D = 10^{15}$ donors/cm^3. Measurements by Faulkner and Buckingham (1968) give exactly the same value for a type 1S922 diode (Texas Instruments, Ltd.) at room temperature. They also obtain a value $\alpha = 1.61$ (Faulkner and Buckingham, 1968) for a Hewlett–Packard step recovery diode, and values ranging from 1.14 to 1.91 for other commercial diodes of unspecified doping profiles (Buckingham and Faulkner, 1969).

It should be noted that only two assumptions have been used to compute the generation–recombination current as a function of applied potential. These are that (1) the electrochemical potentials appear as shown in Fig. 10 and that (2) the contribution to J_R from the heavily doped P region is negligible. The first approximation can be expected to be valid for the low-injection levels involved and the other one is reasonable because almost all of the space-charge region lies on the N side of the junction. Although the elimination of the depletion approximation appears to be the principal change in the SNS theory required to match theory with experiment for forward characteristics, this might not always be necessary. For example, Calzolari and Graffi (1972) show that for reverse-biased silicon diodes, difficulties with capacitance versus voltage measurements can be resolved by taking into account the effective width of the region for which R has a maximum, constant value.

Let us consider in more detail the imperfections which we have loosely called recombination centers or traps. We shall use the elementary classification scheme of Sah (1967a) and of Stockmann (1973), which is graphically represented in Fig. 23. The individual rates r_{CT}, g_{TC}, r_{TV}, and g_{VT} of Eqs. (132) and (134) are expressed in the abbreviated notation CT, TC, TV, and VT, respectively; the associated transition processes are then shown on the energy-band diagram in terms of arrows joining the traps with the appropriate band. The relative magnitudes of CT as compared to TV and VT as compared to TC are indicated by the widths of the arrows. The first possibility is that a trap is more likely to receive an electron from the conduction band than from the valence band (this latter transition is the same as sending a hole from trap to valence band) and, once the electron is in the trap, it is more likely to return to the conduction band than drop to the valence band. The pair of inequalities then specifies that the imperfection level is acting like an electron trap; it is taking carriers from the conduction band, holding them for a finite time,

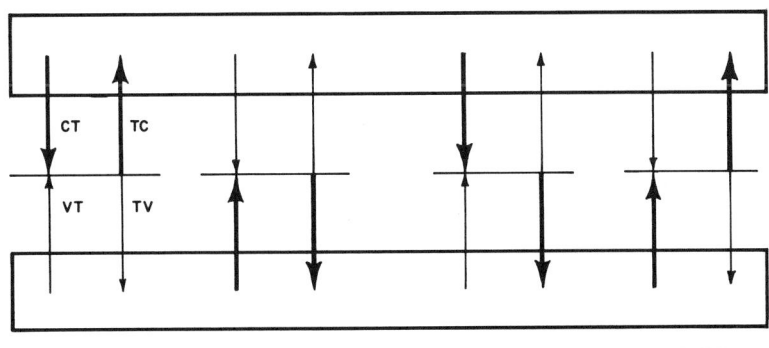

FIG. 23. Sah–Stöckmann classification scheme.

and returning them to their source. This effectively reduces the number of free electrons and—since (98) requires that the *pn* product be fixed for a given bias—there is a corresponding rise in the hole concentration. Thus, electron traps have acceptorlike properties, even though they tend to lie in the upper part of the forbidden band. In the same way, a reversal of both of the inequalities produces a hole trapping level, which has donor properties. The third possibility—electrons which are initially trapped are more likely to go all the way down to the valence band than back to the conduction band—is the recombination process, the imperfection being designated as a recombination center. This classification scheme then clearly shows the quantitative distinction between an electron or hole trap and a recombination center. It should be realized that these are dynamic definitions; it is possible under transient conditions for the role of an imperfection to change from trapping to recombination or vice versa (Nussbaum *et al.*, 1977).

Another way in which the behavior of the traps can change is in connection with a reversal of bias polarity. When excess carriers of amount Δn and Δp are created by injection, optical excitation or any other customary way, the boundary condition (98) is equivalent to

$$pn > n_i^2$$

and the excess carriers decay with the imperfections acting as recombination centers. It is possible, however, to actually reduce n below n_0 (and p below p_0) in the space-charge region of a *PN* junction under reverse bias; the applied field will force out both kinds of carriers, and there is no replenishment by injection. This corresponds to the situation for which (98) becomes

$$pn < n_i^2$$

and the reduction in free carriers encourages upward transitions from valence band to imperfection or imperfection to conduction band. This is the generation center situation, shown as the last alternative in Fig. 23, and represents the inherent physical difference between the forward and reverse behavior of a silicon diode.

This type of calculation is readily extended to more complex situations. As an example, gold in silicon produces two trapping levels (Bullis, 1967). There is a donorlike level at 0.35 eV above the top of the valence band and one which acts like an acceptor at 0.57 eV from the same reference point. (The level closer to the valence band has donor properties because it can function as a hole trap.) If the concentration of each of the two kinds of traps is low enough, then they are noninteracting; free carriers can go from each trap to the two bands but the traps do not communicate with one another. Under these conditions, it is simple to show that (Sah, 1967b) the steady-state net recombination rate is the sum of two terms like (143), one for each level.

This expression was used in the integral of Eq. (161) to compute the relative value of J_R for the silicon junction whose doping profile is shown in Fig. 24. Although this profile is analytically expressed in terms of the complementary error function erfc(x) (Gandhi, 1968), the assumption was made that the junction is abrupt, in order to use (174) to compute ψ_J, which is not unreasonable in view of the steepness of the curve at 3.8 μm from the diffusion face. On the other hand, the variation with position of the quantities a_N and a_P was incorporated in the calculation (Oprea and Mandt, 1974). Figure 25 shows the numerical evaluation of J_R for the two-level trapping model with a total trap concentration of about 10^{14}/cm^3. Qualitatively, it differs from the single-level model in the fact that the transition from $\alpha = 1.0$ to $\alpha = 1.5$ occurs at a higher recombination current, and this can be explained on the basis of the requirement that both trap levels be saturated before the slope shows a significant

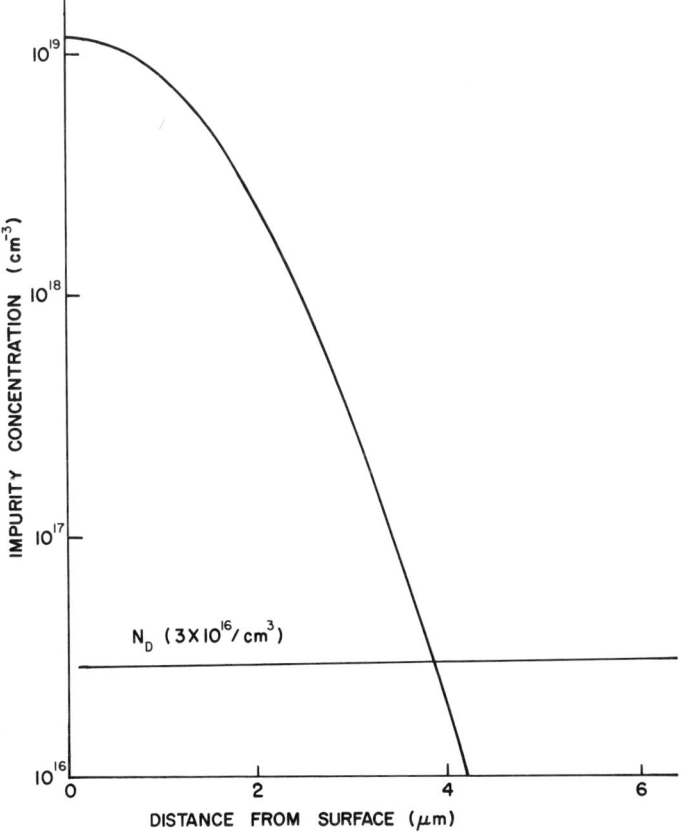

FIG. 24. Doping profile of diode with two gold trapping levels.

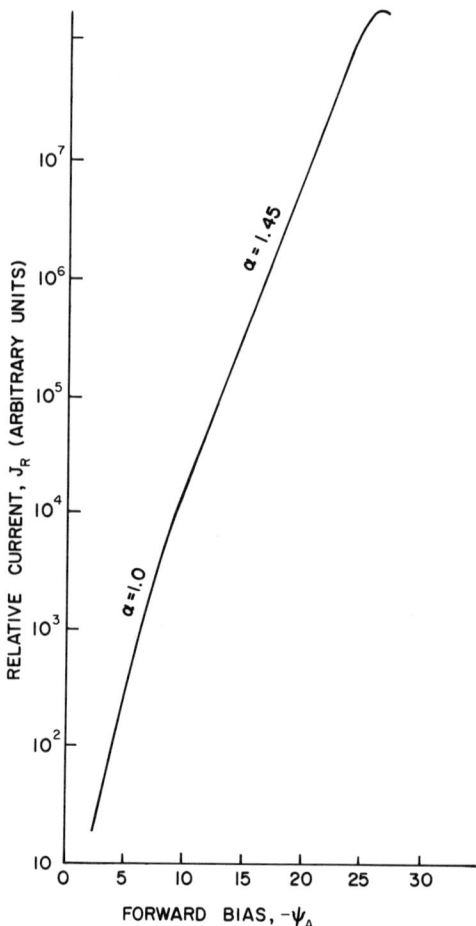

FIG. 25. Current–voltage characteristic of a diode with two independent gold trapping levels.

break. Otherwise, the diode with two noninteracting traps appears to be similar to the original SNS model.

Those junctions which show a constant value of α to within $\pm 10\%$ over four decades of current have been called logarithmic diodes by Buckingham and Faulkner (1969); a discussion of alternative explanations of their behavior has been given in a prior publication (Nussbaum, 1973). One of these is the semiempirical treatment of Barber (1969) which we shall cover briefly in a later section.

6. Boundary Conditions

As discussed in Section 3, among the foundations of Shockley's treatment of the PN-junction diode are the boundary conditions which result from the

2. THE THEORY OF SEMICONDUCTING JUNCTIONS

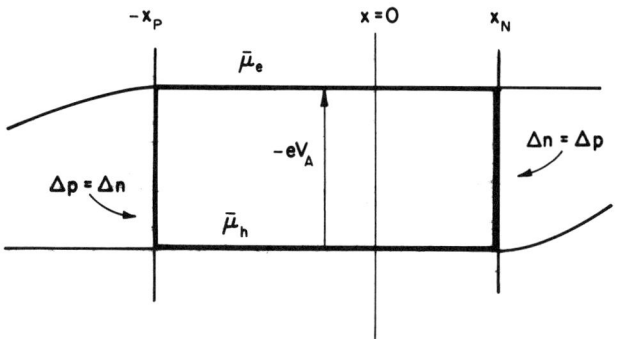

FIG. 26. The Fletcher rectangle.

assumed behavior of the quantities $\bar{\mu}_e$ and $\bar{\mu}_h$ in the space-charge region. These have been generalized by Fletcher (1957) to incorporate space-charge neutrality. To see the physical significance of these conditions, we extract from Fig. 10 those portions of $\bar{\mu}_e$ and $\bar{\mu}_h$ which lie between the edges of the space-charge region (Fig. 26). This diagram shows a rectangle whose upper and lower boundaries correspond to constant values of the electrochemical potentials and whose vertical sides are at the coordinates for which space-charge neutrality is valid.

Using the definitions (69) and (70), the electrochemical potential $\bar{\mu}_{eN}$ for electrons is

$$\bar{\mu}_{eN} = kT \log(n_N/n_i) - eV_N, \tag{176}$$

where the subscript N explicitly designates quantities specified at the edge x_N (see also Fig. 10); similarly

$$\bar{\mu}_{hN} = kT \log(n_i/p_N) - eV_N, \tag{177}$$
$$\bar{\mu}_{eP} = kT \log(n_P/n_i) - eV_P, \tag{178}$$
$$\bar{\mu}_{hP} = kT \log(n_i/p_P) - eV_P. \tag{179}$$

Subtracting (178) from (176) and using the fact that $\bar{\mu}_e$ is a constant across the space-charge region gives

$$n_P = dn_N, \tag{180}$$

where

$$d = \exp[-e(V_A - V_D)/kT] \tag{181}$$

since

$$-e(V_N - V_P) = -e(V_A - V_D).$$

In the same way we obtain

$$p_N = dp_P. \tag{182}$$

The requirement that space-charge neutrality is maintained at the edges x_N and $-x_P$ of the transition region can be expressed as

$$n_N - n_{0N} = p_N - p_{0N}, \tag{183}$$

$$n_P - n_{0P} = p_P - p_{0P}. \tag{184}$$

Combining these two equations with (180) and (182), we obtain the boundary conditions (Fletcher, 1957)

$$\begin{aligned} p_P &= (m_P + m_N d)/(1 - d^2), \\ p_N &= (m_P d + m_N d^2)/(1 - d^2), \\ n_N &= (m_N + m_P d)/(1 - d^2), \\ n_P &= (m_N d + m_P d^2)/(1 - d^2), \end{aligned} \tag{185}$$

where the excess majority-carrier concentrations are defined as

$$m_P = p_{0P} - n_{0P}, \qquad m_N = n_{0N} - p_{0N}. \tag{186}$$

The equilibrium form of the Fletcher boundary conditions is then

$$\begin{aligned} p_{0P} &= (m_P + m_N d_0)/(1 - d_0^2), \\ p_{0N} &= (m_P d_0 + m_N d_0^2)/(1 - d_0^2), \\ n_{0N} &= (m_N + m_P d_0)/(1 - d_0^2), \\ n_{0P} &= (m_N d_0 + m_P d_0^2)/(1 - d_0^2), \end{aligned} \tag{187}$$

where

$$d_0 = \exp(eV_D/kT), \tag{188}$$

Then (180) and (182) reduce to

$$n_{0P} = d_0 n_{0N}, \qquad p_{0N} = d_0 p_{0P}. \tag{189}$$

It is not immediately apparent that the Fletcher conditions are a generalization of those used by Shockley [Eq. (99) plus the corresponding expression for n_P], since space-charge neutrality was not explicitly invoked in Section 3. We can establish the converse relation by a simple, numerical argument. Suppose that we apply a forward bias of 0.307 V to the junction of Fig. 5. This means that d will have a value of 0.1 and the denominators in Eqs. (185) will reduce to unity. The excess majority concentrations by Eq. (186) are p_{0P} and n_{0N}, respectively, to a very good approximation so that Eqs. (185) take on a simpler form. In particular,

$$p_N = d(p_{0P} + n_{0N} d).$$

Furthermore, the term involving n_{0N} is small; the result is that this expression reduces to (99).

On the other hand, consider a junction for which the acceptor concentration has been reduced to $10^{16}/\text{cm}^3$. The barrier becomes 0.250 eV and the forward bias needed to give d a value of 0.1 is now 0.192 V. Although the approximation of replacing $(1 - d^2)$ with unity is still valid, we must keep both terms in the numerator. Hence, for the Fletcher formula to reduce to Shockley's expression, the injection must be at quite a low level. Shockley's low-level boundary conditions, Eq. (99) and its counterpart for electrons, are sometimes called the law of the junction; the Fletcher boundary conditions are therefore a generalization of this law.

Having established this relation between the Shockley and the Fletcher conditions, we can now appreciate that there is an inherent contradiction in Fig. 26 and Eqs. (185). The difficulty is that the electrochemical potentials corresponding to majority carriers are constant throughout the bulk regions. Since the electrostatic potentials are also constant in these regions (Fig. 10), the chemical potentials must show the same behavior. Further, the majority carrier concentrations must equal the respective equilibrium values at the contacts. All of these facts combine to show that the electron concentration at x_N has to be n_{0N} and at $-x_P$ the hole concentration is p_{0P}. We have therefore demonstrated that the left-hand side of Eq. (183) and the right-hand side of Eq. (184) must vanish, which further means that there can be no injection at all. Figures 10 and 26 depict physical situations which cannot be rigorously correct (although they can be taken as perfectly acceptable low-level approximations). This represents just one of several interesting anomalies which we shall consider and resolve. For the time being, however, let us ignore this problem concerning the validity of space-charge neutrality. It should be observed that Fletcher (1957) regarded his results as being primarily applicable to high-level injection, for which the majority carriers deviate from their equilibrium concentrations. He specifically mentions that changes in majority-carrier concentrations in the bulk region will result in large changes in the resistance of these regions, producing what is known as conductivity modulation. This phenomenom shows up in several ways, such as accumulation and exclusion, which will be defined and explained in Section 9. We shall also show shortly that the introduction of conductivity modulation implies that the electrochemical potentials are no longer constant throughout the bulk regions, so that the Fletcher conditions require some modification. First, however, we consider some variations.

The Fletcher conditions have been expressed in an alternate form by Harrick (1962), who solves for the injected concentrations at the boundaries. For example, at the left-hand edge of Fig. 25, we have

$$\Delta p_P = p_P - p_{0P} = n_P - n_{0P} = \Delta n_P = \frac{m_N d + m_P d^2}{1 - d^2} - n_{0P}$$

$$= \frac{(n_{0N} - p_{0N})d + p_{0P}d^2 - n_{0P}}{1 - d^2}. \tag{190}$$

Using
$$d = d_0 \exp(-eV_A/kT) = d_0 \exp(-\psi_A)$$
and (189) gives
$$d = (n_{OP}/n_{ON}) \exp(-\psi_A). \qquad (191)$$
Then
$$\Delta p_P = \frac{[n_{OP} - (p_{ON}n_{OP}/n_{ON})] \exp(-\psi_A) + p_{OP}(n_{OP}/n_{ON})^2 \exp(-2\psi_A) - n_{OP}}{1 - (n_{OP}/n_{ON})^2 \exp(-2\psi_A)}$$
$$= n_{OP}[\exp(-\psi_A) - 1] \frac{(p_{ON}/n_{ON}) \exp(-\psi_A) + 1}{1 - (n_{OP}p_{ON}/n_{ON}p_{OP}) \exp(-2\psi_A)}. \qquad (192)$$

In a similar way
$$\Delta n_N = \Delta p_N = p_{ON}[\exp(-\psi_A) - 1] \left(\frac{(n_{OP}/p_{OP}) \exp(-\psi_A) + 1}{1 - (n_{OP}p_{ON}/n_{ON}p_{OP}) \exp(-2\psi_A)} \right). \qquad (193)$$

Equations (192) and (193) are also found in the literature in the form due to Misawa (Lade and Poncelet, 1964; Gaertner, 1960; Misawa, 1955, 1958)
$$\Delta p_P = \frac{P^2[N^2 + \exp(-\psi_A)]}{P^2 N^2 - \exp(-2\psi_A)} n_{OP}[\exp(-\psi_A) - 1],$$
where
$$P = p_{OP}/n_i, \qquad N = n_{ON}/n_i$$
or
$$P^2 = p_{OP}/n_{OP}, \qquad N^2 = n_{ON}/p_{ON}$$

with a similar expression for Δn_N. The coefficient of the terms in square brackets in (192) and (193) will be recognized as extensions of the Shockley boundary conditions. For example, Eq. (99) gives the injected hole concentration as
$$\Delta p_N = p_N - p_{ON} = p_{ON}[\exp(-\psi_A) - 1]. \qquad (194)$$
Harrick identifies the modifying term in (193) as the effect of high-level injection, but this is incorrect; Eq. (193) is simply the consequence of replacing the boundary condition $\Delta n_N = 0$ with the neutrality statement $\Delta n_N = \Delta p_N$. The Fletcher conditions in any of the forms presented here are valid only in the limit of low-level injection, for—as pointed out in connection with Eq. (85)—constant values of $\bar{\mu}_e$ and $\bar{\mu}_h$ mean that the associated currents must vanish.

2. THE THEORY OF SEMICONDUCTING JUNCTIONS

Although the argument just raised represents the objection to the Fletcher conditions, there are other difficulties which can be demonstrated analytically (van Vliet, 1966; Gummel, 1967). The behavior of the electrochemical potentials across the space-charge region can be expressed, from (178) and (177), as

$$\bar{\mu}_{eN} - \bar{\mu}_{hN} = \bar{\mu}_{eP} - \bar{\mu}_{hN} = e(V_N - V_P) + kT \log(n_P p_N / n_i^2)$$
$$= -e(V_D - V_A) + kT \log(n_P p_N / n_i^2) \quad (195)$$

or by (189),

$$\bar{\mu}_{eN} - \bar{\mu}_{hN} = eV_A + kT \log(n_P p_N / n_{0P} p_{0N}). \quad (196)$$

Substituting for the four quantities inside parentheses converts this into

$$\bar{\mu}_{eN} - \bar{\mu}_{hN} = eV_A + kT \log\left[\left(\frac{d}{d_0}\right)^2 \left(\frac{m_P d + m_N}{m_P d_0 + m_N}\right)\left(\frac{m_P + m_N d}{m_P + m_N d_0}\right)\left(\frac{1 - d_0^2}{1 - d^2}\right)^2\right]. \quad (197)$$

The first factor in square brackets is

$$kT \log(d/d_0)^2 = -2eV_A.$$

As van Vliet (1966) points out, the remaining factors in the argument of the logarithm are all larger than unity for forward bias. This leads to the result that

$$\bar{\mu}_{eN} - \bar{\mu}_{hN} > -eV_A, \quad (198)$$

which contradicts the assumption that the electrochemical potentials are separated by an amount determined by the forward potential. This inconsistency has been established in an alternate way by Gummel (1967). Multiplication of the expressions for p_N and n_N of (185) gives

$$p_N n_N = \frac{d m_P m_N [1 + (m_N/m_P + m_P/m_N) d + d^2]}{(1 - d^2)^2}. \quad (199)$$

The term in square brackets has to be positive and greater than unity and the denominator is less than unity. Hence, (199) reduces to

$$p_N n_N > d m_P m_N.$$

This is equivalent, with very good accuracy, to

$$p_N n_N > d p_{0P} n_{0N}$$

or

$$p_N n_N > n_i^2 \exp(-eV_A/kT), \quad (201)$$

which disagrees with (98).

Other versions of the contradictions demonstrated by (198) and (201) continue to show up in the literature because of the failure to recognize that the gradients of the electrochemical potential cannot vanish. The most recent is that of Guckel *et al.* (1977) and Nussbaum (1978) which we shall reexpress in our own notation. By multiplying Eqs. (185) together, two at a time, we obtain

$$p_P n_P = p_N n_N = \frac{m_P m_N d(1 + d^2) + (m_P^2 + m_N^2) d^2}{(1 - d^2)^2}. \tag{202}$$

By (181),

$$d = \exp[-e(V_A - V_D)/kT] = \exp(\psi_D - \psi_A) = \exp(\psi_P - \psi_N). \tag{203}$$

Then

$$\cosh(\psi_P - \psi_N) = \frac{d + (1/d)}{2} = \frac{d^2 + 1}{2d}, \qquad \sinh(\psi_P - \psi_N) = \frac{d^2 - 1}{2d},$$

and (202) becomes

$$p_P n_P = p_N n_N = \frac{2 m_P m_N \cosh(\psi_P - \psi_N) + m_P^2 + m_N^2}{4 \sinh^2(\psi_P - \psi_N)}. \tag{204}$$

This result is equivalent to Eq. (9) of Guckel *et al.* (1977) (they identify m_P as N_A, the acceptor doping in the P region, and m_N similarly as N_D, but these approximations do not affect the argument). This paper bases its conclusions on (204); however, it is simpler to go back to (202). For example, let

$$m = m_P = m_N$$

for a symmetric junction. Then, at either edge of the space-charge region

$$p_P n_P = p_N n_N = \frac{m^2 d}{(1 - d)^2} = \frac{m^2 \exp(\psi_D) \exp(-\psi_A)}{(1 - d)^2} = \frac{(m^2 n_i^2 / p_{0P} n_{0N}) \exp(-\psi_A)}{(1 - d)^2}. \tag{205}$$

Neglecting minority carriers, this simplifies to

$$p_P n_P = p_N n_N = n_i^2 \exp(-\psi_A)/(1 - d)^2, \tag{206}$$

and since the denominator is always less than unity, Eq. (206) is equivalent to Eq. (201). This development shows that Eq. (204) is valid only in the limit of zero injection. Actually, $1/(1 - d)^2$ can be shown numerically to be quite close to unity until the forward bias is substantially close to the equilibrium barrier potential, so that—considering just boundary conditions—the approximations involved do not seem to be too bad. However, the question of when low-level assumptions affect the nature of the differential equations must also be examined, and this we shall do in Section 8.

The assumed behavior of the electrochemical potentials in the space-charge region has thus generated inconsistencies; the resolution proposed by Gummel is to postulate the existence of a finite slope for both $\bar{\mu}_e$ and $\bar{\mu}_h$. As an initial guess, we shall assume that this slope, taken to be the same for both electrochemical potentials, is specified by a change of magnitude ε over a region of width w (Fig. 27). There will also be gradients in the bulk regions, specified by δ_P and δ_N, so that the bands will tilt in a fashion indicated by Fig. 28. A key feature of this diagram is that the applied bias V_A, which determines the separation of the Fermi levels at the contacts (assumed to be well beyond the injection regions), is broken down into individual contributions in accordance with the relation

$$-eV_A = -eV_J + \delta_N + \delta_P. \tag{207}$$

That portion V_J which is seen by the junction is thus defined as the applied potential less the bulk-region drops, where we are regarding δ_N and δ_P as positive numbers. Having made this correction, let us now show that it still leads to inconsistencies (Nussbaum, 1969). We start (Fig. 28) with

$$\begin{aligned}\varepsilon &= \bar{\mu}_{eN} - \bar{\mu}_{eP} = kT \log(n_N/n_P) - eV_N + eV_P \\ &= kT \log(n_N/n_P) + eV_D - eV_J\end{aligned} \tag{208}$$

from which we obtain

$$n_P = d'n_N, \tag{209}$$

where

$$d' = \exp[(eV_D - eV_J - \varepsilon)/kT]. \tag{210}$$

Equation (210) is a generalization of (181) and we also have that

$$p_N = d'p_P. \tag{211}$$

Equations (209) and (211) then lead to modified Fletcher boundary conditions, which we obtain by simply replacing d with d'.

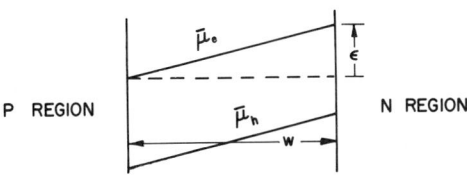

FIG. 27. Electrochemical potentials in the space-charge region with a common gradient ε/w.

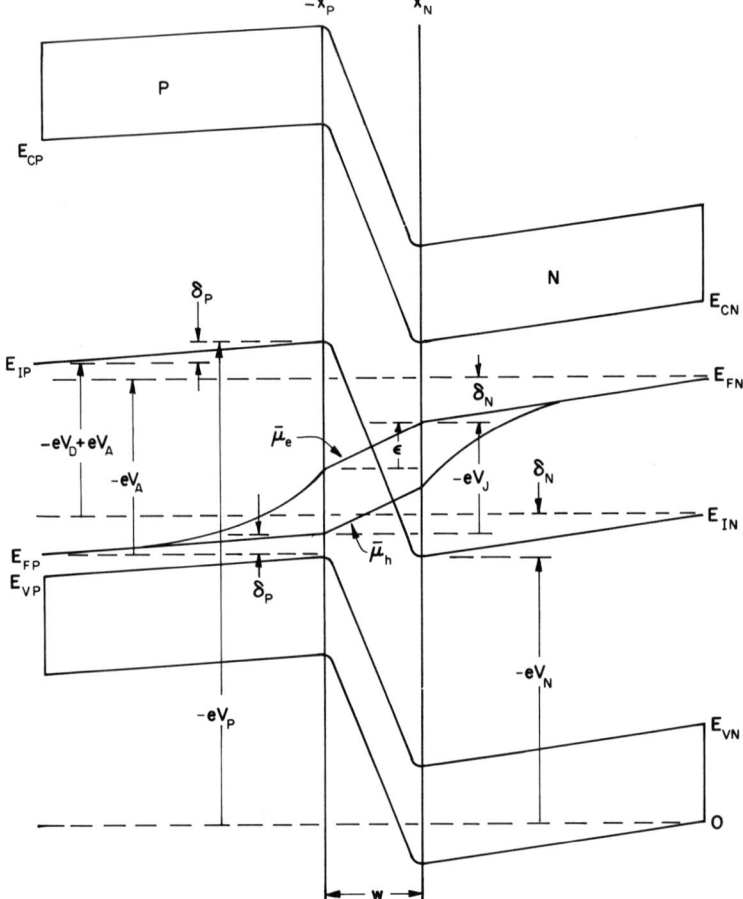

FIG. 28. High-level energy-band diagram for a junction under forward bias.

We may also write

$$\begin{aligned}\bar{\mu}_{eN} - \bar{\mu}_{hN} &= (\bar{\mu}_{eP} + \varepsilon) - \bar{\mu}_{hN} \\ &= kT \log(n_P p_N / n_i^2) + eV_J - eV_D + \varepsilon \\ &= kT \log(n_P p_N / n_{0P} p_{0N}) + eV_J + \varepsilon \end{aligned} \quad (212)$$

using

$$V_D = (kT/e) \log(n_{0P}/n_{0N}). \quad (212a)$$

From (209) and (211), plus the corresponding relations involving d_0, this becomes

$$\bar{\mu}_{eN} - \bar{\mu}_{hN} = kT \log(n_N p_P / n_{0N} p_{0P}) - eV_J - \varepsilon. \quad (213)$$

(Note that it is not necessary to define a quantity d'_0, since δ_N, δ_P, and ε vanish at equilibrium.)

The interpretation of this result comes from Fig. 28. Taking the potential at the right-hand contact as fixed, the applied potential V_A depresses all the energy levels at the left-hand contact by an amount $-eV_A$ below their original positions, which is why E_{FP}, for example, is this amount below E_{FN}. The quantities δ_P and δ_N represent the potential drops in the two bulk regions, so that the difference between the applied potential and these two quantities is the potential across the junction. In energy units

$$-eV_A = \delta_N + \delta_P + (-eV_J). \qquad (214)$$

The figure then shows that

$$\bar{\mu}_{eN} - \bar{\mu}_{hN} = -\varepsilon - eV_J, \qquad (215)$$

which contradicts (213) because of the presence of the term

$$kT \log(n_N p_P / n_{0N} p_{0P}) = kT \log(n_N/n_{0N}) + kT \log(p_P/p_{0P}). \qquad (216)$$

It is seen that the two ratios on the right are deviations in *majority*-carrier concentrations. The phenomenon of majority-carrier injection has been ignored up until now, and this is what has produced these logical difficulties. The terms on the right of (216) are clearly shifts in the majority-carrier electrochemical potentials at the boundaries of the space-charge region. To account for these, we must further alter the band structure, as shown in Fig. 29. At $-x_P$, for example, the quantity $\bar{\mu}_h$ must have a slope which is

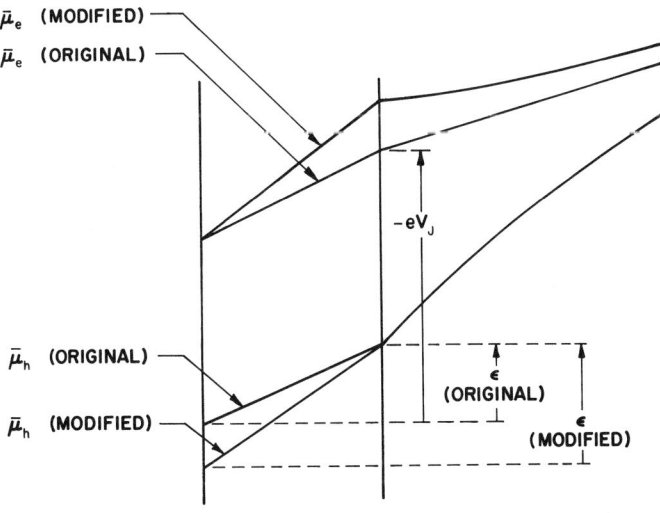

FIG. 29. Modification of electrochemical potentials to incorporate the effects of majority-carrier injection.

actually greater than ε/w; we indicate this by showing both the original and the modified values of ε. Similarly, $\bar{\mu}_e$ at x_N increases by an amount $kT\log(n_N/n_{0N})$ and $\bar{\mu}_e$ is no longer represented by a straight line in the N region. Instead it corresponds to the curve shown; the shape is guesswork but we do know that the slope must be everywhere positive, since otherwise the electron current would flow in the wrong direction. We also realize that the modified position of $\bar{\mu}_{eN}$ is separated by an amount greater than $-eV_J$ from $\bar{\mu}_{hN}$. Therefore, we can no longer identify the separation of the electrochemical potentials at any given point in the space-charge region with either $-eV_A$ or $-eV_J$, for at the edges, the separation is somewhere in between these two quantities.

The necessity for incorporating majority-carrier injection in high-level theories was recognized by Harrick (1962), who states (without proof) that

$$\bar{\mu}_{eN} - \bar{\mu}_{hN} = -eV_J + kT\log\left(1 + \frac{\Delta p_P}{p_{0P}} + \frac{\Delta p_N}{n_{0N}} + \frac{\Delta p_P \Delta n_N}{p_{0P}n_{0N}}\right). \quad (217)$$

This is a combination of (213) and (216) for $\varepsilon = 0$. There have also been other ways (Klein, 1976; Hauser, 1971) proposed for specifying boundary conditions, but these are either equivalent to the original approach of Fletcher or they incorporate the slope ε, as proposed by Gummel. Hauser (1971) and Klein (1976) write (76) as

$$dn/dx - (e/kT)(dV/dx)n = J_e/eD_e. \quad (218)$$

This is of the standard form

$$dy/dx + P(x)y = Q(x)$$

with a solution

$$y = \exp\left(-\int P\,dx\right)\int Q\exp\left(\int P\,dx\right)dx + C\exp\left(-\int P\,dx\right).$$

If x_1 is some arbitrary value of x, then

$$n(x) = \exp[eV(x)/kT]\int_{x_1}^{x}\exp[-eV(x')/kT](J_e/eD_e)\,dx'$$
$$+ C\exp\{[eV(x) - eV(x_1)]/kT\},$$

where x' is a dummy variable. The integral vanishes for $x = x_1$, so that $C = n(x_1)$ and

$$n(x) = \exp[eV(x)/kT]\left(\int_{x_1}^{x}(J_e/eD_e)\exp[-eV(x')/kT]\,dx'\right.$$
$$\left. + n(x_1)\exp[-eV(x_1)/kT]\right). \quad (219)$$

2. THE THEORY OF SEMICONDUCTING JUNCTIONS

Hauser assumes that the integral is small within the space-charge region, and (219) reduces to

$$n(x) = n(x_1)\exp\{(e/kT)[V(x) - V(x_1)]\}, \tag{220}$$

which will be recognized as equivalent to (180); there is, of course, a corresponding relation for holes. This result is not surprising, since the neglect of J_e implies that $d\bar{\mu}_e/dx$ is small, or that $\bar{\mu}_e$ is constant.

Rather than use this approximation, van der Ziel (1973) takes the currents as constant in the space-charge region, i.e., he is neglecting generation–recombination. Integrating (77), this time, from $-x_P$ to x_N and using the fact that the change in V corresponds to a barrier height of $-eV_D - (-eV_J)$ gives

$$p_N = p_P \exp[(eV_D - eV_J - J_h R_{hs})/kT], \tag{221}$$

where

$$R_{hs} = \int_{-x_P}^{x_N} \frac{dx}{\mu_h e p} \tag{222}$$

is the hole resistance per unit area of space-charge layer. Equation (221) has the form of (211) if we identify $J_h R_{hs}$ with ε in (210).

Before using this result, let us differentiate (101) to obtain

$$dp/dx|_N = -(p - p_{0N})/L_h.$$

At medium- or high-injection levels, $p_N \gg p_{0N}$, and this simplifies to

$$dp/dx|_N = -p_N/L_h. \tag{222a}$$

Using (222a) plus the corresponding relation for electrons

$$dn/dx|_N = n_N/L_e \tag{222b}$$

in (76) and (77), along with space-charge neutrality at the boundaries expressed as $dn/dx = dp/dx$, then eliminating the field E_N (the value of E at x_N) gives

$$J_h = eD_h p_N \left[\frac{[1 + (p_N/n_N)]/L_h + [1 + (n_P/p_P)](n_P/n_N)/L_e}{1 - p_N n_P/p_P n_N}\right] \tag{223}$$

with a similar expression for J_e. Equation (223) appears to indicate that J_h is independent of the applied field E; we emphasize that E was eliminated between (76) and (77) for convenience and that this expression in any case assumes high injection.

Returning to (222), van der Ziel points out that $p(x)$ decreases monotonically in going from $-x_P$ to x_N, so that

$$R_{hs} < \int_{-x_P}^{x_N} \frac{dx}{e\mu_e p_N} = \frac{w}{e\mu_h p_N}$$

and

$$J_h R_{hs} < (w/e\mu_h) J_h/p_N. \tag{224}$$

Using J_h/p_N from (223) gives

$$J_h R_{hs} < \frac{kT}{e} \left[\frac{(1 + p_N/n_N)w/L_p + (1 + n_P/p_P)(n_P/n_N)w/L_n}{1 - p_N n_P/p_P n_N} \right], \tag{225}$$

which we recognize as a valid approximation because the ratios w/L_p, w/L_n, p_N/n_N, and n_P/p_P are all small. Hence $\varepsilon(=J_h R_{hs})$ is also small and the assumption that $\bar{\mu}_e$ and $\bar{\mu}_h$ are constant and parallel appears to be verified.

We notice, however, that (225) is based on assumptions which may be expressed as

$$J_{eN} = J_{eP}, \qquad J_{hN} = J_{hP}, \tag{226}$$

whereas (209), (211), and (77) require that

$$J_{hP} = p_P \mu_h (d\bar{\mu}_h/dx)_P \tag{227}$$

and

$$J_{hN} = p_N \mu_h (d\bar{\mu}_h/dx)_N = d' J_{hP} \frac{(d\bar{\mu}_h/dx)_N}{(d\bar{\mu}_h/dx)_P}. \tag{228}$$

Without a priori knowledge of the slopes of $\bar{\mu}_h$ at the two boundaries, we have no justification for taking their ratio such that (228) will reduce to the second of (226).

An alternate approach (Nussbaum, 1975) is to generalize van der Ziel's procedure by writing (77) and (76) at x_N as

$$J_{hN} = e\mu_h p_N E_N - eD_h(dp/dx)_N, \tag{229}$$
$$J_{eN} = e\mu_e n_N E_N + eD_e(dn/dx)_N, \tag{230}$$

with a similar set at $-x_P$. Going through the same procedure as before gives

$$J_{hN} = \frac{\mu_h}{\bar{\mu}_e} \frac{p_N}{n_N} J_{eN} - eD_h \left(\frac{dp}{dx}\right)_N \left(1 + \frac{p_N}{n_N}\right) \tag{231}$$

and

$$J_{eP} = \frac{\mu_e}{\mu_h} \frac{n_P}{p_P} J_{hP} + eD_e \left(\frac{dn}{dx}\right)_P \left(1 + \frac{n_P}{p_P}\right). \tag{232}$$

Expressing the total current as

$$J = J_{eP} + J_{hP} = J_{eN} + J_{hN}, \tag{233}$$

we then obtain

$$J_{hN} = \frac{(\mu_h/\mu_e)(p_N/n_N)J - eD_h(dp/dx)_N[1 + (p_N/n_N)]}{1 + (\mu_h/\mu_e)p_N/n_N}, \quad (234)$$

$$J_{eP} = \frac{(\mu_e/\mu_h)(n_P/p_P)J + eD_e(dn/dx)_P[1 + (n_P/p_P)]}{1 + (\mu_e/\mu_h)n_P/p_P}. \quad (235)$$

Using (185) and (186) in modified form, the term p_N/n_N is

$$p_N/n_N = [(m_P + m_N d')/(m_N + m_P d')]d'. \quad (236)$$

For the silicon diode of Fig. 18, the diffusion potential $-eV_D$ has a magnitude of 27.6 in units of kT. If we assume a forward bias of 16 units and a value for ε of about 20% of the bias, we find that d' is approximately 4.6×10^{-7} and (236) gives $p_N/n_N = 4.5 \times 10^{-3}$. Letting μ_e be approximately equal to μ_h, Eq. (234) will then reduce to

$$\frac{J_{hN}}{p_N} = \frac{4.5 \times 10^{-3} J}{p_N} + \frac{eD_h}{L_h}. \quad (237)$$

Inserting this into (224), the inequality to be verified now becomes

$$J_{hN}R_{hs} < \left(\frac{4.5 \times 10^{-3} Jw}{e\mu_h p_N} + \frac{kT}{e}\frac{w}{L_h}\right). \quad (238)$$

The second term on the right is small, leaving

$$R_{hs} < \frac{4.5 \times 10^{-3} Jw}{e\mu_h p_N J_{hN}}. \quad (239)$$

Since the injected hole current which reaches the N side of the space-charge region should be a significant fraction of the total current, the right-hand side of (239) is like the right-hand side of (224), except for the numerical factor which now invalidates the inequality. On the other, a similar calculation applied to (235) shows that J_{eP}/n_P is small and van der Ziel's conclusion is correct.

This apparently contradictory result is in fact confirmed by an exact numerical calculation performed by Stone (1971), who obtained the behavior shown in Fig. 46 for a diode with the doping specified above. (We shall consider the details of this calculation in a later section.) We note that about 15% of the total applied potential appears in the form of a change in $\bar{\mu}_h$ in the transition region, and the rest goes into the bulk-region potential difference. It is thus apparent that the original Fletcher boundary conditions are not universally valid and must be modified to suit the device under study. This conclusion also underlines the fact that it is necessary to be fairly careful about justifying the approximations used in any specific calculation.

For example, Sah (1966) established the constant behavior of the electrochemical potentials by basing a numerical calculation on the depletion approximation; as we have already seen in connection with generation–recombination theory, this approach may lead to difficulties.

In this analysis of the boundary conditions, we have indicated the various difficulties that can arise because of restrictive conditions placed on the behavior of the electrochemical potentials and the failure to take into account majority-carrier injection. As already stated, Fletcher's original presentation incorporates conductivity modulation, so that he was effectively letting the electrochemical potentials have finite slopes in the bulk regions. However, although it was not explicitly stated, his use of (180) and (182), when combined with (181), imply that he was taking $\bar{\mu}_e$ and $\bar{\mu}_h$ to behave as shown in Fig. 26, in the space-charge region only. This assumption can be justified with the argument that this region is so thin that any change is relatively minor (a look ahead to Fig. 46 indicates the danger of this assumption). If we accept this argument, then the conductivity modulation removes the difficulty about satisfying space-charge neutrality, as brought out at the beginning of this section, but other difficulties still exist. For example, there is the question of how to define the actual junction potential $-eV_J$ and how it is related to the applied potential difference $-eV_A$ across the junction terminals and to the behavior of the majority- and minority-carrier electrochemical potentials in the bulk and space-charge regions. It has in fact been pointed out by van Vliet (1971) that the Fletcher conditions do not require that we identify $-eV_J$ as the separation of the electrochemical potentials; it suffices that they be constant and parallel across the space-charge region, and, in addition, $-eV_J$ must be specified as the difference between the height of the barrier under injection and at equilibrium. We should also recognize that the Fletcher conditions appear to have the same form at low and at high injection, but the latter still involves a neglect of majority-carrier injection, which leads to contradictions.

In this review of the work published on boundary conditions, we have concentrated on the extensions of Shockley's low-level treatment. Some of the earlier attempts to improve his approximations are due to Webster (1954), Hall (1954), Rittner (1954), Misawa (1958), Matz (1960), and others; this work has been briefly discussed by van Vliet (1966).

IV. The *PN* Injection under High Injection

7. Approximate and Semiempirical High-Injection Theories

Treatments of junction theory which have eliminated some or all of Shockley's assumptions are so numerous that it is not really possible to give any complete, coherent survey of this literature. As typical examples of papers that have been cited by subsequent authors, we may mention Graham and Hauser (1972), Armstrong et al. (1956), Lieb et al. (1962) and Herlet (1968). There is also an extensive Soviet literature, available in English in Soviet Physics—Semiconductors, Soviet Physics—Technical Physics, and in Radio Engineering and Electronic Physics.

The fact that we cannot readily summarize all of this material is of no special consequence, for there is a way around this difficulty. As we shall discuss shortly, the steady-state behavior of the electrons and holes at any point in a one-dimensional PN-junction device is governed by a set of simultaneous, nonlinear differential equations which have already been given. In particular, there are two generalized Ohm's laws in the form

$$J_e = \mu_e kT\, dn/dx - ne\mu_e\, dV/dx, \tag{76'}$$

$$J_h = -\mu_h kT\, dp/dx - pe\mu_h\, dV/dx, \tag{77'}$$

two continuity equations

$$(1/e)\, dJ_h/dx + R = 0, \tag{11'}$$

$$-(1/e)\, dJ_e/dx + R = 0, \tag{12'}$$

and Poisson's equation

$$d^2\psi/dr^2 = [(n - p)/2n_i] - a, \tag{110'}$$

where ψ is defined by (109) and R is given by (142). This group of five simultaneous nonlinear equations in the unknown quantities n, p, J_e, J_h, and V comprise what is known as the van Roosbroeck (1950) model. [See also their development as Eqs. (9)–(14), (74)–(82), (110), and (135)–(142).] Some specific numerical procedures will be presented, but first we shall consider diodes operating at high levels in a general, intuitive way.

Solving (76) and (77) for E and equating the two expressions gives

$$E = \frac{J_h + eD_h\, dp/dx}{pe\mu_h}$$

$$= \frac{J_e - eD_e\, dn/dx}{ne\mu_e}. \tag{240}$$

By conservation of charge

$$p - n + N_D - N_A = 0 \tag{241}$$

and differentiating

$$dn/dx = dp/dx.$$

Putting this into (240) and solving for dp/dx gives

$$dp/dx = (J_e p - J_h nc)/ecD_h(n + p), \tag{242}$$

where

$$c = \mu_e/\mu_h \qquad (243)$$

is the mobility ratio. For an asymmetric junction like those previously considered, the injected hole current into the N region will be considerably larger than the electron current going the other way, and we can assume that $J_e \ll J_h$. Further, the number of free electrons n or free holes p is larger than the number of ionized donors, so that (241) simplifies to

$$n \sim p \qquad (243)$$

and (242), with these approximations, becomes

$$J_h = -2eD_h\,dp/dx, \qquad (244)$$

which indicates that the high-injection current should be exactly twice the low-injection value of (93); that is, the sum of the drift and diffusion current is equal to twice the low-level diffusion current alone. To express this conclusion in another way, at low levels, only the diffusion current is significant, but at high levels the drift contribution increases, with a maximum equal to the diffusion current.

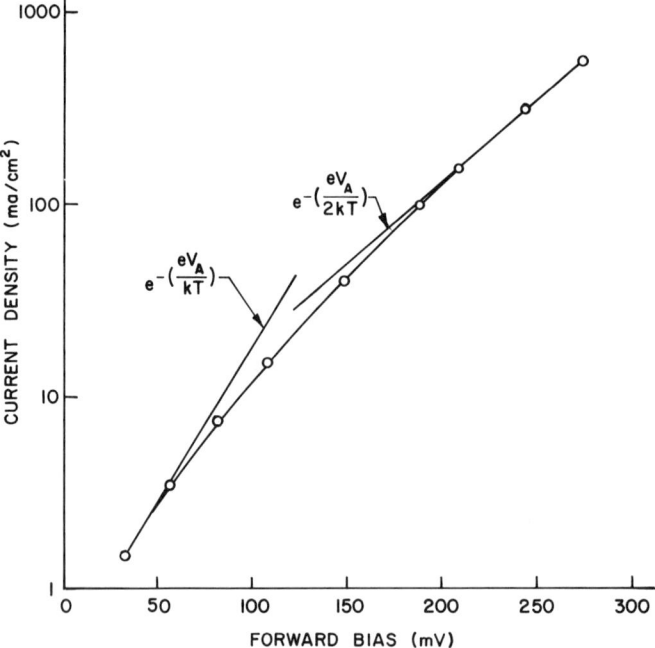

FIG. 30. The current–voltage characteristic for a germanium alloyed diode.

This result also affects the boundary conditions, for (243) in (98) is then
$$p = n = n_i \exp(-eV_A/2kT), \tag{245}$$
which leads to a characteristic of the form [cf. Eq. (107)]
$$J = J_s[\exp(-eV_A/2kT) - 1]. \tag{246}$$
This change in the slope of the J–V curve has been observed in germanium diodes; Fig. 30, taken from Nussbaum (1962), is an early alloyed device. This change also shows up in silicon diodes, but is obscured by other contributions (Fig. 22).

A great deal of work has been done in an attempt to justify approximations that would make the van Roosbroeck model more tractable, while still retaining the necessary rigor. Because the numerical approach we shall discuss in the following section simply bypasses this problem, we shall consider only a single example of this type of analytic approach. The one we have chosen, that of Barber (1969), is interesting for two reasons; it brings in some important concepts which are used in high-injection–level theory and it shows remarkable agreement with experiment for one particular class of diodes (see Section 8). We start with the steady-state form of the continuity equations (11) and (12), and combine them with the definitions (76) and (77) to obtain
$$e\mu_h(p\,dE/dx + E\,dp/dx) - eD_h\,d^2p/dx^2 = -eR, \tag{247}$$
$$e\mu_e(n\,dE/dx + E\,dn/dx) + eD_e\,d^2n/dx^2 = eR. \tag{248}$$
Space-charge neutrality implies that
$$dn/dx = dp/dx, \quad d^2n/dx^2 = d^2p/dx^2. \tag{249}$$
Multiply (247) by $n\mu_e$, (248) by $p\mu_h$, and subtract the first equation from the second to obtain, with the use of (249),
$$e(n\mu_e D_h + p\mu_h D_e)\,d^2p/dx^2 + \mu_h\mu_e E(n - p)\,dp/dx = eR(n\mu_e + p\mu_h). \tag{250}$$
This equation may be written as
$$eD^*\,d^2p/dx^2 - \mu^* E\,dp/dx = eR, \tag{251}$$
where the ambipolar diffusivity D^* is defined by
$$\begin{aligned}D^* &= (n\mu_e D_h + p\mu_h D_e)/(n\mu_e + p\mu_h) \\ &= (n + p)D_e D_h/(nD_e + pD_h)\end{aligned} \tag{252}$$
and μ^* is the ambipolar mobility whose definition is
$$\mu^* = \mu_e\mu_h(n - p)/(n\mu_e + p\mu_h). \tag{253}$$
These definitions indicate that D^* reduces to D_p and μ^* to μ_p when $n \gg p$, with corresponding relations for electronic quantities.

We note that these definitions lead to
$$\mu^*/D^* = (e/kT)(n-p)/(n+p), \quad (254)$$
which may be regarded as a generalization of the Einstein relation (79); this, in turn, gives an alternate form of (247), namely
$$d^2p/dx^2 - [(e/kT)(n-p)E/(n+p)]\,dp/dx + (dJ_h/dx)/eD^* = 0. \quad (255)$$
Define a hole injection coefficient Γ as
$$\Gamma = p/(n-p) \quad (256)$$
so that
$$2\Gamma + 1 = (n+p)/(n-p),$$
and by (254),
$$D^* = (2\Gamma + 1)D_e D_h/[(\Gamma + 1)D_e + \Gamma D_h]. \quad (257)$$
The physical significance of Γ can be obtained by considering, for example, the injection of holes into the N region of a diode. For low levels, (256) may be written as
$$\Gamma = (p_{0N} + \Delta p)/(n_{0N} + \Delta n - p_{0N} - \Delta p) \sim \Delta p/n_{0N} \quad (258)$$
so that Γ is small and, in fact, goes to zero in the limit of no injection. Intermediate injection is defined by the condition $p = n/2$, so that $\Gamma = 1$ and high injection corresponds to $p = n$, making Γ infinite.

For the same type of situation, the Hall–Shockley–Read recombination rate equation (142) using the conditions $\Delta n = \Delta p$, $n_0^* = p_0^* = n_i$, $n_{0N} > n_i$, and $n_{0N} \gg p_{0N}$ gives
$$R = \frac{\Delta p(n_{0N} + \Delta p)}{(n_{0N} + \Delta p)\tau_{h0} + \Delta p\, \tau_{e0}}$$
or by (258),
$$R = \frac{\Delta p(1 + \Gamma)}{(1 + \Gamma)\tau_{h0} + \Gamma \tau_{e0}}. \quad (259)$$
This may be written as
$$R = \Delta p/\tau^*,$$
where the injection-dependent lifetime τ^* is
$$\tau^* = [(1 + \Gamma)\tau_{h0} + \Gamma \tau_{e0}]/(1 + \Gamma). \quad (260)$$
This concept can be incorporated into (251), which, assuming the term in E is small, becomes
$$D^* d^2p/dx^2 = R = (p - p_{0N})/\tau^*$$

2. THE THEORY OF SEMICONDUCTING JUNCTIONS

or

$$d^2p/dx^2 = (p - p_{0N})/L^{*2}, \qquad (261)$$

where the injection-dependent diffusion length is

$$L^* = \sqrt{D^*\tau^*}. \qquad (262)$$

Since both D^* and τ^* are bilinear functions of Γ, it is reasonable to assume that L^* is independent of Δp, so that (261) can be integrated as if it were the Shockley low-level equation with the boundary conditions specified by (98) at $x = 0$ and (100) at $x = \infty$. The solution (101) is written as

$$p_N - p_{0N} = p_{0N}[\exp(-eV_A/kT) - 1]\exp(-x/L^*). \qquad (263)$$

When Γ is small, both D^* and τ^* reduce to their customary low-level values, and (263) leads to (105); we designate this low-injection current as

$$J_{hN}^{(L)} = (eD_h p_{0N}/L_h)[\exp(-eV_A/kT) - 1]. \qquad (264)$$

For the high-injection situation, we use (245) in the form

$$p = n_i \exp(-eV_A/2kT) = p_{0N} + \Delta p$$

or to reasonable accuracy

$$\Delta p = n_i \exp(-eV_A/2kT). \qquad (265)$$

This boundary condition at $x = 0$, when combined with the appropriate form of (93), gives a high-level current

$$\begin{aligned} J_{hN}^{(H)} &= -eD^*(dp/dx)_{x=0} \\ &= n_i e[2D_e D_h/(D_e + D_h)(\tau_{e0} + \tau_{h0})]^{1/2} \exp(-eV_A/2kT), \end{aligned} \qquad (266)$$

which identifies J_S in (246).

Returning to (255), we incorporate (240), (256), and (79) to obtain

$$\frac{dJ_h}{dx} = \frac{eD^*}{1 + 2\Gamma}\left(\frac{J_h + eD_h\, dp_N/dx}{ep_N D_h}\right)\frac{dp_N}{dx} - eD^*\frac{d^2 p_N}{dx^2}. \qquad (267)$$

By (263),

$$dp/dx = d\,\Delta p/dx = -\Delta p/L^* \qquad (268)$$

and

$$d^2 p/dx^2 = \Delta p/L^{*2}. \qquad (269)$$

We also make use of the fact that for high injection

$$J_h = -eD^*\, dp/dx = eD^*\,\Delta p/L^* \qquad (270)$$

and
$$\Delta p = p_N - p_{0N} \sim p_N \tag{271}$$

to convert (267) into the relation

$$\frac{dJ_h}{dx} = -\frac{e(p_N - p_{0N})}{\tau^*}\left[1 + \frac{D^* - D_h}{D_h(1 + 2\Gamma)}\right]. \tag{272}$$

This is like (90), with the net recombination term having the form of (91). However, not only has τ_h been generalized to τ^*, but the second term in brackets alters the value of R. Since D^* reduces to D_h at low-injection levels and since Γ is infinite at high levels, this term is of significance only in the intermediate range, where it was shown by Barber to represent a correction of the order of 20% for germanium, silicon, and indium antimonide diodes. However, rather than use this result directly, the hole contribution to the total current was regarded as due to injection at two distinct junctions: the boundary between the P^+ and the space-charge regions will always operate at low levels because of the magnitude of the potential that can be sustained there, while the other edge of the space-charge region can operate at high levels. The conductances of these two effective junctions are in series so that the currents combine in accordance with the relation

$$J_{hN} = J_{hN}^{(L)} J_{hN}^{(H)} / (J_{hN}^{(L)} + J_{hN}^{(H)}), \tag{273}$$

where $J_{hN}^{(L)}$ is given by (264) and $J_{hN}^{(H)}$ by (266). Since (273) is an explicit expression in terms of material parameters, whereas (272) requires integration, it is easier to use the latter. It was found, however, that the behavior of (273) in the intermediate range corresponded quite closely to the predictions of (272).

The total current in the diode can now be expressed as

$$J = J_{hN} + J_{eP} + J_R, \tag{274}$$

where the injected low-level electron current J_{eP} is

$$J_{eP} = (eD_e n_{0P}/L_e)[\exp(-eV_A/kT) - 1] \tag{275}$$

and the generation–recombination contribution J_R was taken as a generalization of (164). The complete expression comes by adding (275), (164) as modified by (166), and (273) to obtain the equation given by Barber (1969)

$$J = J_{e0}^{(L)}[\exp(-\psi_A) - 1] + J_{R0}\frac{\exp(-\psi_A/2)}{(\psi_A - \psi_0)^{1/2}} + \frac{J_{h0}^{(L)} J_{h0}^{(H)} \exp(-\psi_A/2)[\exp(-\psi_A) - 1]}{J_{h0}^{(L)}[\exp(-\psi_A) - 1] J_{h0}^{(H)} \exp(-\psi_A/2)}, \tag{276}$$

where

$$\psi_A = eV_A/kT,$$
$$J_{e0}^{(L)} = eD_e n_{0P}/L_e,$$
$$J_{h0}^{(L)} = eD_h p_{0N}/L_h,$$
$$J_{h0}^{(H)} = n_i e \left(\frac{2D_e D_h}{(D_e + D_h)(\tau_{e0} + \tau_{h0})}\right)^{1/2},$$
$$J_{R0} = \pi n_i kT (2\varepsilon/e\tau_{e0}\tau_{h0})^{1/2}[(1/N_A) + (1/N_D)]^{1/2}.$$

Equation (276), with some of the constants used as adjustable parameters, gives the solid curves of Figs. 20 and 22. The experimental points show the remarkable ability of this semiempirical theory to explain the behavior of these two diodes. (The flattening of the characteristic at the very highest currents is due to the inclusion of the bulk ohmic drop in the lightly doped N region.) Although this agreement is really very good, it should be noted that Eq. (276) is a combination of several different kinds of dependence on V_A. We shall see in the next section that this type of fit does not occur for diodes with medium and long N regions; the results shown are possible only for short diodes.

8. The Numerical Analysis of the van Roosbroeck Model

a. Background

Quite a few schemes have been proposed for treating diodes in a numerical fashion, each with its own advantages and disadvantages. We may cite those of van der Maesen *et al.* (1962), Fulkerson and Nussbaum (1966), Sanchez (1967a,b,c,d, 1968), de Mari (1968), Gummel (1964), Arandjelovic (1970), Vandorpe and Xuong (1971), Choo (1971b, 1972), and Graham and Hauser (1970). Some of these are based on the approach used by de Mari (1968), who extended the iterative method of Gummel (1964). The de Mari treatment uses a trial solution for the electrostatic potential to estimate the carrier densities. These carrier densities are then used in a linearized form of Poisson's equation to correct the trial potential. The new trial potential is used to recalculate the carrier densities and the procedure is repeated until the corrections to the potential become small. This method has the advantages that the device is solved as a unit without applying boundary conditions at internal interfaces and it is probably not as sensitive to changes in the initial conditions as an iterated numerical integration for a boundary value problem would be. There are several drawbacks to this method, however. It requires a large amount of computer memory to store the data pertaining to each discrete point within the device, especially for a long device; the step size must be chosen in advance and is quite difficult to change once it is chosen; and the solution requires the inversion of a tridiagonal matrix, which may require considerable computer time. In addition, if recombination is not neglected it is necessary to perform a separate numerical integration for each point in the diode at which the solution is evaluated.

Other methods, such as from Sanchez (1967a,b,c,d, 1968), use direct numerical integration of the equations. The primary advantages of the direct integration methods are that there is no need to store the entire point-by-point solution in the computer, the solution does not require the manipulation of large arrays of data, and the step size of the numerical integration can be conveniently changed at any point in the integration if accuracy considerations should require it. Direct numerical integration, however, has one serious disadvantage; it leads to a two-point boundary value problem. This difficulty is overcome by estimating the values of the unknown initial conditions at one boundary, integrating across the diode to the other boundary, comparing the calculated values at this boundary with the known boundary conditions there, and then modifying the estimates of the unknown initial conditions. The integration is then repeated, the comparison is made, and the process is repeated until the calculated values agree with the known boundary values.

b. Description of Stone's Method

The solution which we shall describe in detail is that of Stone (1971). This differs from those described above in that the diode is divided into three parts: a heavily doped P^+ region, a space-charge layer, and a moderately doped N region. This division—previously used by Fulkerson and Nussbaum (1966) for germanium junctions—is rather arbitrary as is the specification of boundaries, but this viewpoint permits some very useful simplifications which will be explained and justified. The consequence is that it turns out that it is reasonably easy to observe the effect of changes in crucial parameters, such as lifetime and impurity concentrations. Hence, a series of diodes was studied and compared.

For purposes of calculation, it is convenient to have the differential equation system in dimensionless form. Hence, the van Roosbroeck (1950) model, whose full specification requires Eqs. (11), (12), (76), (77), (109), (110), and (142), is restated as

$$dn'/dr = J'_e(D_{h0}/D_e) - (n' + n'_0)(E'_0 + E') + n'_0 E'_0, \tag{277a}$$

$$dp'/dr = (J'_e - J')(D_{h0}/D_e) + (p' + p'_0)(E'_0 + E') - p'_0 E'_0, \tag{277b}$$

$$dJ'_e/dr = \frac{(L_D/L_h)^2[(p'_0 + p')n' + n'_0 p']}{n'_0 + n' + n_0^{*'} + \tau(p'_0 + p' + p_0^{*'})}, \tag{277c}$$

$$dE'/dr = (p' - n')/2, \tag{277d}$$

$$dV'/dr = -E', \tag{277e}$$

2. THE THEORY OF SEMICONDUCTING JUNCTIONS

TABLE I
Definition of Dimensionless Symbols

Name	Definition
Excess electric field	$E' = e(E - E_0)L_D/kT$
	$= -d(\psi - \psi_0)/dr$
Equilibrium electric field	$E'_0 = eE_0 L_D/kT = -d\psi_0/dr$
Current density	$J' = JL_D/eD_{h0}n_i$
Electron current density	$J'_e = J_e L_D/eD_{h0}n_i$
Excess electron concentration	$n' = (n - n_0)/n_i = \Delta n/n_i$
Equilibrium electron concentration	$n'_0 = n_0/n_i = e^{\psi_0}$
Electron recombination probability	$n_0^{*\prime} = n_0^*/n_i$
Excess hole concentration	$p' = (p - p_0)/n_i = \Delta p/n_i$
Equilibrium hole concentration	$p'_0 = p_0/n_i = e^{-\psi_0}$
Hole recombination probability	$p_0^{*\prime} = p_0^*/n_i$
Distance	$r = x/L_D$
Reduced potential	$\psi = eV/kT$
Lifetime ratio	$\tau = \tau_{e0}/\tau_{h0}$
Excess potential	$V' = \psi - \psi_0$
Debye length	$L_D = (\varepsilon kT/2n_i e^2)^{1/2}$
Reduced barrier potential	ψ_D

where the symbols are defined in Table I. Equations (277a) and (277b) are Ohm's law for electrons and holes, respectively, Eq. (277c) is the continuity relation for electrons—the corresponding relation for holes is not needed, since we can eliminate J_h with the condition

$$J_e = J - J_h, \qquad (278)$$

where J is the total current in the junction—incorporating Eq. (142) (the HSR expression) and (277d) is Poisson's equation expressed in terms of an excess electric field intensity E'. The diffusion constants enter as ratios so that the effects of field and doping variation on these parameters can be studied; the normalization was specified in terms of a constant value D_{h0}, whereas D_e and D_h are variable. We also note that the Poisson equation (277d) does not contain N_D and N_A because E', n', and p' represent excess values and the equilibrium terms drop out.

The complete model thus involves five first-order, simultaneous differential equations in n', p', J'_e, E', and V'. The boundary conditions are specified by requiring the excess carrier concentrations at the contacts to vanish (this is known as the infinite recombination-rate condition). Using the geometry of Fig. 31, this means that

$$\begin{aligned} p'(-W_P) = n'(-W_P) = 0, \\ p'(W_N) = n'(W_N) = 0. \end{aligned} \qquad (279)$$

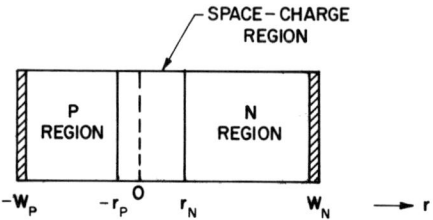

FIG. 31. Abrupt diode geometry.

The remaining boundary condition will involve choosing the reference for the potential V' at one of the contacts. In order to solve these equations, we must also know the behavior of the equilibrium quantities E'_0, V'_0, n'_0, and p'_0. As discussed in Section 2, the first integral of the Poisson–Boltzmann equation, by (41), is

$$\frac{d\psi_0}{dr} = \begin{cases} \{2[\cosh \psi_0 - \cosh u_{0P} + a_P(u_{0P} - \psi_0)]\}^{1/2}, & r \leq 0 \\ \{2[\cosh \psi_0 - \cosh u_{0N} + a_N(u_{0N} - \psi_0)]\}^{1/2}, & r \geq 0 \end{cases} \quad (280)$$

subject to the boundary condition (46) at $r = 0$. We thus have six equations and six boundary conditions, specified at three different places. To restore this to a two-point boundary value problem of the form of (277), we use the rule

$$df/dr = (df/d\psi_0) d\psi_0/dr \quad (281)$$

and divide the set (277) by $d\psi_0/dr = -E_0$ to obtain

$$\frac{dn'}{d\psi_0} = \frac{(D_{h0}/D_e)J'_e - (n' + n'_0)E'}{-E'_0} + n', \quad (282a)$$

$$\frac{dp'}{d\psi_0} = \frac{(D_{h0}/D_h)(J'_e - J') + (p' + p'_0)E'}{-E_0} - p', \quad (282b)$$

$$\frac{dJ'_e}{d\psi_0} = \frac{(L_D/L_h)^2[(p'_0 + p')n' + n'_0 p']}{-E'_0[(n'_0 + n_0^{*'} + n') + \tau(p'_0 + p_0^{*'} + p')]}, \quad (282c)$$

$$\frac{dE'}{d\psi_0} = \frac{p' - n'}{-2E'_0}, \quad (282d)$$

$$\frac{dV'}{d\psi_0} = \frac{E'}{E'_0}. \quad (282e)$$

The arbitrary division of the diode into three separate regions is now explicitly introduced, and each region will be analyzed separately. Starting

2. THE THEORY OF SEMICONDUCTING JUNCTIONS

with the N region, the neutrality condition $\Delta n = \Delta p$ previously used with the Fletcher boundary conditions is invoked here, so that $n' = p'$ in the set (282), and furthermore

$$dn'/dr = dp'/dr. \tag{283}$$

If we multiply Eq. (277a) by $(p' + p'_0)$ and use (277b) to find $(p' + p'_0)(E' + E'_0)$, we may eliminate p' in favor of n'. Similar manipulation with the remaining equations in (277) reduces them to

$$\frac{dp'}{dr} = \frac{(D_{h0}/D_e)(p'_{0N} + p')J'_e + (J'_e - J')(n'_{0N} + p')(D_{h0}/D_h)}{(p'_{0N} + n'_{0N} + 2p')}, \tag{284a}$$

$$\frac{dJ'_e}{dr} = \frac{p'(p'_{0N} + n'_{0N} + p')(L_D/L_h)^2}{n'_{0N} + n^{*'}_0 + p' + \tau(p'_{0N} + p^{*'}_0 + p')}, \tag{284b}$$

$$\frac{dV'}{dr} = -E' = \frac{(D_{h0}/D_h)(J'_e - J') - J'_e(D_{h0}/D_e)}{(p'_{0N} + n'_{0N} + 2p')}. \tag{284c}$$

In obtaining these equations, the fact that E_0 is zero in the bulk regions was used. (As discussed in Section 2, we use this criterion to establish the edges of the space-charge region for the equilibrium diode. However, the actual field E was not taken to vanish, as is done in the usual low-level treatment.)

The P-region equations can be simplified in exactly the same way and yield the same set, with the appropriate values of the constants n'_0, p'_0, D_e, and D_h. In addition, n' is replaced by p' for convenience to obtain

$$\frac{dn'}{dr} = \frac{(D_{h0}/D_e)(p'_{0P} + n')J'_e + (J'_e - J')(n'_{0P} + n')(D_{h0}/D_h)}{(p'_{0P} + n'_{0P} + 2n')}, \tag{285a}$$

$$\frac{dJ'_e}{dr} = \frac{n'(p'_{0P} + n'_{0P} + n')(L_D/L_h)^2}{n'_{0P} + n^{*'}_0 + n' + \tau(p'_{0P} + p^{*'}_0 + n')}, \tag{285b}$$

$$\frac{dV'}{dr} = -E' = \frac{(D_{h0}/D_h)(J'_e - J') - J'_e(D_{h0}/D_e)}{(p'_{0P} + n'_{0P} + 2n')}. \tag{285c}$$

For most of the diodes investigated the P region was much more heavily doped than the N region. This implies that even when the N region is at high injection ($p' \gg n'_{0N}$ at the N edge of the space-charge region), the P region will still be at low injection ($n' \ll p'_{0P}$) as discussed in the preceding section. It is useful, therefore, to develop an approximate analytic solution for the P region, which can be used as a first approximation to find the boundary conditions at the interface. Low-level injection in the P region implies two conditions:

(1) The excess carrier densities at $x = -x_P$ are much less than the equilibrium hole concentration.

(2) The electric field term can be neglected when considering minority-carrier current flow.

Applying these two conditions together with space-charge neutrality to Eqs. (277a) and (277c) gives

$$dn'/dr = J'_e(D_{h0}/D_e), \qquad (286a)$$
$$dJ'_e/dr = (L_D/L_h)^2 n'/\tau. \qquad (286b)$$

Differentiating the first equation and substituting to eliminate dJ'_e/dr gives

$$d^2n'/dr^2 = (L_D/L_e)^2 n'. \qquad (287)$$

Setting $n' = 0$ at $r = -W_P$, the solution of Eq. (287) is found to be

$$n'(r) = \frac{n'(-r_P)\sinh[(L_D/L_e)(W_P + r)]}{\sinh[(L_D/L_e)(W_P - r_P)]} \qquad (288a)$$

so that

$$J'_e(r) = \frac{n'(-r_P)(L_D/L_e)(D_e/D_{h0})\cosh[(L_D/L_e)(W_P + r)]}{\sinh[(L_D/L_e)(W_P - r_P)]}. \qquad (288b)$$

Then $E'(r)$ can be calculated from Eq. (285c) to be

$$E'(r) = \frac{J'_e(D_{h0}/D_e) - (J'_e - J')(D_{h0}/D_h)}{(p'_{0P} + n'_{0P} + 2n')}. \qquad (288c)$$

We have now placed in normalized form three sets of differential equations as follows:

(a) Eqs. (282) for the transition region,
(b) Eqs. (284) for the N region,
(c) Eqs. (285) for the P region.

To solve them for a particular device, we specify the physical dimensions and electrochemical properties (doping, minority-carrier lifetime, diffusion constants, etc.). The normalized diode current density J' is then chosen and an initial estimate is made of the electron current density $J'_e(-r_P)$ at the interface. The P-region equations are then solved to determine values of $n(-r_P)$ and $E(-r_P)$ to use as initial conditions for the transition-region integration. Depending on the current density and the relative doping of the P region, either the numerical solution from Eqs. (285) or the analytic solution from Eqs. (288) may be used. (For most of the diodes studied, the P region is much more heavily doped than the N region and thus the analytic and numerical solutions are practically identical, except at very high forward currents.)

Using the initial estimate of $J'_e(-r_P)$, the values of $n'(-r_P)$ and $E'(-r_P)$ from the P-region solution, and letting $V'(-r_P)$ be zero (since the reference for a potential is arbitrary), the transition-region solution is found by adjusting the value of $p(-r_P)$ until the integration converges and makes $p'(r_N) \sim n'(r_N)$, since neutrality requires that $p' - n'$ at both edges be small. Equations (282) are integrated from $u_{0P} + \delta_1$ to $u_{0N} - \delta_2$ where δ_1 and δ_2 are small positive numbers, and where u_{0N} and u_{0P} are the relative chemical potentials. As δ_1 and δ_2 are made smaller, $[p'(-r_P) - n'(-r_P)]$ and $[p'(r_N) - n'(r_N)]$, respectively, become smaller. [Note, however, that δ_1 and δ_2 can never be zero since $E'(u_{0N}) = E'(u_{0P}) = 0$ would cause a singularity in the transition-region equations.] The value of $J'_e(r_N)$ obtained from this solution can be used as an initial condition for the N region, and Eqs. (284) are solved to find a value of $p'(r_N)$ which will give $p'(W_N) = 0$ at the N-region contact. The value of $p'(r_N)$ obtained from the N-region solution is then compared to $p'(r_N)$ obtained from the transition-region solution. If these two values are not the same, the initial value chosen for $J'_e(-r_P)$ is modified and the entire solution is repeated until the two values obtained for $p'(r_N)$ are sufficiently close, as specified by the condition

$$|p'(r_N)_{SC} - p'(r_N)_N|/|p'(r_N)_N| \leq 10^{-2},$$

where the subscript SC refers to the value of p' obtained from the space-charge region solution. Now the boundary conditions are all matched and the three solutions are redone using these corrected values.

One drawback of this solution method is that the electric field at the N-region interface is not matched, and no attempt was made to compare the two values of $E'(r_N)$. In general, they can be quite different, leading to some inaccuracy near the interface at r_N. The value of $E'(r_N)$ can be adjusted by changing δ_2; generally $E(r_N)$ decreases as δ_2 is decreased. It was found that a mismatch of $E'(r_N)$ does not seriously affect the other variables; that is, the carrier concentrations and electron and hole current densities exhibit only small changes (less than one percent), even for large changes in δ_2 and $E'(r_N)$.

c. Specification of Diode Parameters

The procedure just described was applied to the thirteen diodes of Table II. As indicated, an attempt was made to study the effect of varying the key parameters and, in addition, a few cases were arranged to match commercially available diodes, for which the $J-V$ curves could actually be measured. All of the diodes except D11 and D12 were solved using both the numerical method and an approximate method known as the recombination–diffusion

TABLE II
DIODE PARAMETERS

Diode	u_{0N}	u_{0P}	Doping N_D (cm^{-3})	N_A (cm^{-3})	Length	Trap position	Lifetime μsec τ_{e0}	τ_{h0}	Description
D1	−18.4	9.2	1.5×10^{14}	1.5×10^{18}	L[d]	I[e]	1.1	0.4	"Standard" diode[f]
D2	−18.4	9.2	1.5×10^{14}	1.5×10^{18}	L	I	1.1	11.0	Long lifetime
D3	−18.4	9.2	1.5×10^{14}	1.5×10^{18}	L	P	1.1	0.4	P-type traps
D4	−18.4	9.2	1.5×10^{14}	1.5×10^{18}	L	N	1.1	0.4	N-type traps
D5[a]	−18.4	9.2	1.5×10^{14}	1.5×10^{18}	L	I	1.1	0.4	Linear field
D6	−18.4	9.2	1.5×10^{14}	1.5×10^{18}	M	I	1.1	0.4	Medium length
D7	−18.4	9.2	1.5×10^{14}	1.5×10^{18}	S	I	1.1	0.4	Short
D8	−18.7	14.1	2.0×10^{16}	2.0×10^{18}	S	I	1.1	0.4	Short, heavily doped
D9	−18.4	9.2	1.5×10^{14}	1.5×10^{18}	S	I	0.011	0.004	Very low lifetime

D10	−18.4	9.2	1.5×10^{14}	2.0×10^{18}	S	I	0.11	0.04	Low lifetime
D11[b]	−18.7	14.1	2.0×10^{16}	2.0×10^{18}	S	I	0.011	0.004	Very low lifetime, heavily doped
D12[b]	−9.2	18.4	1.5×10^{18}	1.5×10^{14}	S	I	0.11	0.04	Reversed doping
D13[c]	−18.4	9.2	1.5×10^{14}	1.5×10^{18}	L	I	1.1	0.4	Field-dependent diffusion coefficients

[a] Linear field diode.
[b] Diffusion-recombination solution only.
[c] Field-dependent diffusion coefficients.

[d] Bulk-region length (μm)

	N	P
Short (S)	22.36	2.504
Medium (M)	22.36	22.36
Long (L)	223.6	44.72

[e] I: $n_0^* = p_0^* = n_i$;
N: $n_0^* = 10^2 n_i$;
P: $p_0^* = 10^2 n_i$.
[f] "Standard" in the sense that it is used as a reference to which all other diodes are compared.

solution. The recombination–diffusion solution consists of adding the results of the simple diode equation to a simplified numerical integration of the recombination integral across the transition region, and will be discussed in more detail shortly. Only this solution was considered for D11 and D12.

All of the diodes except D5 are abrupt; this one was assumed to have an equilibrium field which varies linearly with distance (Fig. 32) in the transition region. The doping profile needed to produce such a field can be calculated from Poisson's equation. The equilibrium electrostatic potential may be expressed as

$$\psi_0(r) = \psi_0(-A) - \int_{-A}^{r} E'_0(r')\,dr', \qquad (289)$$

where r' is a dummy variable and the reduced variables of Table I are used. As the figure shows,

$$E'_0 = \begin{cases} -E'_{min}\left(\dfrac{r}{A} + 1\right), & -A \leq r \leq 0 \\[2mm] E'_{min}\left(\dfrac{r}{B} - 1\right), & 0 \leq r \leq B. \end{cases} \qquad (290)$$

Substituting and integrating gives

$$\psi_0(r) = \begin{cases} u_{OP} + [2\psi_D/(A+B)][(r^2/2A) + r + (A/2)], \\ u_{OP} + [A\psi_D/(A+B)] - [2\psi_D/(A+B)][(r^2/2B) - r], \end{cases} \qquad (291)$$

where the diffusion barrier is

$$\psi_D = \psi_0(B) - \psi_0(-A)$$

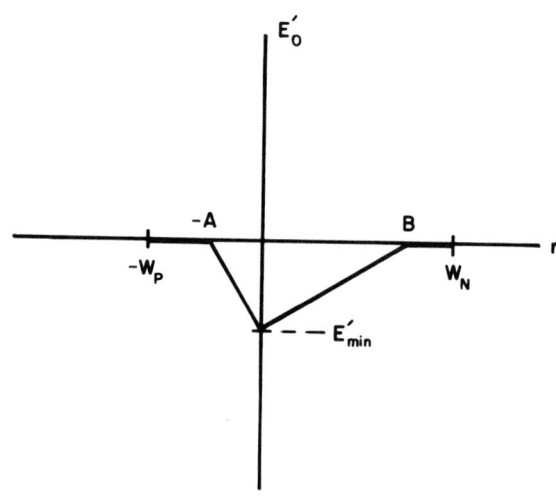

Fig. 32. Linear-field diode geometry.

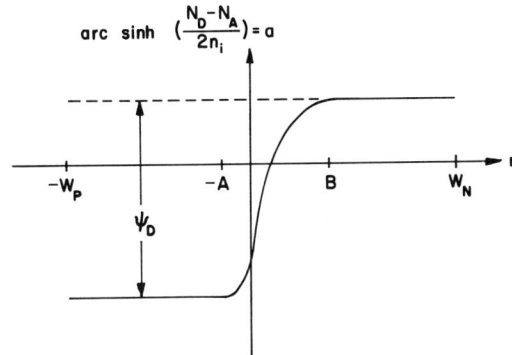

FIG. 33. Doping profile which leads to a linear field.

so that
$$E'_{min} = -2\psi_D/(A + B). \tag{292}$$

By (291),
$$\psi_0(0) = [A\psi_D/(A + B)] + u_{OP}. \tag{293}$$

Putting $\psi_0(r)$ from (291) into Poisson's equation (34), written as
$$d^2\psi_0/dr^2 = \sinh \psi_0 - a(r), \tag{294}$$

where a now depends on distance and—for the purpose of simplifying this specific procedure—we have shifted the energy reference level to the position of E_F, we obtain

$$a(r) = \begin{cases} \sinh\{u_{OP} + [2\psi_D/(A + B)][(r^2/2A) + r + A/2]\} \\ \qquad - 2\psi_D/A(A + B), \quad -A \leq r \leq 0 \\ \sinh\{u_{OP} + [A\psi_D/(A + B)] - [2\psi_D/(A + B)][(r^2/2B) - r]\} \\ \qquad + 2\psi_D/B(A + B), \quad 0 \leq r \leq B \end{cases} \tag{295}$$

and this is plotted in Fig. 33. It is also possible to find E_0 as a function of ψ_D—this is needed for the integration—but we shall omit the details here. Again we note that Poisson's equation is rather difficult to handle analytically; a linear field corresponds to a doping profile which is a hyperbolic function of a quadratic.

d. Effect of Parameters on Electrical Behavior

We now consider some of the correlations between junction parameters and the forward I–V characteristics. The length of the diode is the variable in Fig. 34, whereas Figs. 35 and 36 show the effect of lifetime changes on two different groups. The major effect of increasing the bulk-region length of a diode is to cause the I–V characteristic to rise less rapidly at higher

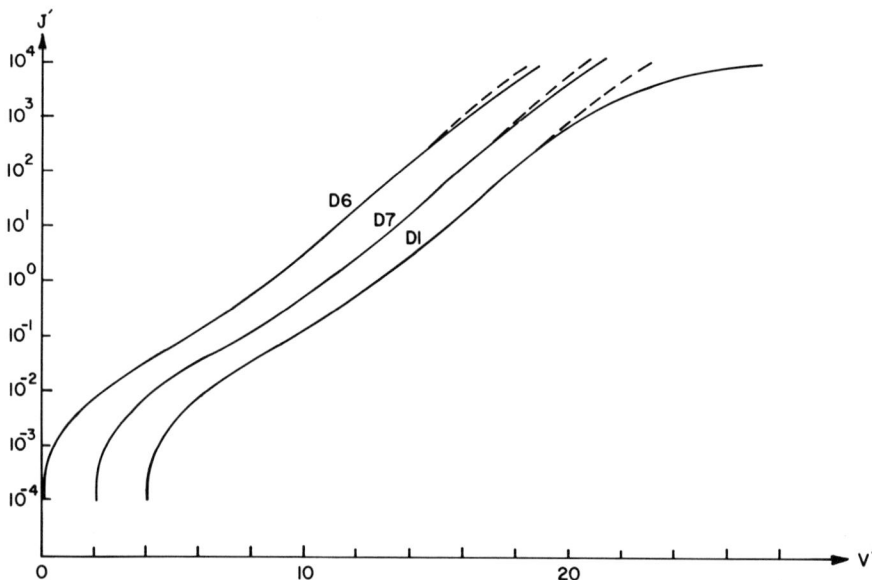

FIG. 34. Effect of bulk-region length on diode characteristics.

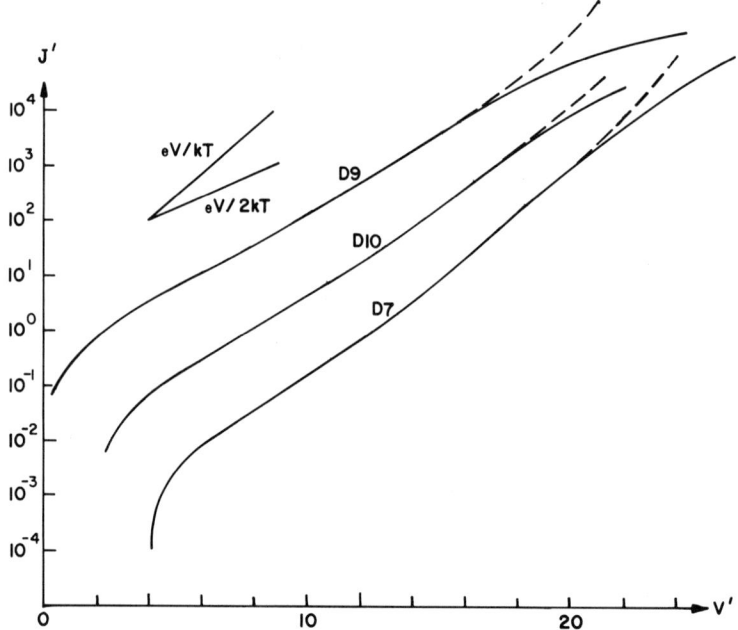

FIG. 35. The effect of lifetime changes on diode characteristics.

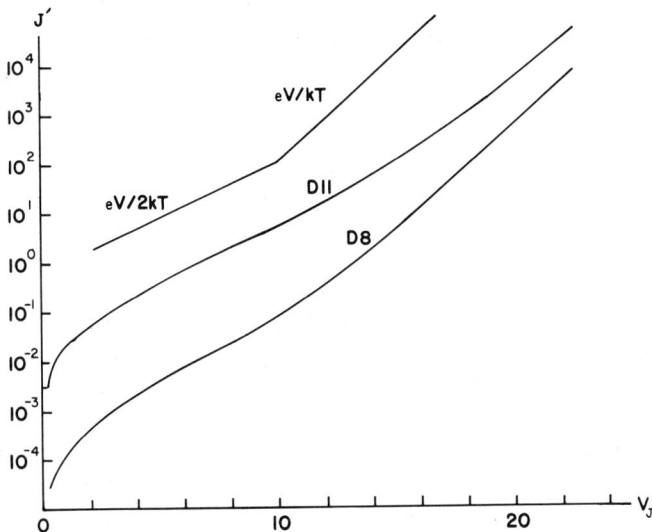

FIG. 36. The effect of lifetime changes on heavily doped diodes.

current densities; in the high current region, the bulk voltage drop goes up with length. Figure 34 shows that the characteristic for the longest diode (D1) bends over quite noticeably at higher current densities, and the shorter diodes (D6 and D7) show much less bending, as expected. Below a normalized current density of 10^2, however, all of the curves are very similar. Notice also that the transition-region voltage drop (dashed curve) is nearly identical for all three diodes. The total voltage drops for D6 and D7 are almost identical because the lightly doped N region in both diodes is the same length. The heavily doped P regions have different lengths but, because of the heavy doping, the voltage drop across the P regions is negligible even at high current densities

Decreasing the minority-carrier lifetime in a silicon diode can be expected to have three distinct effects, as follows:

(i) Recombination in the transition region will increase, causing an increase in diode current for a given voltage.

(ii) Decreasing the lifetime will cause a decrease in the minority-carrier diffusion length which will in turn cause an increase in the saturation current J_S of (108). This larger saturation current will then cause an increase in diode current for a given voltage.

(iii) A shorter diffusion length will also cause an increase in the bulk-region voltage drop causing the I–V curve to bend over faster at high current densities.

Figure 35 shows the numerical I–V characteristics for three abrupt silicon diodes with different lifetimes (all other parameters are the same). All three of the expected characteristics are readily apparent in these curves. Note that the slope of the curves lies between eV/kT and $eV/2kT$ and that although the transition-region voltage drops (dashed curves) begin to bend upward toward a slope of eV/kT at high current densities, the total voltage drops (which are what would be measured externally) actually bend downward due to the increasing bulk-region voltage drops. Figure 36 shows the effects of changing the lifetime for two heavily doped diodes. These curves show particularly well how recombination influences the I–V characteristics at low currents. Note how the curves show a slope less than eV/kT in the low current region where transition-region recombination dominates the I–V characteristics. As the current increases, the curves approach a slope of eV/kT as predicted by the diode equation (107). Note also that decreasing the lifetime causes an increase in the saturation current (the curve for D11 lies above the curve for D8), and the recombination is increased as well (the bump in the I–V curve at low currents is more pronounced for D11 than for D8). The curves shown in Fig. 36 were obtained using the recombination–diffusion solution, and consequently, potential drops across the

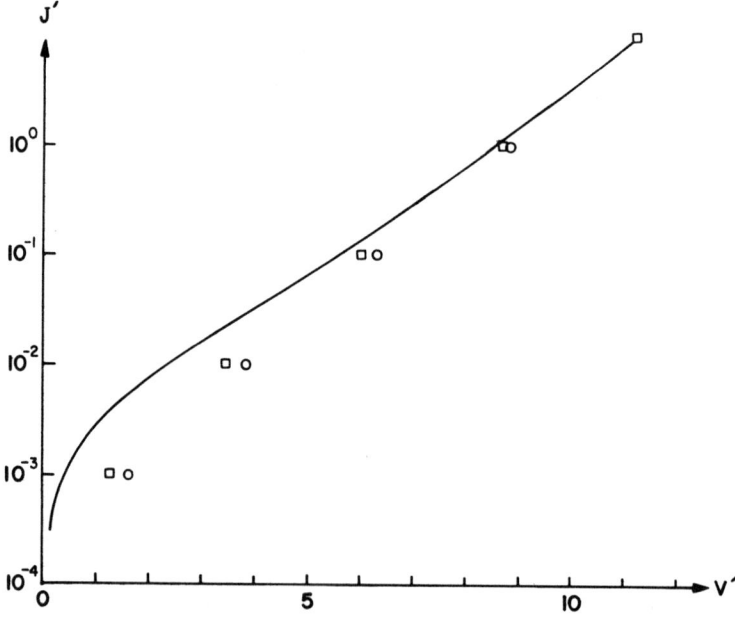

FIG. 37. Effect of trap position on diode characteristics.

bulk regions were neglected. Hence, the plot uses the normalized junction voltage V'_J as the abscissa.

Associated with the lifetime is the trap position E_T, since this affects the recombination rate via the quantities n_0^* and p_0^*. Table II specifies diode D1 with traps at the intrinsic level, whereas D4 has its traps above E_I and D3 has the traps shifted to an equal distance below the intrinsic level. The effect of these changes in trap position is shown in Fig. 37. The current is the greatest and the recombination rate the highest for traps at the center of the forbidden gap.

Figure 38 shows the effects on the abrupt diode characteristic of doping. Note that for D7, the more lightly doped diode, the curve bends down noticeably at the top of the graph due to the bulk-region voltage drop. The heavily doped diode D8, on the other hand, approaches the eV/kT straight-line characteristic at high currents. The larger bump in the curve of D8 at low currents indicates that recombination in the transition region becomes more significant as the doping is increased. Recombination current should decrease with an increase in doping because the width of the transition region will decrease. Apparently the recombination current is not as strong a function of doping as the saturation current, and consequently the recombination current will become more significant in the low-level characteristic as doping is increased.

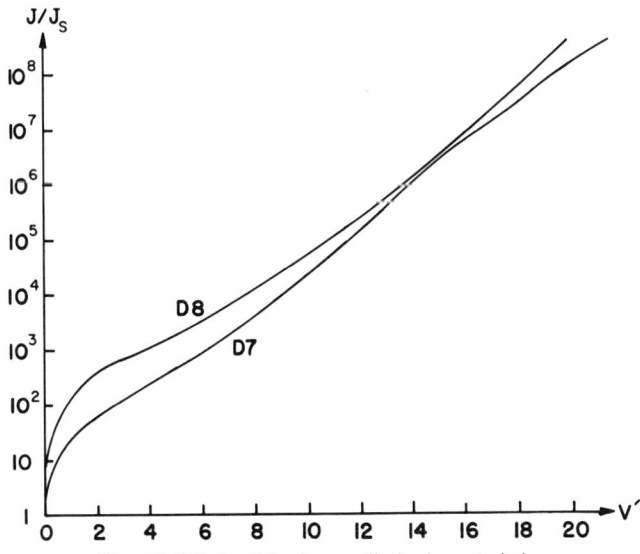

FIG. 38. Effects of doping on diode characteristics.

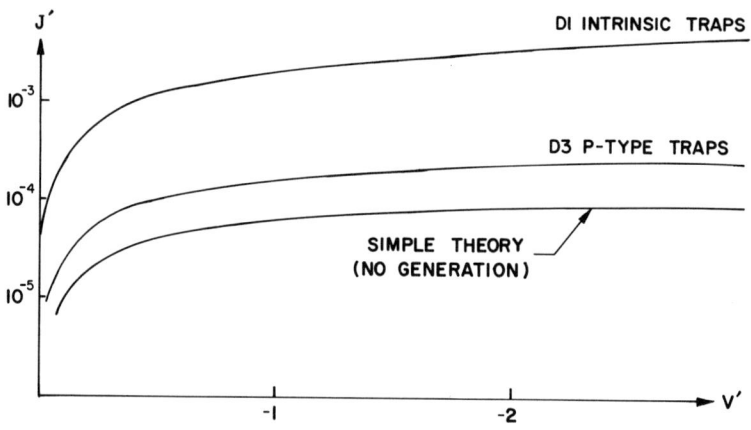

FIG. 39. Reverse current for two diodes.

For silicon diodes biased in the reverse direction, generation of carriers in the transition region is the dominant effect in the I–V characteristic, causing the reverse current in a silicon diode to be as much as several orders of magnitude greater than the value predicted by the simple low-level theory. We have seen this in an approximate way in Section 5, and Fig. 39 compares the numerically calculated reverse currents of two different diodes with the reverse current predicted by the simple diode equation. It is clear that generation can play a major role in determining the reverse characteristics of a silicon diode.

The diode D5 has a doping profile which yields an equilibrium electric field in the transition region that is linear. The region over which the field is nonzero was chosen to give a transition-region width comparable to that of the abrupt diode D1. The constants of Fig. 32 were chosen to be

$$A = 1.0 \times 10^{-3}, \quad B = 1.0 \times 10^{-1}.$$

Some care must be exercised in choosing these constants; it was found that with the numerical techniques used to solve the abrupt diodes, the solution of the linear field diode diverged for two values of A and B slightly larger than those given above because of roundoff difficulties in the differential equation for electron concentration (282a). The characteristic of diode D5 is nearly identical to that of D1. Figure 40 shows that the recombination current for D5 falls below the recombination current for D1 at higher current densities. The characteristics of both diodes are almost identical, however, because the diffusion current (the part of the current predicted by the simple diode equa-

2. THE THEORY OF SEMICONDUCTING JUNCTIONS 127

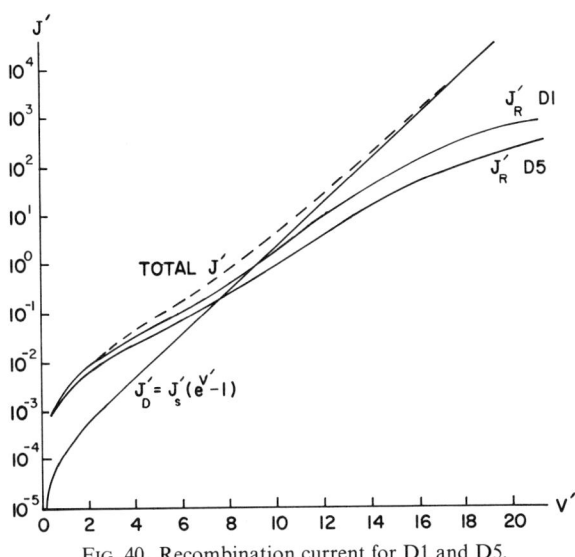

FIG. 40. Recombination current for D1 and D5.

tion), which is the same for both diodes, dominates the solutions in the region where the recombination currents are different.

An abrupt diode solution was also obtained using the field and doping-dependent diffusion coefficients derived from measurements in silicon by Caughey and Thomas (1967). In spite of the fact that these nonconstant diffusion coefficients tend to be substantially smaller than the constant coefficients used for the other diodes, the curve for D13 (not shown) is almost identical to that for D1 over the entire current range solved. The recombination and diffusion components of the current are also nearly identical.

e. Generation–Recombination Considerations

The theory given in Section 5 was used to make some predictions about the behavior of a silicon diode in the generation–recombination regime, whereas the present section covers all the details and difficulties involved in a complete solution of the van Roosbroeck system of equations. It is worthwhile to examine if some compromise is available which gives acceptable accuracy with a reasonable amount of effort. The method which Stone devised, which he calls the recombination–diffusion solution, was to simply write the total current as the sum of a recombination and a diffusion contribution. The former was calculated by evaluating an integral similar to those of Section 5, and the latter was the standard low-level solution. A comparison of this approach to the previous treatment is shown in Fig. 41.

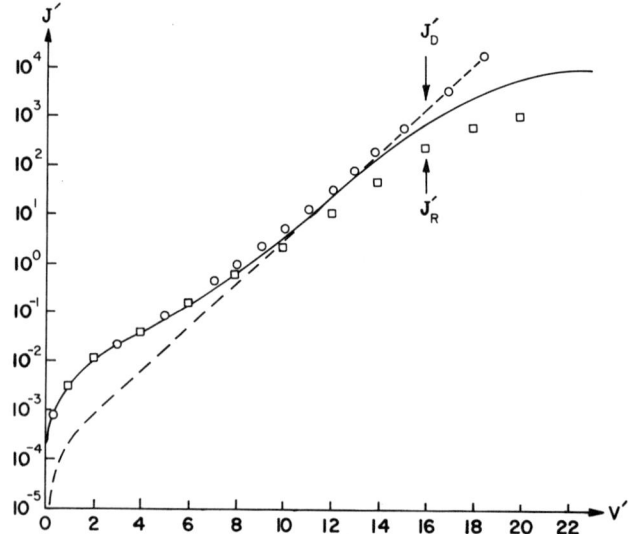

FIG. 41. Comparison of numerical and diffusion–recombination solutions for D1.

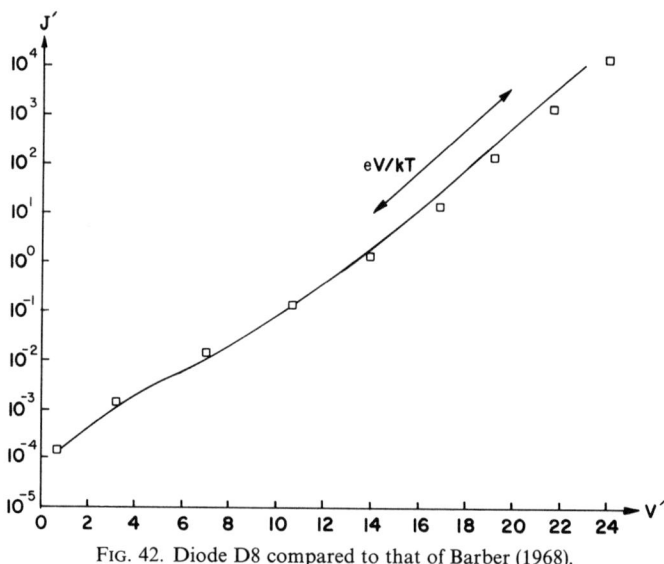

FIG. 42. Diode D8 compared to that of Barber (1968).

It is seen that this approximation matches the exact solution nicely at low and medium injection levels, but deviates sharply at high levels. Figure 42 compares the same quantities for diode D8, taken to be identical to that of Barber (1968), and here the agreement is perfect over the full injection range. The explanation lies in the comparative lengths; D1 is a long diode with a

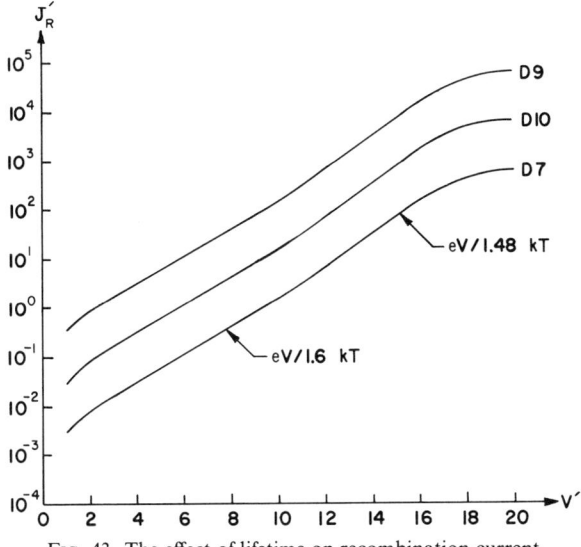

FIG. 43. The effect of lifetime on recombination current.

220-μm-bulk N region whereas the corresponding length for D8 is only 20 μm.

The effect of lifetime was also studied and this again agrees with our earlier work. As Fig. 43 shows, a rise in lifetime produces the expected decrease in recombination current but has no effect on the slope of the characteristic. Note that α is of the order of 1.5–1.6 for traps at the intrinsic level in a short diode, as previously predicted.

f. Boundary Conditions

The assumption concerning the constant and parallel property of the electrochemical potentials is one that has generated some controversy in the literature (see Section 6); the calculation of the behavior of $\bar{\mu}_e$ and $\bar{\mu}_h$ for D1 (a long diode) is shown in Fig. 44 with a forward bias of about two units,

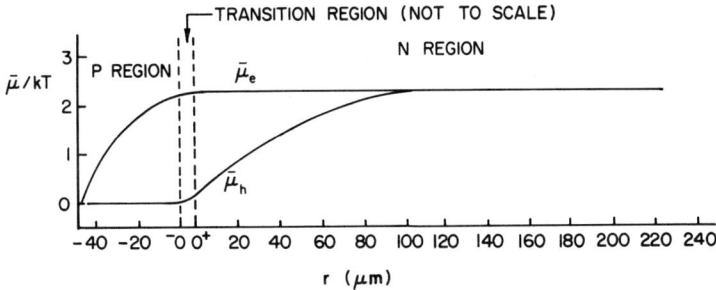

FIG. 44. Electrochemical potentials for D1 under forward bias.

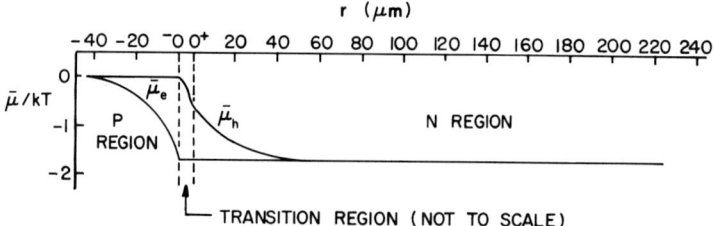

FIG. 45. Electrochemical potentials for D1 under reverse bias.

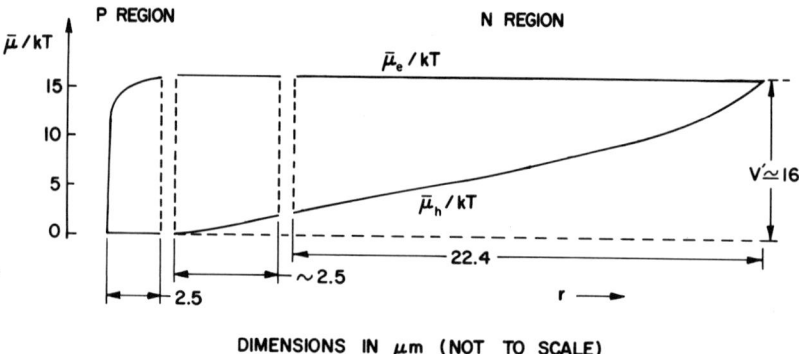

FIG. 46. Electrochemical potentials for D9 under forward bias.

which is in the very low injection range (Fig. 34). The potentials are definitely not flat in the transition region, although the relative changes are small. For the reverse-biased case (Fig. 45), $\bar{\mu}_h$ changes significantly. Note that the slope of the electrochemical potentials is always positive (Fig. 44) under forward bias, and the opposite under reverse bias (Fig. 45), as required by the associated Ohm's law. The contrasting situation for a short diode under an intermediate bias is shown in Fig. 46, which indicates that $\bar{\mu}_e$ is roughly constant across the space-charge region whereas $\bar{\mu}_h$ slopes upward by an amount equal to about 2.0 units. Now the relation between the electrons at the two edges of the space-charge region, as given by (180) and (181), is

$$n(-r_P) = n(r_N)\exp[-e(V_A - V_D)/kT] \qquad (296)$$

and we have shown in Eq. (211) that this should now be altered for the holes to

$$p(r_N) = p(-r_P)\exp[-e(V_A - V_D + \varepsilon)] \qquad (297)$$

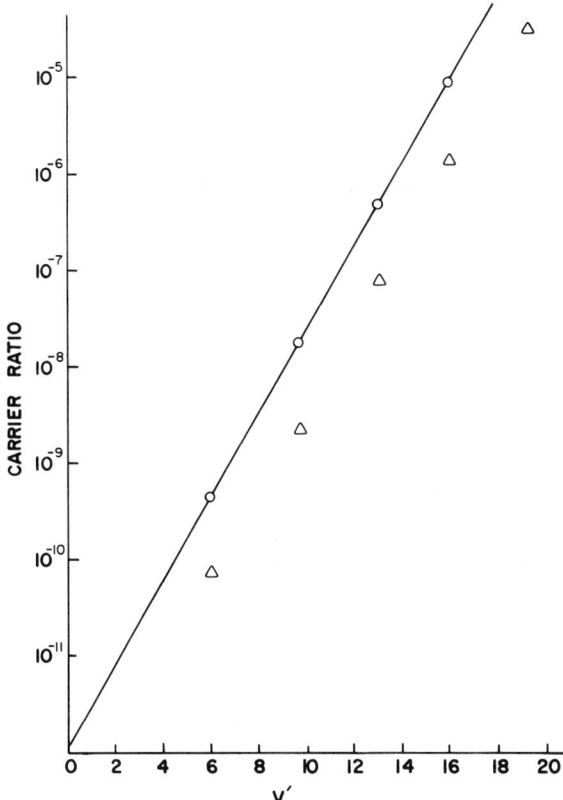

FIG. 47. Transition-region boundary ratios for D9.

for the case $\delta_N = \delta_P = 0$. In normalized units, this means that the plot of $p'(r_N)/p'(-r_P)$ should be shifted 2.0 units to the right of the curve for $n'(-r_P)/n'(r_N)$, and Fig. 47—which is a plot of the carrier ratios at the space-charge region edges for D9—verifies this. That is, the Fletcher conditions are verified for $\bar{\mu}_e$, which is flat, and must be modified for $\bar{\mu}_h$, as previously discussed.

g. *Experimental Results*

Figures 34 through 47 represent the results of the computer calculations done by Stone (1971); in addition, two commercial diodes had their characteristics measured as part of his study (Fig 48): a type 1N759, 12-V zener diode, and a type 1N914, high-speed gold-doped switching diode. The zener diode shows the expected slope upward at higher forward currents as diffusion begins to dominate over recombination. The switching diode curve,

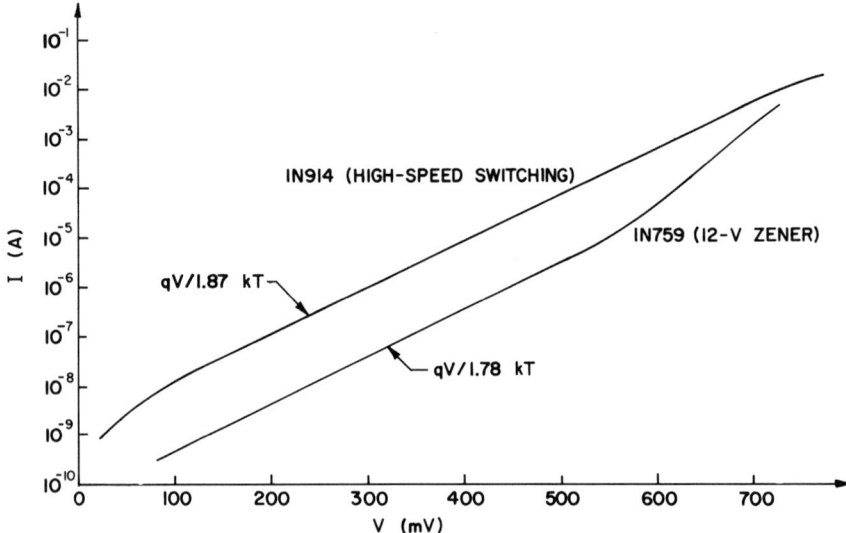

FIG. 48. Characteristic curves for two commercial diodes.

however, bends downward because the low lifetime caused by the gold doping also causes the bulk-region voltage drops to become significant before the diffusion current can dominate. Notice the similarities between the curves for the 1N914 and D9, and the 1N759 and D7 (Fig. 35).

The measured properties of the type 1N914 correspond to the assumed parameters in Table II. We have already pointed out that D8 closely matches the diode measured by Barber; it has the characteristic curve of Fig. 22, which agrees with both Barber's theory and the exact numerical calculations, for reasons already stated.

A similar sort of behavior in an early alloyed-junction silicon diode was observed by Shields (1959). The diode was designed to have a P^+NN^- configuration, and he explained the changes in slope from $\alpha = 2.0$ (the upper part of the generation–recombination regime) to $\alpha = 1.0$ (where the diffusion dominates) and then to $\alpha = 1.9$ by successive transitions from a P^+NN^- structure to P^+N, PIN^-, and P^+N^- as the injection level increases. This type of approach leads to approximate expressions for the current involving the voltage to various powers, and their combination is very much like Eq. (276).

We may thus summarize Stone's work by noting that the arbitrary division of the diode into three distinct regions is not as rash an approximation as might initially appear. The solutions correspond to real diodes in spite of the fact that we are using a one-dimensional model and thus implicitly ignoring the effect of surface recombination, crystalline imperfections, and other related items.

h. Comparison with Other Calculations

The work of Graham and Hauser (1970) involves the quasi-linearization technique of Bellman and Kalaba (1965); they had originally tried deMari's method, but found that the HSR recombination-rate expression produced diverging solutions. They studied three P^+NN^- and three P^+IN^- diodes. Their diode number 6 is about the closest to one of those in Table II; it has a P^+ region which is 250 μm long and has 10^{18} acceptors/cm^3, an N region which is 35 μm long with 2.5×10^{15} cm^3 donors, and an N^- region which can be regarded as providing an ohmic contact. (No details are given about trap positions or lifetimes.) As Fig. 49 shows, the numberical solution bears

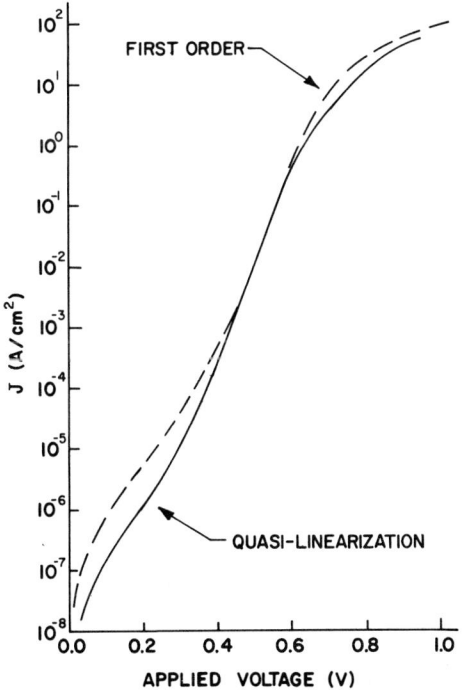

FIG. 49. Characteristics of Graham and Hauser diode no. 6.

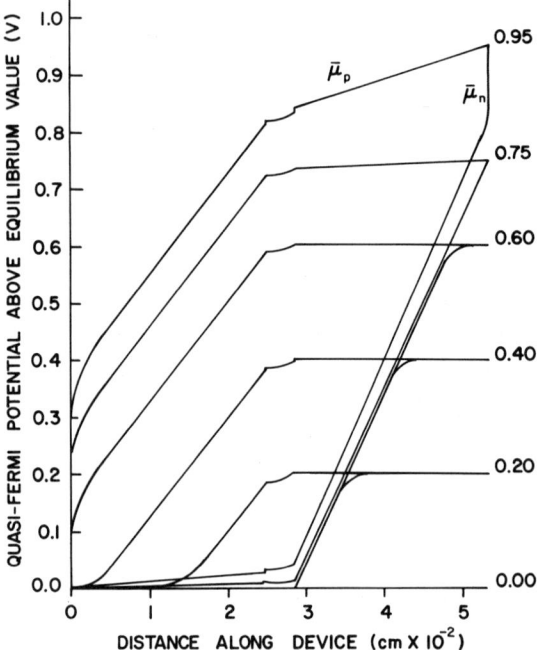

FIG. 50. Electrochemical potentials for the diode of Fig. 49.

some resemblance to the short diodes already discussed; we might infer that the length of the heavily doped P region does not have much effect on the characteristic properties. We note that the upper part of the generation–recombination part ($\alpha = 2$) is barely one decade, which contrasts with Fig. 22. Graham and Hauser have also computed and plotted the electrochemical potentials (Fig. 50). On the scale shown, it is not possible to tell much about their behavior in the space-charge regions, but the discontinuities in $\bar{\mu}_p$ would argue against the original Fletcher boundary conditions, in the fashion of Figs. 44 and 46.

The combined result of the theory and the numerical modeling is that a silicon diode is a much more complicated device than Shockley's original theory envisaged. Over its operating range of many decades, it shows a value of α which depends explicitly on the nature of the recombination processes and of the importance of their contribution to the total current. In some instances, α will meet the logarithmic diode criterion of Buckingham and Faulkner (1969); these devices have some useful applications.

V. Heterojunctions and Contacts

9. IDEAL HETEROJUNCTIONS

The concept of a *PN* junction with different materials in the two regions was suggested prior to 1960. For example, Kroemer (1957) showed that such an arrangement—which he called a wide-gap emitter—should enhance the injection efficiency of a transistor. The first comprehensive theory of ideal heterojunction diodes is that of Anderson (1962); his model has served as a basis for the interpretation of almost all subsequent experimental work. It has recently been pointed out (Adams and Nussbaum, 1979), however, that this theory involves boundary conditions which do not meet the requirements imposed by electromagnetic field theory. When these boundary conditions are corrected, they confirm a band structure which had been previously proposed (Chang, 1965).

A simple way of presenting the Anderson model is to use the numerical treatment of Milnes and Feucht (1972). Consider a heterojunction composed of *P*-type germanium and *N*-type GaAs, with material constants as specified in Table III. These parameters, in the order listed, are gap width E_G, electron affinity χ, donor or acceptor concentration N_D or N_A, position of Fermi level $(E_C - E_F)$ or $(E_F - E_V)$, dielectric coefficient ε_r, lattice constant a, and mobility ratio c. The electron affinity χ may be defined as the energy required to take an electron at the bottom E_C of the conduction band and place it at rest outside the semiconductor. Its final energy, measured with respect to the arbitrary reference level of Fig. 1, is known as the vacuum level E_{vac}. (We shall consider the implications of this definition in great detail in Section 10.) A significant feature of the two materials is that their lattice constants match to

TABLE III

HETEROJUNCTION PARAMETERS

Symbol	*P*-Ge	*N*-GaAs	Name
E_G (eV)	0.70	1.45	Energy gap
χ (eV)	4.13	4.07	Electron affinity
N_D (cm^{-3})	—	10^{16}	Donor concentration
N_A (cm^{-3})	3×10^{16}	—	Acceptor concentration
$(E_C - E_F)$ (eV)	—	0.1	Position of Fermi level
$(E_F - E_V)$ (eV)	0.14	—	Position of Fermi level
ε_r	16.0	11.5	Dielectric coefficient
a (nm)	0.5658	0.5654	Lattice constant
$c = \mu_e/\mu_h$	2.0	21.67	Mobility ratio

within 0.1%. This implies that there are no interface states, so that the Anderson model deals only with ideal heterojunctions.

When the depletion approximation is applied to such a heterojunction, Eqs. (53) and (54) are altered to

$$V_N(x) = -(eN_D/2\varepsilon_N)(x - x_N)^2 + V_{ON} \tag{298}$$

and

$$V_P(x) = (eN_A/2\varepsilon_P)(x + x_P)^2 + V_{OP}, \tag{299}$$

where the meaning of the subscripts on ε is obvious. Then at the junction ($x = 0$) we obtain

$$V_N(0) - V_{ON} \equiv V_{DN} = -(eN_D/2\varepsilon_N)x_N^2, \tag{300}$$

$$V_P(0) - V_{OP} \equiv V_{DP} = (eN_A/2\varepsilon_P)x_P^2, \tag{301}$$

where V_{DN} and V_{DP} are diffusion or barrier potentials which represent the band bending on each side of the junction. [These quantities, which indicate how the total diffusion potential is divided into two parts between the two regions of the junction, should not be confused with the equilibrium potentials V_N and V_P as used in Eq. (34a).] Dividing (300) by (301) gives

$$V_{DN}/V_{DP} = -N_D \varepsilon_P x_N^2 / N_A \varepsilon_N x_P^2. \tag{302}$$

To eliminate $(x_N/x_P)^2$ from (302), we differentiate Eqs. (298) and (299) to obtain the normal components of $\mathbf{D} = \varepsilon_r \mathbf{E}$ and use these derivatives in the boundary condition

$$\varepsilon_N \left.\frac{dV}{dx}\right|_N = \varepsilon_P \left.\frac{dV}{dx}\right|_P \tag{303}$$

to obtain

$$N_D x_N = -N_A x_P, \tag{304}$$

which is the same as (55). Combining this with (302) shows that

$$V_{DN}/V_{DP} = -N_A \varepsilon_P / N_D \varepsilon_N. \tag{305}$$

Equation (300) indicates that V_{DN} is inherently negative, so that the total barrier is

$$V_D = V_{DP} + |V_{DN}|. \tag{306}$$

Since we are going to convert these potentials into energies, it is simpler to work with magnitudes and we rewrite the above equations as

$$\rho = |V_{DN}/V_{DP}| = N_A \varepsilon_P / N_D \varepsilon_N \tag{307}$$

and
$$V_D = V_{DP} + V_{DN} = (1 + \rho)V_{DP}. \tag{308}$$

Substituting the numerical values of Table III into (307) and (308) leads to Fig. 51, which is roughly to scale. The construction procedure is as follows: starting with the common Fermi level E_F, the positions of E_C and E_V in each region come from a knowledge of the doping and the gap width. The vacuum level positions are set by the affinities and continuity condition (303) is applied to E_{vac}; i.e., V_D is identified as the shift in E_{vac} across the junction, with V_{DP} and V_{DN} representing the distribution between the two regions. These quantities are then simply transferred to E_C, E_V, and E_I to produce the discontinuities shown. The P side of the conduction band has what is known as a "notch" and the right-hand side contains a "spike" (Sharma and Purohit, 1972). The valence band, on the other hand, has a

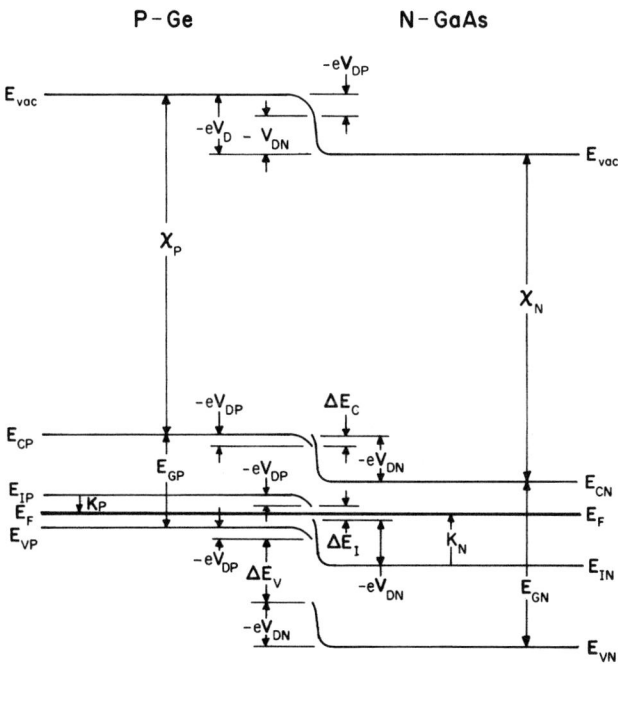

FIG. 51. Equilibrium energy-band diagram for a P-Ge/N-GaAs heterojunction, assuming a continuous vacuum level.

simple discontinuity, as does the intrinsic level. The magnitude ΔE_I of the discontinuity shown for the intrinsic level in Fig. 51 can be calculated by starting at the lower edge of this break in E_I and adding energies in accordance with the relation

$$\Delta E_I = -(-eV_{DN}) + K_N + (-K_P) - (-eV_{DP}) \tag{308a}$$

or

$$\Delta E_I = eV_D + kT \log_e(n_{ON}/n_{OP})(n_{iP}/n_{iN}). \tag{309}$$

For a homojunction, with $n_{iP} = n_{iN}$, this indicates that $\Delta E_I = 0$, since the right-hand side is equivalent to (189). It is also possible to have these discontinuities when both materials are either N-type or P-type; this is the isotype heterojunction, whereas Fig. 51 illustrates an anisotype junction. From this diagram we see that

$$\Delta E_C = -eV_{DN} - \chi_N - (-eV_D) + \chi_P - eV_{DP} = \chi_P - \chi_N \tag{310}$$

and

$$-\Delta E_V = -eV_{DN} - E_{GN} - \chi_N - (eV_D) + \chi_P + E_{GP} - eV_{DP}$$

or

$$\Delta E_V = (E_{GN} - E_{GP}) - (\chi_P - \chi_N). \tag{311}$$

Equations (310) and (311) comprise what is known as the affinity rule. For the example in question they give

$$\Delta E_C = 0.06 \quad \text{eV}, \quad \Delta E_V = 0.69 \quad \text{eV}, \tag{311a}$$

so that the scale is violated to some extent in Fig. 51.

The original objection to the affinity rule came from Chang (1965), who extended the Fletcher boundary conditions to heterojunctions. However, a simpler and more fundamental approach can be based on a generalization of (34a). The energy $-eV_D$ associated with the diffusion potential should have a contribution from each of the two regions of the junction which depends on the position of the Fermi level above or below its intrinsic position. Hence, we should define the diffusion barrier via an extension of (34a), namely

$$\begin{aligned}-eV_D &= kT \log(n_{ON}/n_{iN}) + kT \log(n_{iP}/n_{OP}) \\ &= \tfrac{1}{2}kT \log(n_{ON}/n_{iN})^2 + \tfrac{1}{2}kT \log(n_{iP}/n_{OP})^2 \\ &= \tfrac{1}{2}kT \log(n_{ON}/p_{ON}) + \tfrac{1}{2}kT \log(p_{OP}/n_{OP}) \end{aligned} \tag{312}$$

2. THE THEORY OF SEMICONDUCTING JUNCTIONS

using Boltzmann nondegenerate statistics and the usual symbols, or, for a heterojunction

$$-eV_D = \tfrac{1}{2}kT \log(n_{0N}p_{0P}/n_{0P}p_{0N}). \tag{313}$$

Equation (313) reduces, as it should, to (34a) for a homojunction. Using Eqs. (1)–(4), the argument of the logarithm is

$$\frac{n_{0N}p_{0P}}{n_{0P}p_{0N}} = \left(\frac{m_{eN}m_{hP}}{m_{eP}m_{hN}}\right)^{3/2} \frac{\exp[(E_{CP} - E_{CN})/kT]}{\exp[(E_{VN} - E_{VP})/kT]}. \tag{314}$$

If we accept the conditions that make (8b) valid, then (314) can be written in terms of mobility ratios, so that (313) becomes

$$-eV_D = \tfrac{3}{4}kT \log(c_P/c_N) + \tfrac{1}{2}[E_{CP} - E_{CN}) + (E_{VP} - E_{VN})], \tag{315}$$

where c_P and c_N are the mobility ratios in each of the two regions of the heterojunction. But the barrier should also be a measure of what is happening in the conduction band; in fact, it should be $E_{CP} - E_{CN}$ corrected by the discontinuity ΔE_C, or

$$-eV_D = (E_{CP} - E_{CN}) + \Delta E_C. \tag{316}$$

If these two expressions for the diffusion potential are to be consistent, it then follows that

$$\begin{aligned}\Delta E_C &= \tfrac{1}{2}[(E_{VP} - E_{VN}) + (E_{CN} - E_{CP})] - \tfrac{3}{4}kT \log(c_N/c_P) \\ &= \tfrac{1}{2}(E_{GN} - E_{GP}) - \tfrac{3}{4}kT \log(c_N/c_P).\end{aligned} \tag{317}$$

In the same way, there should be an expression

$$-eV_D = (E_{VP} - E_{VN}) - \Delta E_V \tag{318}$$

involving the discontinuity in the valence band and this, when combined with the original expression, leads to

$$\Delta E_V = \tfrac{1}{2}(E_{GN} - E_{GP}) + \tfrac{3}{4}kT \log(c_N/c_P), \tag{319}$$

and these relations replace the previous affinity rule, Eqs. (310) and (311). The logarithmic term for many material combinations is small, since c_N is roughly the same as c_P; the consequence is that the discontinuities are then identical. However, for the example of Table III, substitution into Eqs. (317) and (319) gives

$$\Delta E_C = 0.33 \quad \text{eV}, \qquad \Delta E_V = 0.42 \quad \text{eV},$$

and these values differ significantly from those of (311a). We thus see that if three different possible definitions of the barrier potential are to be mutually consistent, then the discontinuities in the valence and conduction bands must be determined as indicated.

The form of Eqs. (316) and (318) requires some detailed consideration, as does the physical basis on which they rest. In view of the fact that the conduction band of Fig. 51 contains the spike–notch discontinuity, it might have been more reasonable to have explicitly designated ΔE_C in (316) as a negative quantity, and conversely in (318) [which would simply have reversed the signs in (317) and (319), respectively]. We can justify this procedure to some extent by considering a hypothetical heterojunction for which the conduction band has a continuous edge. In such a situation, it is quite evident that the barrier for electrons is simply the quantity $E_{CP} - E_{CN}$, with the sign dependent on the relative energies of the two sides. Now suppose that the introduction of impurities restores the spike–notch discontinuity. As Fig. 51 indicates, there is an energy overlap between states, making the barrier easier to surmount. The barrier should be lowered by an amount equal to the discontinuity in band edges at the junction, and similarly, the corresponding simple discontinuity in the valence band increases the barrier for holes. These statements are expressed in quantitative form by Eqs. (316) and (318), where it is assumed that the sign of the discontinuity is chosen as just discussed. Admittedly, this justification is very intuitive, and we shall consider other arguments—both theoretical and experimental—which bear on the validity of the expressions given here.

Although nothing was stipulated about the specific nature of E_I at the junction, we can deduce an important conclusion about the behavior of this level from the fact that we have used it as a potential reference. It will be shown just below (Adams and Nussbaum, 1979) that the proposed new form of the affinity rule follows from the relation

$$-eV_D = -eV_{DN} + (-eV_{DP}). \tag{320}$$

Thus, *if* E_I plays the role of stipulating an absolute electrostatic potential, then the arguments just given imply that it is also continuous. It should be explicitly noted, however, that *this conclusion is highly controversial* and may possibly be settled only by very careful experimental work. The principal objection is that the intrinsic level in some sense may be regarded as fictitious and no special properties should be assigned to it.

Another way of comparing the Anderson affinity rule with the one that we are advancing is based on the following physical argument: consider a homojunction with the band structure of Fig. 3 and apply tension to the N region only. This process will lower the symmetry from cubic to tetragonal, there will be an associated change in the size of the energy gap, and we will have created a heterojunction with a virtually perfect lattice match at the interface. Some internal level (not necessarily the intrinsic level) can be chosen to specify the electrostatic potential and it must be continuous; a discontinuity implies that work can be done with zero displacement, and this is possible only if the field is infinite. Furthermore, the final arrangement of the energy levels and of the inevitable discontinuities will depend only on the internal rearrangements in the distorted lattice, so that—as we are predicting—the new band structure depends only on inherent parameters of the two parts of the junction and not on work functions. On the other hand, the change in the surface dipole layer should result in a discontinuity in the vacuum level, and this effect is also a feature of our model.

Some questions about the meaning of affinity have been raised by Kroemer (1975). His approach is based on the concept that the potential energy of an electron in a solid involves two distinct types of forces: the electrostatic part, representing the potential of the electron if its own presence does not affect any of the other electrons, and the polarization part, which is the change in potential due to the effect of a particular electron on all the other electrons. In particular, the electron affinity is

$$\chi = \chi^{(e)} + \chi^{(p)}, \tag{321}$$

where $\chi^{(e)}$ is the electrostatic term and $\chi^{(p)}$ is the polarization contribution. This idea is then applied to a very elementary model: a heterojunction whose doping is such that there is no band bending (his arguments do not affect this assumption in any case). As Fig. 52 indicates, there is a discontinuity $\Delta\chi^{(p)}$ at the interface which may be calculated from the relation

$$\chi_1 - (\chi_2 + \Delta E_C) = \chi_1^{(p)} - (\chi_2^{(p)} + \Delta\chi^{(p)})$$

or

$$\Delta E_C = \chi_1 - \chi_2 - [\chi_1^{(p)} - \chi_2^{(p)}] + \Delta\chi^{(p)}. \tag{322}$$

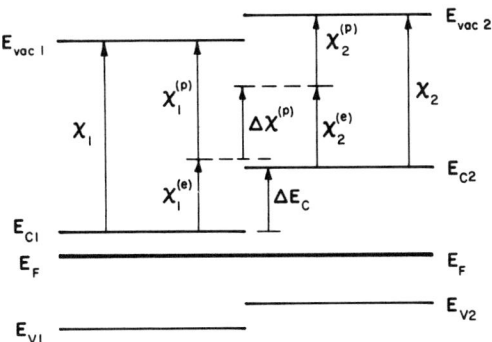

FIG. 52. Decomposition of affinities for a heterojunction.

Now this is consistent with (310) only if

$$\Delta\chi^{(p)} = \chi_1^{(p)} - \chi_2^{(p)}. \tag{323}$$

Kroemer points out that (323) implies that the two surface dipole layers would simply superimpose themselves on one another linearly as they gradually come into contact—this means that they are not permitted to interact, which is certainly not true for a properly fabricated heterojunction. He therefore proposed discarding the heterojunction affinity rule and attempting to calculate ΔE_C from a detailed knowledge of the band structures of each of the two materials. [Note: we have reversed the roles of $\chi^{(e)}$ and $\chi^{(p)}$ from Kroemer's original assignment; their position on the diagram is arbitrary and it seems physically reasonable that (324) should involve the discontinuity in the polarization contribution.]

Since then, Frensley and Kroemer (1976) have performed this calculation and believe that they have in fact confirmed the Anderson form of the affinity rule. However, their approach involves a number of assumptions—plus the approximations which are inherent in the pseudopotential method—and there is no really satisfactory way of accurately determining the validity of this calculation. Another type of calculation, involving LCAO theory, was done by Harrison (1977) to determine the valence-band maxima for a large number of elements and compounds. One of the conclusions of this work is that the Frensley–Kroemer approach appears to be reasonable for heterojunctions with a good lattice match, but not otherwise.

We now return to the question of boundary conditions, as used in junction theory. We have already mentioned that the normal component of **D** must be continuous at $x = 0$, giving the intrinsic level E_I a continuous slope, but this places no requirements on the value of E_I. In principle, the band structure of a junction could show a discontinuity which would be common to the levels E_V, E_C, and E_I. It is interesting to note that this question has not been considered by Shockley (1949, 1950) in his original or subsequent treatment and it has apparently been overlooked by others (including the present author) (Nussbaum, 1962). The reason why the bands in the transition region have the customary form illustrated in Fig. 3, which represents a symmetric junction, is that those levels which bend and join smoothly at $x = 0$ are all subject not only to the two conditions stated above but to a third one as well: it is well known (Lorrain and Corson, 1970) that the electrostatic potential V is continuous for a wide variety of charge distributions. A discontinuity in V implies an infinite field, which is ruled out on physical grounds. We can therefore require that E_I, which we have taken to be a measure of the potential, to be continuous. All the other levels in Fig. 3 must then behave in the same way, and this includes E_{vac}, which is not shown and whose specific properties we shall discuss in connection with metal–semiconductor contacts.

2. THE THEORY OF SEMICONDUCTING JUNCTIONS

The boundary condition which we shall use is then

$$\varepsilon_P \left.\frac{dV}{dx}\right|_P = \varepsilon_N \left.\frac{dV}{dx}\right|_N \tag{324}$$

or in terms of the normalized variables of Section 2, we have

$$\frac{dV}{dx} = \frac{1}{e}\frac{dK}{dx} = \frac{1}{e}\frac{kT}{L_D}\frac{du}{dr} = \left(\frac{2n_i kT}{\varepsilon}\right)^{1/2}\frac{du}{dr}$$

so that (324) becomes

$$\sqrt{\varepsilon_P n_{iP}}\left.\frac{du}{dr}\right|_P = \sqrt{\varepsilon_N n_{iN}}\left.\frac{du}{dr}\right|_N.$$

Squaring and using (41) gives

$$\varepsilon_P n_{iP}[\cosh u(0) - a_P u(0) + C_P] = \varepsilon_N n_{iN}[\cosh u(0) - a_N u(0) + C_N], \tag{325}$$

where, as in (47), $u(0)$ is the value of u at the junction ($r = 0$). Rearranging, we obtain

$$(R - 1)\cosh u(0) + (a_N - a_P R)u(0) + (RC_P - C_N) = 0, \tag{326}$$

where

$$R = \varepsilon_P n_{iP}/\varepsilon_N n_{iN}. \tag{327}$$

We recognize that for $R = 1$, (326) reduces to (47), as it should for a homojunction. To solve the transcendental equation (326) numerically, we obtain the equilibrium values of u from the information given in Table III concerning the values of the mobilities, the magnitude of the gap widths, and the positions of the bands with respect to the common Fermi level. The results are

$$u_{0P} = -8.66, \qquad u_{0N} = 22.44.$$

Then by (35),

$$a_P = -2.89 \times 10^3, \qquad a_N = 2.79 \times 10^9.$$

The definition (29) gives the intrinsic concentrations as

$$n_{iP} = 5.2 \times 10^{12}, \qquad n_{iN} = 1.8 \times 10^6,$$

from which

$$R = 4.03 \times 10^6.$$

Although the intrinsic concentration for germanium as calculated in this fashion differs somewhat from that used in Eq. (34a), we shall use the figures just above because our purpose here is merely to illustrate the consequences of the assumption that the intrinsic level is continuous. We also note that our value of R differs by about the same amount from the value 1.49×10^6, as determined by Table 2.3 of Sze (1969).

To find the constants of integration, we use (43), obtaining

$$C_P = a_P u_{0P} - \cosh u_{0P} = 2.21 \times 10^4,$$
$$C_N = a_N u_{0N} - \cosh u_{0N} = 5.97 \times 10^{10},$$

and (326) becomes

$$4.03 \times 10^6 \cosh u(0) + 1.44 \times 10^{10} u(0) + 2.94 \times 10^{10} = 0. \quad (328)$$

Using the Newton–Raphson method, the solutions to this equation are found to be

$$u(0) = \begin{cases} -11.075, \\ -2.042. \end{cases} \quad (329)$$

This double root is inherent in Eq. (326), as we can see by solving it for R to obtain

$$R = \frac{\cosh u(0) - \cosh u_{0N} + \sinh u_{0N}[u_{0N} - u(0)]}{\cosh u(0) - \cosh u_{0P} + \sinh u_{0P}[u_{0P} - u(0)]}. \quad (330)$$

Then the numerator vanishes for $u(0) = u_{0N}$ and the denominator does likewise for $u(0) = u_{0P}$. Hence $R = 0$ corresponds to $u(0) = u_{0N}$ and $R = \infty$ to $u(0) = u_{0P}$. Also, when $|u(0)|$ is very large, then only $\cosh u(0)$ is significant in (330), making $R = 1$. If we plot R as a function of $u(0)$, as shown in Fig. 53, these features are evident and we also see that the curve has three distinct branches, separated by the values $u(0) = u_{0N}$ or u_{0P}. In fact, two roots are possible for any value of R; they are both negative for $R > 2.7 \times 10^6$, they have opposite signs when R lies in the range $1.0 < R < 2.7 \times 10^6$, and for $R < 1.0$, they are both positive. To resolve the ambiguity about which root corresponds to the actual heterojunction, we realize that $u(0)$ must obey

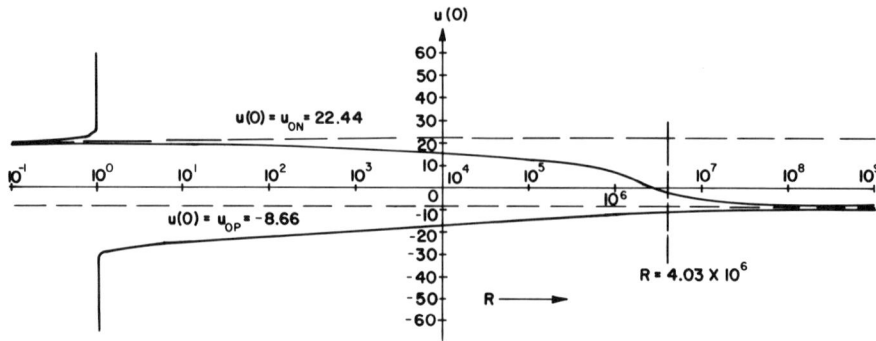

FIG. 53. Graphical determination of $u(0)$ for a heterojunction.

2. THE THEORY OF SEMICONDUCTING JUNCTIONS

the inequality $u_{0P} < u(0) < u_{0N}$, so that only the central branch has physical meaning. Otherwise the bands would bend away from one another and there would be a discontinuity in E_I, violating the boundary condition. Thus, from (329),

$$u(0) = -2.042, \qquad K(0) = -0.05 \quad \text{eV},$$

and this solution enables us to construct the energy-band diagram (again, approximately to scale) of Fig. 54a.

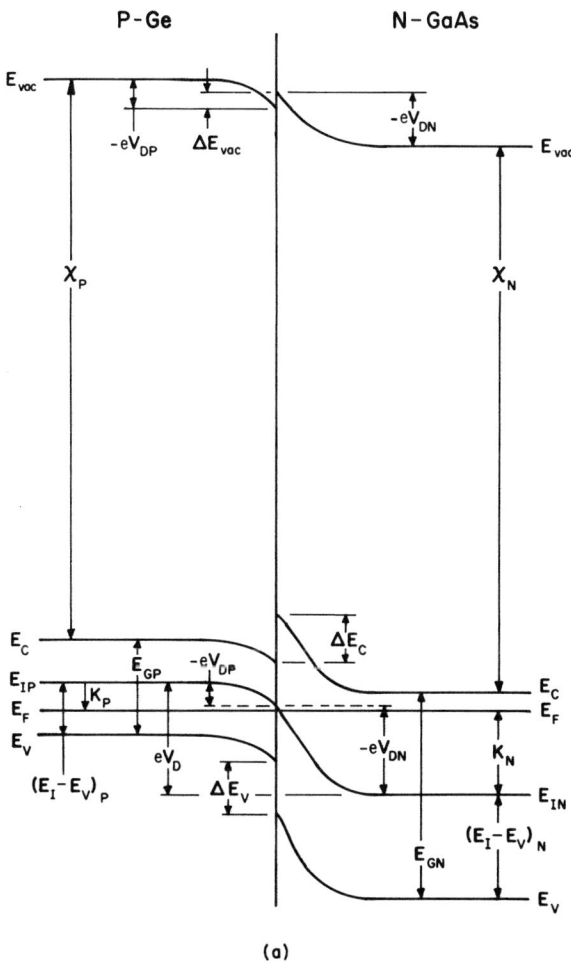

(a)

FIG. 54. (a) Equilibrium energy-band diagram for a p-Ge/N-GaAs heterojunction assuming a continuous intrinsic level. (b) Energy-band diagram for a doubly intrinsic Ge/GaAs heterojunction, assuming a continuous vacuum level. (c) Energy-band diagram for a doubly intrinsic Ge/GaAs heterojunction, assuming a continuous intrinsic level.

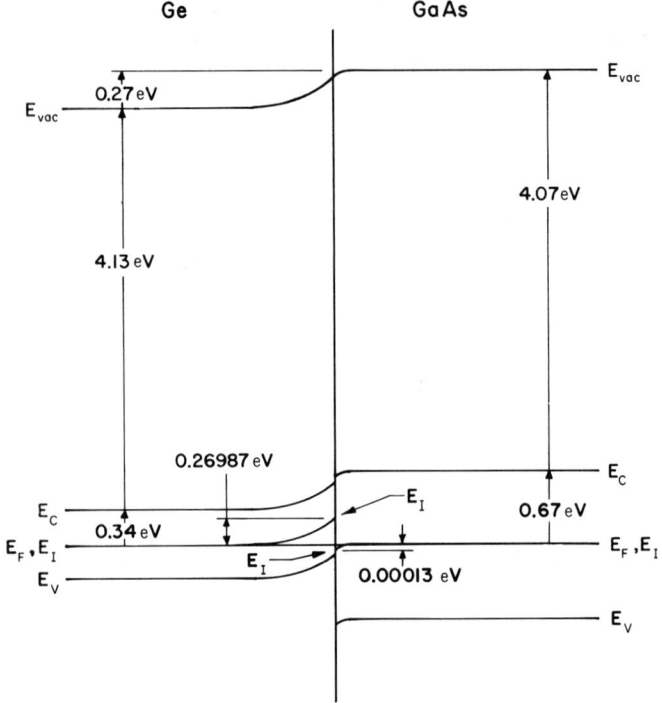

(b)

FIG. 54. (*Continued*)

Note here that we have explicitly taken E_I to be continuous. This figure shows that

$$\Delta E_C = -E_{GP} + (-eV_{DP}) + (E_I - E_V)_P + eV_D - (E_I - E_V)_N \\ + E_{GN} + (-eV_{DN}) \\ = [E_{GN} - (E_I - E_V)_N] - [E_{GP} - (E_I - E_V)_P].$$

Combining the procedure of Eq. (316) with the definition (5) gives

$$\Delta E_C = \tfrac{1}{2}(E_{GN} - E_{GP}) - \tfrac{3}{4}kT \log(c_N/c_P) \qquad (331)$$

and similarly

$$\Delta E_V = \tfrac{1}{2}(E_{GN} - E_{GP}) + \tfrac{3}{4}kT \log(c_N/c_P). \qquad (332)$$

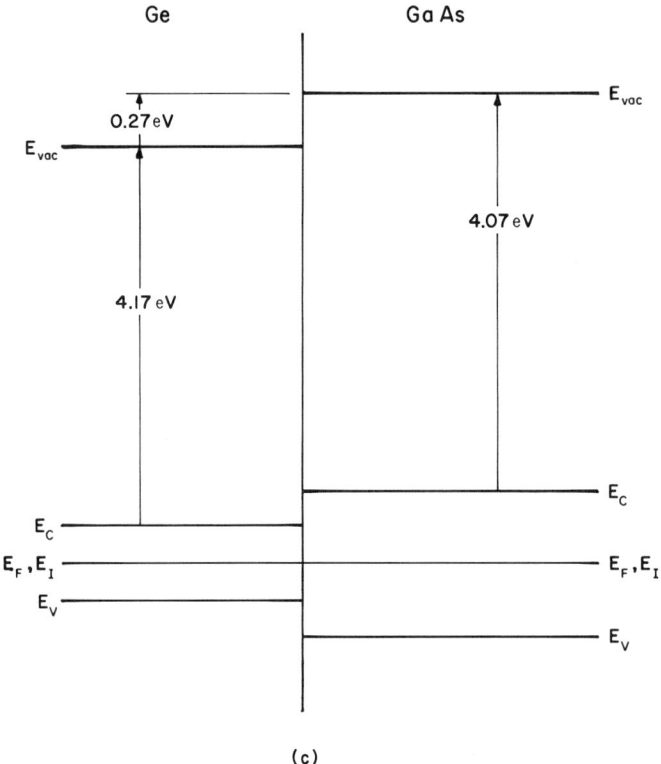

(c)

FIG. 54. (*Continued*)

These are identical to (317) and (319), respectively. However, the discussion associated with Eq. (320) shows why the conditions used with this model justify the conclusion just reached.

A very specialized, but explicit, example of how the two models under consideration lead to rather different predictions is the doubly intrinsic heterojunction. If we consider intrinsic Ge joined in an ideal structure to intrinsic GaAs, the constants of Table III permit us to immediately construct the portions of Fig. 54b which lie outside the space-charge region. The central section, however, cannot be handled in the same way as Fig. 51, for the depletion approximation is no longer valid. Instead, we use the complete Poisson–Boltzmann equation, and (34) assumes the form

$$d^2u/dr^2 = \sinh u \tag{332a}$$

with a first integral

$$du/dr = [2(\cosh u + C)]^{1/2}. \qquad (332b)$$

To evaluate the arbitrary constant C, we use the boundary condition that $u = 0$ for intrinsic material in the equilibrium regions of the junction, where du/dr vanishes, so that

$$C = -1.$$

Hence

$$du/dr = \pm[2(\cosh u - 1)]^{1/2} = 2\sinh(u/2). \qquad (332c)$$

There is an ambiguity of sign in this result which we shall discuss shortly. Rearranging (332c),

$$dr = du/[2\sinh(u/2)] = \tfrac{1}{2}\operatorname{csch}(u/2)\,du$$

and integrating again

$$r = \log_e \tanh(u/4) + C'.$$

To evaluate the second arbitrary constant C', let

$$u = u(0) \quad \text{at} \quad r = 0.$$

Then

$$C' = -\log \tanh[u(0)/4]$$

and

$$r = \log\left(\frac{\tanh(u/4)}{\tanh[u(0)/4]}\right)$$

or

$$u = 4\operatorname{arctanh}\{\exp r \tanh[u(0)/4]\}. \qquad (332d)$$

This expression for u is equivalent to one obtained for the potential in the semiconductor part of a metal–intrinsic Schottky diode by McKelvey (1966).

To find the positions of the intrinsic level just at the junction in Fig. 54b, we use the form of (330) which is applicable to this doubly intrinsic junction, namely,

$$R = [\cosh u_1(0) - 1]/[\cosh u_2(0) - 1], \qquad (332e)$$

where the subscripts refer to GaAs and Ge, respectively. The figure shows that the barrier, as determined by the relative positions of the vacuum level

2. THE THEORY OF SEMICONDUCTING JUNCTIONS

in the two neutral regions, is 0.27 eV. It follows that (converting from the normalized chemical potential u to the actual value K)

$$|K_1(0)| + |K_2(0)| = 0.27. \tag{332f}$$

Using $R = 4.03 \times 10^6$, as indicated in Fig. 53, we must solve (332e) and (332f) simultaneously by a numerical method; the Newton–Raphson procedure (which turned out to be somewhat difficult) gives magnitudes of

$$|K_1(0)| = 0.00013 \quad \text{eV}, \qquad |K_2(0)| = 0.26987 \quad \text{eV},$$

and these results are incorporated into Fig. 54b.

Because of the tremendous difference between the two values of the chemical potential at the junction, Fig. 54b cannot show the band bending with any reasonable scale. Furthermore, it is also difficult to show the continuity of the normal component of the displacement vector across the junction, since the band diagram indicates du/dr, which is a measure E_n rather than D_n. It is simple to show, however, that in the normalized units of Eq. (34), we may write

$$D_n = -(2n_i \varepsilon kT)^{1/2} \, du/dr$$

so that there is a direct connection between the sign of the slope du/dr which is common to the band edges in each of the two regions of the junction and the value of D_n at any point in these regions. Thus, D_n can be continuous only if du/dr possesses this property at the junction. Now the assumed behavior of the vacuum level in the Anderson-model energy-band diagram of Fig. 54b results in $K_2(0)$ being positive and $K_1(0)$ being negative, so that when we compute du/dr from (332c) by evaluating the expression $2 \sinh(K/kT)$, we find a change in the sign of the slope at the junction. The presence of the double sign before the radical does not alter this conclusion, for we shall now show that once having determined the particular sign to be used in (332c), it is invariant across the junction. Looking back at the Ge–GaAs junction of Fig. 51, it is quite clear that du/dr is negative (and continuous) at the junction. Let us (in principle) alter the impurity concentration so as to achieve the doubly intrinsic junction of Fig. 54a.

During the course of this process, the magnitude of du/dr may change, but the signs will always be identical on either side of the junction. This is equivalent to saying that once we have chosen a sign for the radical in (44) to match du/dr, *it is the same for both regions*; that is, it is inconceivable that when the heterojunction goes from being lightly doped to a doubly intrinsic structure, one set of bands abruptly undergoes a reversal in the sign of the slope.

We can use a related argument in connection with the final solution given by Eq. (332d). In the equilibrium part of the Ge region, u can go to zero as required because of the term $\exp(r)$, where r is now large and negative. This same term, however, will prevent the boundary condition from being satisfied in the GaAs equilibrium region, and again we reach the conclusion that the resolution of this difficulty is for $u(0)$ to be zero.

The energy-band structure based on these considerations is shown in Fig. 54c, and it indicates a sharp discontinuity in the vacuum level. Although this may appear to be an objection to the model on which Fig. 54c is based, it is believed that the concepts of work function and of vacuum level need some discussion, and this will be given in Section 10. In addition, the question of whether or not the vacuum level can have structure is somewhat controversial as well. For example, Moore (1969) shows the vacuum level associated with the two sides of a heterojunction as everywhere constant, with the affinities determining the position of the conduction band for the two materials just at the junction. This idea, also, we shall discuss later.

When we consider the experimental confirmation of either of these models, we find that not much has been published. Consider first the capacity versus voltage measurements of Shay et al. (1976); the parameters of the InP/CdS junction they reported on [as taken from their paper or from Milnes and Feucht (1972)] are listed in Table IV. Using the same procedure as before, we find from (307) and (308) that

$$-eV_D = 0.69 \quad \text{eV}$$

with

$$-eV_{DN} = 0.02 \quad \text{eV}, \qquad -eV_{DP} = 0.67 \quad \text{eV}.$$

The fact that V_{DN} is almost negligible explains the remark of Wagner et al. (1975) that there is no interfacial spike in the conduction band. The band structure (Fig. 55) differs from the previous example in both this feature and in the fact that the discontinuities at the two band edges are similar. As is evident from the figure, $\Delta E_C = 0.52$ eV. Applying Eq. (310) combined with the data from Table IV gives

$$|\Delta E_C| = |\chi_P - \chi_N| = 0.52 \quad \text{eV},$$
$$\Delta E_V = E_{GN} - E_{GP} - |\chi_P - \chi_N| = 1.60 \quad \text{eV}.$$

In addition, Shay et al. (1976) have found the magnitude of the diffusion potential to be 0.65 eV from measurements of $1/C^2$ versus reverse voltage, so that the properties of this junction conform to the predictions of the affinity rule. It should be noted that the large discontinuity in the valence band implies a large barrier for holes; conduction by electrons dominates.

2. THE THEORY OF SEMICONDUCTING JUNCTIONS

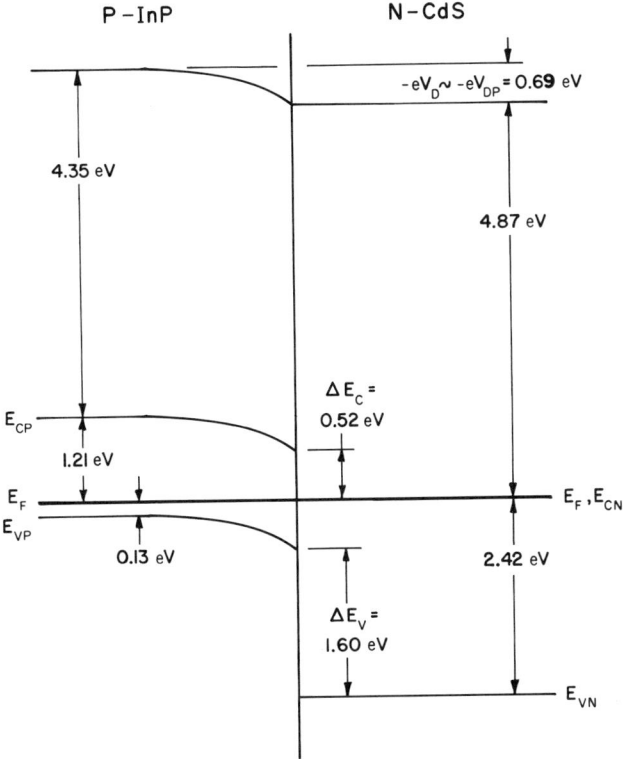

FIG. 55. Equilibrium energy-band diagram for a P-InP/N-CdS heterojunction.

TABLE IV

P-InP/N-CdS Heterojunction Parameters

	P-InP	N-CdS
E_G (eV)	1.34	2.42
χ (eV)	4.35	4.87
N_D (cm^{-3})	—	2×10^{18}
N_A (cm^{-3})	5×10^{16}	—
$(E_C - E_F)$ (eV)	—	0
$(E_F - E_V)$ (eV)	0.13	—
ε_r	12.1	9.7
a (nm)	0.5869	0.5850
$c = \mu_e/\mu_n$	45	(not known)

On the other hand, Eq. (331) gives

$$\Delta E_C = 0.54 - \tfrac{3}{4}kT \log(c_N/45),$$

where the approximation is justified by the fact that almost any value of the mobility ratio for CdS within reason will have little effect on the contribution from the logarithmic term. The discontinuity in the valence band is, of course, identical. Figure 55 indicates that the conduction-band barrier will be the one that primarily determines the electrical properties of the heterojunction, regardless of the specific nature of the discontinuities. Since ΔE_C will have a significant effect on the value of the barrier potential and since it has virtually the same magnitude for either of the two possible models, we conclude that we cannot distinguish between them for this particular example.

Another convenient measurement on junctions is current I versus applied voltage V_A. Anderson's theory for the current–voltage characteristic of the type of heterojunction illustrated in Fig. 51 takes advantage of the fact that the spike–notch discontinuity in the conduction band is not too much of a barrier to electron current, whereas the jump discontinuity in the valence band is a substantial obstacle for holes. We shall apply this argument to a heterojunction composed of N-type Ge and P-type GaAs (the doping is opposite to that of the Milnes–Feucht example), so that the spike–notch discontinuity is in the valence band. This implies that it is the hole barrier which is now smaller and that the electron current can be neglected. Low-level theory indicates that an applied potential V_A will cause a hole current of magnitude

$$J_h = (eD_h p_{0P}/L_h) \exp[e(V_A - V_D)/kT] \tag{333}$$

using (105) and (189).

Let the valence band have the structure shown in Fig. 56. Then holes going from GaAs to Ge are impeded by the barrier of magnitude eV_{DP} and those going in the opposite direction see a smaller barrier $\Delta E_V - (-eV_{DN})$. Let V_A

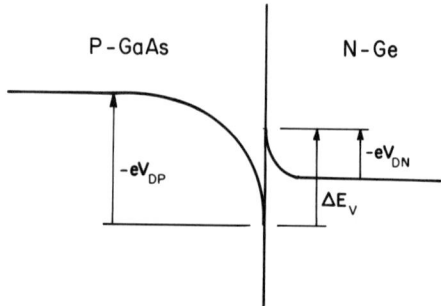

FIG. 56. A spike–notch discontinuity in the valence band.

2. THE THEORY OF SEMICONDUCTING JUNCTIONS

be the applied forward bias, which divides itself such that

$$V_A = V_{AP} + V_{AN}. \tag{334}$$

Each part of the bias decreases its respective barrier and we obtain the net current as the difference of two terms like (333), or

$$J_h = A_P \exp[e(V_{DP} - V_{AP})/kT] \\ - A_N \exp\{-[\Delta E_V + e(V_{DN} - V_{AN})]/kT\}, \tag{335}$$

where A_P and A_N involve material parameters. The two terms are equal at equilibrium, so that—with $V_{AP} = V_{AN} = 0$—we obtain

$$A_N \exp[-(\Delta E_V + eV_{DN})/kT] = A_P \exp(eV_{DP}/kT)$$

or

$$A_N \exp(-\Delta E_V/kT) = A_P \exp(eV_D/kT). \tag{336}$$

Letting $A_P = A$ and replacing J_h with J, since the electron current can be neglected, (335) becomes

$$J = A\{\exp[e(V_{DP} - V_{AP})/kT] - \exp[e(V_D - V_{DN} + V_{AN})/kT]\}. \tag{337}$$

Subtracting (334) from (308), we obtain

$$V_D - V_A = (V_{DP} - V_{AP}) + (V_{DN} - V_{AN}). \tag{338}$$

Also, (307) under bias is

$$\rho = (V_{DN} - V_{AN})/(V_{DP} - V_{AP}). \tag{339}$$

Combining these last two equations gives

$$V_{DN} - V_{AN} = \rho(V_D - V_A)/(\rho + 1), \\ V_{DP} - V_{AP} = (V_D - V_A)/(\rho + 1), \tag{340}$$

and then (337) is converted into

$$J = A\left[\exp\frac{e(V_D - V_A)}{(\rho + 1)kT} - \exp\frac{eV_D}{kT}\exp\frac{-e\rho(V_D - V_A)}{(\rho + 1)kT}\right] \\ = A\exp\frac{eV_D}{(\rho + 1)kT}\left[\exp\frac{-eV_A}{(\rho + 1)kT} - \exp\frac{e\rho V_A}{(\rho + 1)kT}\right]. \tag{341}$$

For forward bias, the second term can be neglected, so that (341) reduces to

$$J = A\exp[e(V_D - V_A)/(\rho + 1)kT]. \tag{342}$$

An example is that of Riben and Feucht (1966), who measured the voltage–current characteristic of a P-GaAs/N-Ge heterojunction. Their results are shown in Fig. 57, and they explain the very noticeable differences between theory and experiment by invoking recombination-tunneling theory.

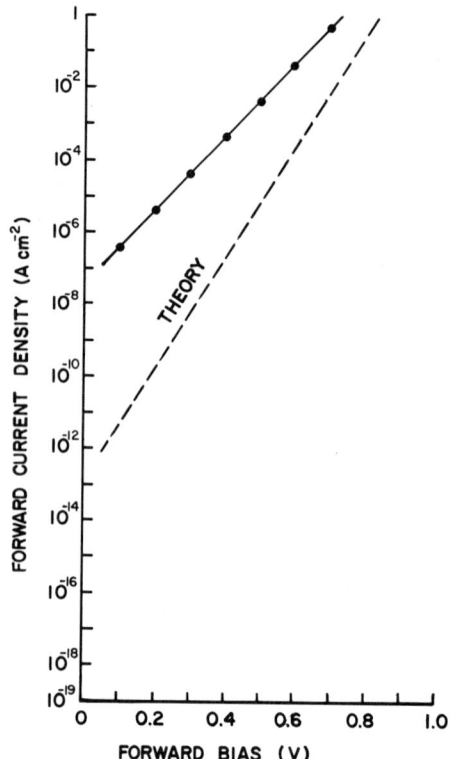

FIG. 57. Voltage–current characteristics for a Ge/GaAs heterojunction. (After Riben and Feucht, 1966.)

Although it is difficult to obtain accurate numbers from a small graph in a published article, we can estimate the intercepts in Fig. 57 as $J = 10^{-13}$ A/cm² for the theoretical curve and $J = 3 \times 10^{-8}$ A/cm² for the measured curve. Using

$$A = eD_h p_{0P}/L_h$$
$$= eN_A \sqrt{D_h/\tau_h},$$

where

$$D_h = 20 \text{ cm}^2/\text{sec}, \qquad \tau_h = 10^{-9} \text{ sec},$$
$$N_A = 1.77 \times 10^{17}/\text{cm}^3,$$
$$N_D = 1.50 \times 10^{18}/\text{cm}^3,$$

we obtain $A = 4.01 \times 10^3$ A/cm².

Riben and Feucht replace (342) with the approximation

$$J = A \exp(eV_{DP}/kT) \exp[-eV_A/(\rho + 1)kT]. \tag{343}$$

2. THE THEORY OF SEMICONDUCTING JUNCTIONS

Letting $V_A = 0$, (343) then gives

$$-eV_{DP} \text{ (observed)} = 0.65 \text{ eV,}$$
$$-eV_{DP} \text{ (theoretical)} = 0.97 \text{ eV.} \quad (344)$$

Using the more exact equation (342) with $V_A = 0$ gives

$$eV_D = kT(1 + \rho) \log_e(J/A). \quad (345)$$

But by (308),

$$eV_{DP} = eV_D/(1 + \rho),$$

and when this is combined with (345) we obtain the same values that have already been specified in Eq. (344), above.

We now wish to compare these results with the predictions for the quantity eV_{DP} that are based on the two different energy-band models. Using $n_{iN} = 2.4 \times 10^{13}/\text{cm}^3$ and $n_{iP} = 1.8 \times 10^6/\text{cm}^3$ in conjunction with the impurity concentrations cited by Riben and Feucht, we find from Eq. (35) that

$$a_N = 3.13 \times 10^4, \quad a_P = -4.92 \times 10^{10},$$

from which

$$E_F - E_{IN} = 0.28 \text{ eV,} \quad E_{IP} - E_F = 0.64 \text{ eV.}$$

These figures, combined with the constants of Table III, enable us to construct the equilibrium portions of the energy-band diagram of Fig. 58, which

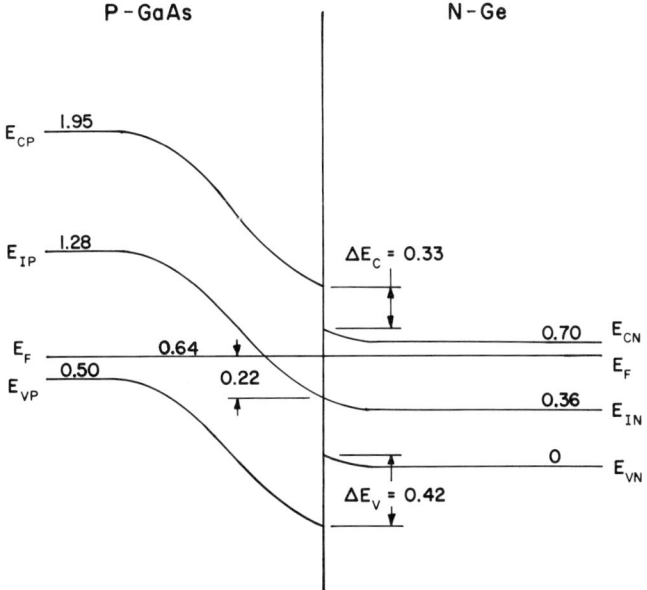

FIG. 58. Energy-band structure for the Ge/GaAs heterojunction of Fig. 57, assuming a continuous intrinsic level. (All energies in units of eV.)

uses the assumption that E_I is continuous; note that in this figure we have arbitrarily chosen the equilibrium position of the valence band for Ge as the reference level. The diffusion potential indicated by this figure is

$$-eV_D = 0.28 + |-0.64| = 0.92 \quad \text{eV}.$$

We compute ρ from (307) as

$$\rho = N_A/N_D = 0.085.$$

Then by (308),

$$eV_{DP} = 0.85 \quad \text{eV}.$$

The model based on Eq. (326), with $R = 2.48 \times 10^{-7}$, produces the transcendental equation

$$-\cosh u(0) + 1.57 \times 10^5 \, u(0) - 1.38 \times 10^6 = 0$$

from which

$$u(0) = \begin{cases} 8.811, \\ 14.378. \end{cases}$$

Since the principal branch is the lower value,

$$K(0) = 0.22 \quad \text{eV}.$$

This crossing point is incorporated into the energy-band diagram of Fig. 58; it can then be seen that

$$-eV_{DP} = 0.86 \quad \text{eV}, \qquad -eV_{DN} = 0.06 \quad \text{eV},$$

and

$$-eV_D = 0.92 \quad \text{eV}.$$

Although we have not provided the corresponding energy-band diagram for the Anderson model, it is simple to see from Fig. 58 and the affinity values of Table III that

$$-eV_D = (1.95 + 4.13) - (0.70 + 4.07) = 1.31 \quad \text{eV},$$

and therefore

$$-eV_{DP} = 1.21 \quad \text{eV}.$$

These calculations are summarized in Table V, which indicates that the band diagram of Fig. 58 appears to come closer to experiment than the original heterojunction model.

2. THE THEORY OF SEMICONDUCTING JUNCTIONS

TABLE V

RESULTS FOR P-GaAs/N-Ge HETEROJUNCTION

	Magnitude of eV_{DP}	
From I–V characteristic	Predicted: 0.97 eV	Observed: 0.65 eV
From energy–band model	Anderson: 1.21 eV	Adams–Nussbaum: 0.86 eV

The most unusual heterojunction that has been described in the literature is one of a series whose composition is P-GaSb$_{1-y}$As$_y$/N-In$_{1-x}$GaAs$_x$. Sakaki et al. (1977) fabricated four such diodes, with x and y values as shown in Table VI. The measured voltage–current characteristics are shown in Fig. 59. These junctions are smooth films which were grown by molecular beam

TABLE VI

COMPOSITION OF THE FOUR
DIODES OF FIG. 59

Designation	x	y
A	0.62	0.64
B	0.52	0.56
C	0.50	0.28
D	0.16	0.10

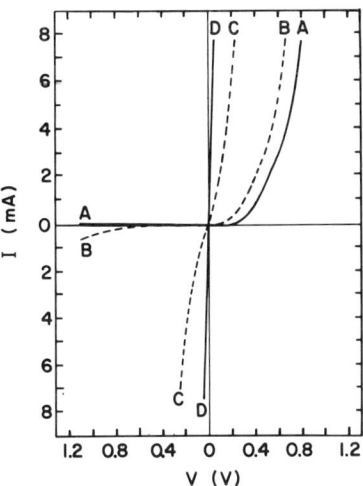

FIG. 59. Voltage–current characteristics of four P-GaSb$_{1-y}$As$_y$/N-In$_{1-x}$Ga$_x$As diodes. (From Sakaki et al., 1977.)

epitaxy, and great care was taken to achieve lattice match within 2.5% through control of x and y. High-energy electron diffraction monitoring during growth indicated the junctions were both ideal and abrupt. The reason for this particular combination of structure and materials, which represents the culmination of a series of studies (Sai-Halasz et al., 1977), is to have close control over the bandgaps and affinities. The objective was to achieve a band structure involving accumulation layers.

Let us digress briefly to discuss this topic. Suppose it were possible in the structure of Fig. 3 to have the bands bend in the opposite way to that shown. That is, the N-region bands move down rather than up, so that the electron concentration increases near the junction, and correspondingly for the P region. Then the bands, instead of acting as barriers, actually encourage minority-carrier injection and the accumulation process results in a diode which conducts well regardless of the bias polarity. Such junctions are described as "ohmic," but we realize that they are truly linear only for small applied potentials, when the Taylor expansion of the exponential function can be approximated by the first-order term. It is clear that this hypothetical structure violates the boundary condition on the electrostatic potential (but not on the field, since a discontinuous function can still have a continuous derivative).

As Davies (1957) has pointed out, however, there is a way of achieving accumulation in a homojunction; the Anderson model for the Sakaki et al. diode D just above will be seen to lead to accumulation in a heterojunction and in the next section, we shall consider this phenomenon for a metal–semiconductor contact. It is possible to obtain rectification and injection at the interface between regions of light and heavy doping which are entirely P-type or entirely N-type. For example, a contact may be made to P-type germanium by alloying a small region on the surface with indium. When the indium-doped germanium recrystallizes, it forms a region which is so heavily doped, that it is referred to as P^+-type and the contact is said to be a P^+P junction, or a heavy–light (HL) junction. Similarly, tin alloyed to N-type germanium gives an NN^- junction. We can apply the Fletcher boundary conditions (185) to an LH junction by letting the P region correspond to the light doping and the N region be replaced by a heavily doped P region. However, when we change m_N in (186) to m_H, we note that it is now a negative quantity, since holes are the majority carrier in both the light and heavy regions. The boundary conditions are then

$$\begin{aligned}
p_L &= (m_L - m_H d)/(1 - d^2), \\
p_H &= (m_L d - m_H d^2)/(1 - d^2), \\
n_H &= (m_L d - m_H)/(1 - d^2), \\
n_L &= (m_L d^2 - m_H d)/(1 - d^2),
\end{aligned} \tag{346}$$

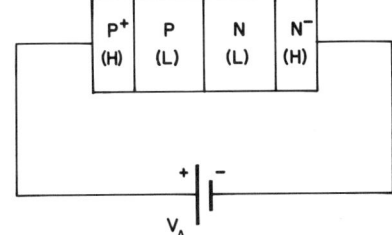

FIG. 60. Light–heavy junctions incorporated into a diode.

and similarly, there is a corresponding set for the equilibrium concentrations, which are denoted p_{0L}, p_{0H}, n_{0L}, and n_{0H}. It may be shown by simple algebra that

$$n_L - n_{0L} = p_L - p_{0L} = (d - d_0)(p_{0L}d + n_{0H})/(1 - d^2),$$
$$n_H - n_{0H} = p_H - p_{0H} = (d - d_0)(n_{0H}d + p_{0L})/(1 - d^2). \quad (347)$$

These equations specify the injected carrier density of each side of the LH junction. The term $(1 - d^2)$ in the denominator is positive, and slightly smaller than unity, for d is small. The quantity $(d - d_0)$ is positive under forward bias. Hence, the entire right-hand side is a positive quantity. This means that $(n_L - n_{0L})$ and $(n_H - n_{0H})$ are both positive, so that the concentration of minority carriers, namely electrons, is increased on both sides of the junction. The increase in the P^+ region is electron injection exactly like the injection of electrons into the P region of a PN junction. In the lightly doped region, however, the situation is an anomalous one. We refer to Fig. 60, which shows a PN junction and the narrow contact regions, which are heavily doped. The P^+P (or HL) junction is biased so that the P^+ region is positive (this is like the normal forward biasing of a PN junction), and the electrons would be expected to be injected from the P region to the P^+ region, just as they are from the N region to the P region. However, the simultaneous increase of electrons on the right-hand side of the P^+P junction is not what we normally expect because the negative bias does not encourage injection. Hence, we have achieved an accumulation condition.

We also realize that when the applied potential is negative, then so is $(d - d_0)$, and there is a decrease in the minority-carrier concentration on each side of the HL junction. The process analogous to injection has been called exclusion by Low (1955), and in place of accumulation we now have extraction. This loss of carriers at the LH junction can occur because the applied field will remove carriers from the junction faster than they can be replenished. The combined effect of these four processes is to assist the junction action with its conducting behavior; that is, injection and accumulation at the LH contacts insure that there is no rectification at these regions under forward bias, while exclusion and extraction make the total impedance slightly greater under reverse bias.

Returning to the diodes of Fig. 59, a band structure based on the Anderson model is shown in Fig. 61 for diode A and in Fig. 62 for diode D. The energies shown on these diagrams were inferred from information given (Sakaki *et al.*, 1977) and may not be completely accurate, but we are going to use them only for qualitative purposes in any case. It is seen that the combination of energy gaps and affinities of Fig. 62 plus the stipulation of a continuous vacuum level leads to a reversal of the normal barrier; that is, there is an accumulation layer on each side of the junction and this is taken to be the explanation of the nonrectifying characteristic for diode D shown in Fig. 59.

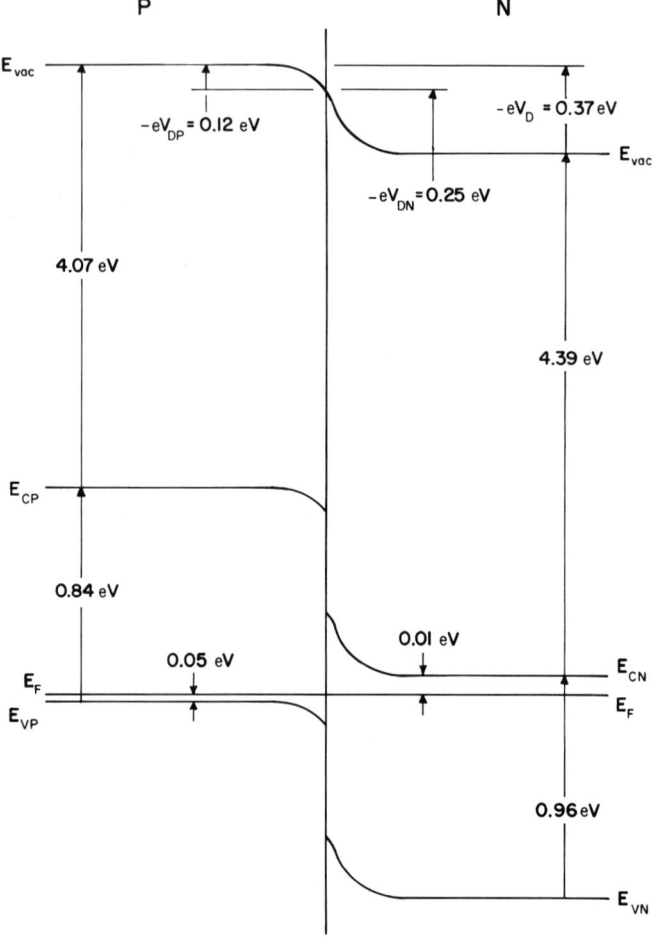

FIG. 61. Band structure of diode A, Fig. 59.

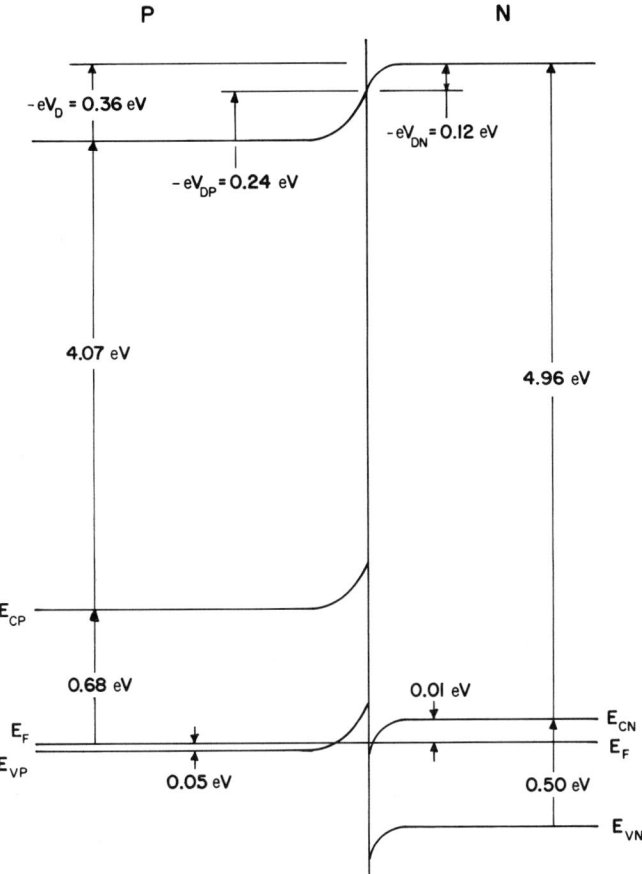

FIG. 62. Band structure of diode D, Fig. 59.

Although we do not have enough information to determine the structure for the same diode with a continuous intrinsic level, it is clear that this condition would produce the barrier-type of band bending and the resulting diagram could show a spike–notch discontinuity in both bands. This is a situation which we have not encountered before and it provides an alternate explanation for the nonrectifying property. As Eq. (341) indicates, the current is exponential under forward bias (neglecting the second term in brackets, and similar, but not identical, under reverse bias, since now the first term can be neglected). For a double spike–notch structure at low voltages, we would expect the linear-looking characteristic of Fig. 59. The converse of this is to have the two normal discontinuities of Fig. 61. Then the barrier is large in both bands and the diode is an extremely good rectifier, as the measurements

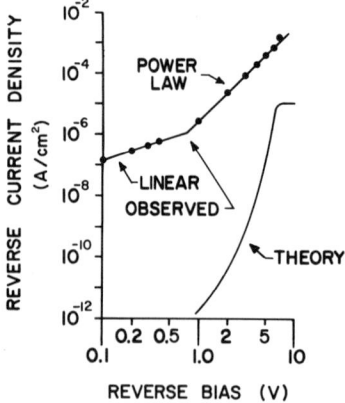

Fig. 63. Reverse characteristic behavior of the diode of Fig. 57.

show. We can predict that shifting the continuity condition from the vacuum to the intrinsic level in Fig. 61 makes very little change in the discontinuities, so that the same explanation holds. The intermediate situation, with a normal discontinuity in one band, and a spike–notch type in the other band, should lead to a "soft" reverse characteristic. For example, the heterojunction of Fig. 57 has the reverse behavior shown in Fig. 63. Note that this is a log–log plot, so that the exponential behavior predicted by (341) is not observed; rather, it is linear at low biases, and then goes over to a power law. The Riben and Feucht band diagram (not shown) is similar in form, and again, either model qualitatively explains the observed behavior. However, Eq. (341) is simply the application of the Shockley low-level theory of Section 3 to heterojunctions; we might speculate that some rigorous calculations of the type discussed in the preceding section might give a better understanding of the electrical characteristics of heterojunctions.

It should be reemphasized that all of the above has been restricted to ideal junctions; a previous article (Tansley, 1971) in this series looks at the nature of the discontinuities when the lattice constants are not matched. This article also considers surface states and the possibility of tunneling through the interface as an explanation for the deviations from the Shockley theory. The most extensive summary of the heterojunction literature is that of Sharma and Purohit (1972), who give a number of tables of materials properties.

10. Unified Theories of Contacts and Junctions

The final topic which we shall review is the equilibrium band structure of various combinations of materials: homojunctions, heterojunctions, metal–semiconductor contacts, and metal–metal contacts. We have come full circle with regard to metal–semiconductor devices, also known as Schottky barriers, because they are the subject of the introductory quotation from Davydov

2. THE THEORY OF SEMICONDUCTING JUNCTIONS

(1938). However, we shall consider their behavior under an applied potential very briefly; we refer the interested reader to fuller accounts elsewhere (Sze, 1969; Padovani, 1971; Rhoderick, 1970, 1974). Our purpose here is to examine the foundations of junction and contact theory with the object of examining the principles common to different kinds of physical systems and of resolving the differences in the explanations given of how they function. It is interesting to note that two almost diametrically opposite approaches are commonly used to analyze the band structure of dissimilar materials in contact, and we shall attempt to correlate these diverging points of view.

One of these unified theories rests on the properties of the vacuum level; the most consistent and clear exposition of this viewpoint is that of van der Ziel (1976), whose main features we shall now summarize.

Consider first the energy bands of a metal (Fig. 64) at room temperature. The lower bands are completely occupied by electrons, the top one is empty, and there is a partially filled intermediate band. At absolute zero, the Fermi level E_F would be the demarkation energy between filled and empty levels in this "conduction band," but for finite temperature, there is a distribution

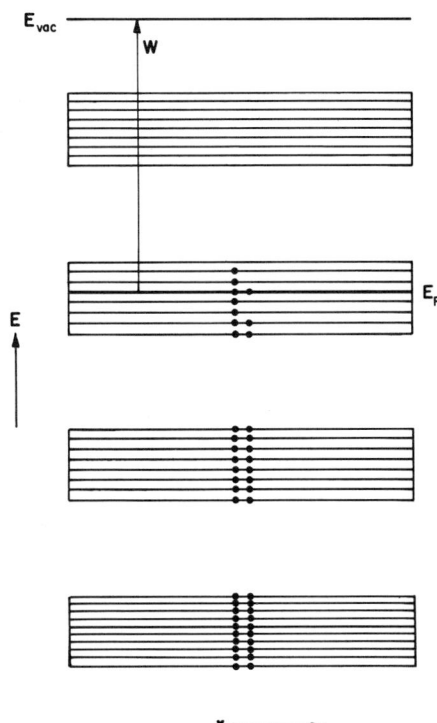

FIG. 64. Energy bands for a metal.

within about $\pm 2kT$ of E_F. The energy difference W between the vacuum level and the Fermi level is the true work function; as Herring and Nichols (1949) indicate, the term "true" is used to differentiate W from quantities such as the photoelectric work function, which represents the energy required to remove an electron from some particular level in a material (say cesium) by absorbing energy from a photon, or the thermionic work function, involving emission due to heating. The nature of the surface is involved in the specification of W; for example, it should take a different amount of work to move along the (100) direction than along the (111) direction of a cubic crystal. Actually, there is no precise agreement on the definition of work function. Herring and Nichols specify it as the difference between the electrochemical potential $\bar{\mu}$ of the electrons just inside the conductor and the electrostatic potential energy $-eV$ of an electron in the vacuum just outside, so that E_F is interpreted as the average value of $\bar{\mu}$ for all the electrons. Quantitatively, this may be expressed as

$$W = \bar{\mu} - (-eV_{\text{vac}}) = K - eV + eV_{\text{vac}},$$

where $-eV_{\text{vac}}$ is the external electrostatic potential and $-eV$ is the internal potential.

Wigner and Bardeen (1935) call it the difference in energy between the original, neutral lattice and the same lattice with one electron removed, assuming that the electron comes from the highest occupied state (i.e., the Fermi level or vicinity). They also recognize that the final position of the electron must be specified, since W is different for different crystallographic planes; in fact, they calculate work functions for a number of alkali metals by the Wigner–Seitz method, assuming that the electron is at a distance from a surface plane which is small compared to the dimensions of this plane but large on the atomic scale. Bleaney and Bleaney (1976) define the work function as the energy required to remove an electron from the top of the energy distribution and take it out to infinity (which actually involves a finite amount of work).

There is a somewhat inconsistent aspect of Fig. 64 which should be mentioned. It is not really possible to show the vacuum level on a one-dimensional diagram; hence having the vacuum level in a position above the internal bands is an attempt to represent a two-dimensional situation on a one-dimensional diagram. Some writers are very meticulous about avoiding this difficulty—van der Ziel (1976), for example—while others use the convention shown. We shall employ whichever is convenient.

An estimate of the work function can be obtained in an elementary way. We shall show below how the solution to the Schrödinger equation corresponding to a potential $V = 0$ inside a cube of edge L and $V = \infty$ outside this cube indicates that, at absolute zero, the maximum energy W_0 (which is related to, but not the same as, the Fermi energy) is obtained from the electron

2. THE THEORY OF SEMICONDUCTING JUNCTIONS

TABLE VII
Work Functions for Various Metals

	W_0 (eV)	Work function Photoelectric (eV)	Thermionic (eV)
Li	4.7	2.2	
Na	3.1	1.9	
K	2.1	1.8	
Cu	7.0	4.1	4.5
Ag	5.5	4.7	4.3
Au	5.5	4.8	4.25
Mo	5.9		4.2
W	5.8	4.49	4.5
Pt	6.0	> 6.2	5.3
Ni	7.4	4.9	4.5

density n in accordance with the formula (Nussbaum, 1962)

$$W_0 = (3n/\pi)^{2/3}(h^2/8m). \tag{348}$$

As an example, $n = 5.9 \times 10^{28}/\text{m}^3$ for silver, giving $W_0 = 5.5$ eV. If we regard the vacuum level as the energy reference at infinity, then W_0 can be taken as a measure of the true work function (ignoring, for the moment, surface effects), since it is specified initially with respect to an arbitrary zero inside the metal; using E_{vac} as the reference simply shifts this point. Some values of W_0 for other metals and a comparison with measured work functions is shown in Table VII, taken from Bleaney and Bleaney (1976).

We shall briefly indicate the origin of (348), since some associated results will be needed later. The solution to the Schrödinger equation for the potential cube leads to the eigenvalues

$$E = (h^2/8mL^2)(n_x^2 + n_y^2 + n_z^2), \tag{349}$$

where $n_x, n_y, n_z = 0, \pm 1, \pm 2, \ldots$. If we construct a sphere of radius $R = (n_x^2 + n_y^2 + n_z^2)^{1/2}$ with the center at the origin of the space defined by these integers, then its volume is a direct measure of the number N of eigenvalues or states specified by (349). Only the positive octant need be considered, so that (349) shows that

$$\tfrac{1}{2}N = \tfrac{1}{8}(\tfrac{4}{3}\pi R_m^3) = \tfrac{1}{6}\pi R_m^3 = (\pi/6)[W_0(8mL^2)/h^2]^{3/2},$$

where $N/2$ indicates that each level holds one electron with spin up and one with spin down, and where W_0 is the value of E for the maximum radius R_m. Using $V = L^3$ and a density of electrons $n = N/V$ gives (348). Furthermore, differentiating (349) shows that the number of states per unit volume in a spherical shell of thickness dR is

$$dG = 4\pi m^{3/2}(2E)^{1/2}\, dE/h^3 \qquad (350)$$

or in terms of momentum

$$dG = 4\pi p^2\, dp/h^3. \qquad (351)$$

This formula combined with Boltzmann statistics leads to Eqs. (3) and (4); we plan to apply it to a more general situation.

A detailed examination of the concept of work function is that of Ashcroft and Mermin (1976). They point out that an infinite, perfectly periodic crystal can be described (in principle) by a Schrödinger equation in which the potential term is the sum of the individual electrostatic potentials due to each unit cell, all of which are identical. Each of these terms can be represented as an integral over some unknown charge distribution $q_v(\mathbf{r})$ of the form (Lorrain and Corson, 1970)

$$V(\mathbf{r}) = -e \int \frac{q_v(\mathbf{r}')\, d\mathbf{r}'}{|\mathbf{r}-\mathbf{r}'|},$$

where $\mathbf{r}$ is the spatial coordinate and $\mathbf{r}'$ is the variable of integration. The inverse distance can be expanded in the usual fashion to produce a monopole, dipole, quadrupole, etc., contribution. The monopole term will vanish, however, since a cyrstal is normally neutral. A lattice with inversion symmetry will have no dipole moment, and for cubic structures with inversion symmetry, the quadrupole moment (the coefficient of $1/r^3$) will be zero, and so will the next term; the lowest-order nonzero term is the one involving $1/r^5$.

Thus, an electron taken from within a few kT of the Fermi level of a metal will have virtually no energy if it is brought to rest at a position which is a few lattice constants beyond the external surface, and the work function W would be expected to have a value

$$W = E_F. \qquad (352)$$

(Ashcroft and Mermin write this as $W = -E_F$, where energies inside the metal are measured with respect to zero at the infinite position; we shall stay with our original convention, as shown in Fig. 1.) This result, however, ignores the termination of the lattice periodicity at the surface.

The electronic charge distribution may not match the Bravais lattice structure of the interior, and there will be a surface dipole sheet, called the double layer; a detailed discussion of the nature of this dipole sheet has been given by Callen (1957). Hence, work W_S must be done in crossing the double layer and (352) becomes

$$W = E_F + W_S. \qquad (353)$$

2. THE THEORY OF SEMICONDUCTING JUNCTIONS

Furthermore, there can even be an additional net surface charge, for if we take an electron out of the crystal via a (100 face with work function W_{100} and return it through a (111) face with work function W_{111}, the total energy in the cycle is conserved. Hence, the quantity $(W_{100} - W_{111})$ is just balanced by the difference in electrostatic potentials between the two faces, and this, in turn, is created by the net surface charge. It is reasonable to expect, however, that the associated field is small compared to the one inside the dipole layer.

The consequence of the above development is that an electron which is reasonably close to the metallic surface experiences a force which is primarily due to the surface double layer, since the lattice contribution falls off extremely rapidly. Hence, we can bring the electron to rest and there will be a level representing vanishing total energy at this point.

Then the electron can be removed still farther away so that when it is at a distance which is large compared to the dimensions of the crystal face, the metal appears to be a continuum and the usual Coulomb image force, with an inverse square dependence, sets up an attraction which must be overcome to continue out to infinity. This implies that there are *two* vacuum levels, and such an idea has actually been proposed by Hagstrum (1976). He calls these simply the near level and the level at infinity. It would carry us too far afield to examine his arguments; we shall merely remark that there is ambiguity and disagreement involved in applying the concept of vacuum level to solids. Hence, theories which are based on its explicit properties have to be carefully examined.

Returning now to the development by van der Ziel, we start with two isolated metals whose band structure is indicated in Fig. 65. The key feature of this diagram is the assumption that the vacuum level is a common reference level for the two metals and all other energies are expressed with respect to it. Let the samples be arranged physically as shown in Fig. 66; then the band structure will be based on a uniform Fermi level when equilibrium is attained

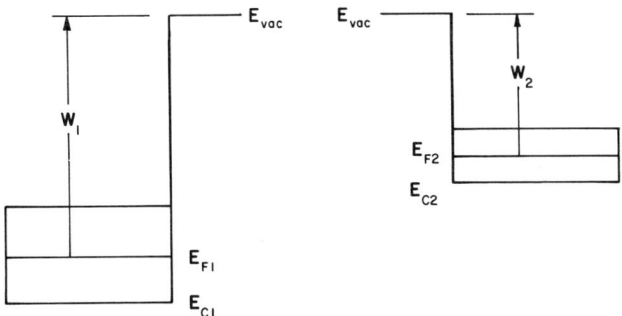

FIG. 65. Energy bands for two widely separated metals.

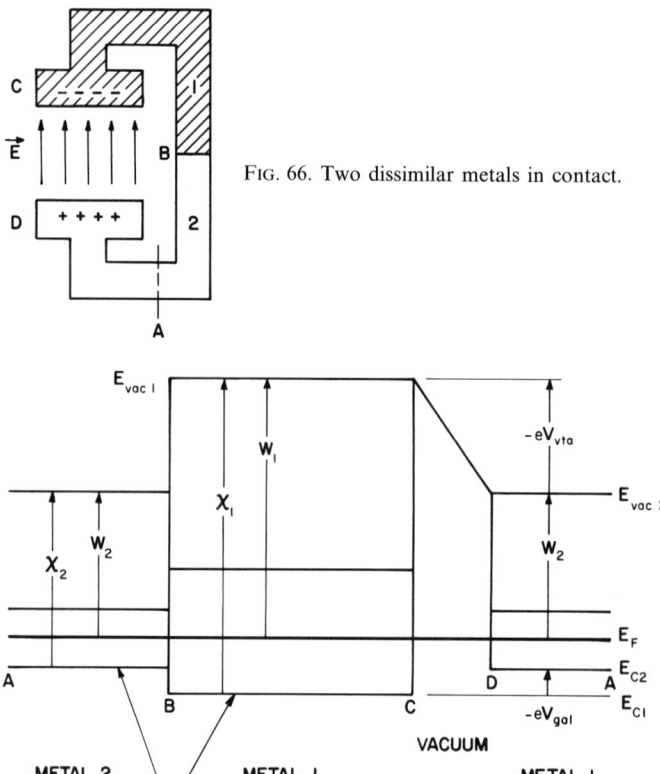

Fig. 66. Two dissimilar metals in contact.

Fig. 67. The energy-band diagram for two metals in contact.

(Fig. 67). This implies that electrons are transferred from the metal with the smaller work function (metal 2) to the other one, and a field **E** is established in the gap CD; the charges responsible for this field lie on the surfaces of the two metals.

Taking the reference level at some point near the bottom of Fig. 65, the assumption of a common vacuum level for the two metals shows that

$$E_{F1} + W_1 = E_{F2} + W_2. \tag{354}$$

After contact and the establishment of equilibrium, all the electrons in metal 2 have lost energy and all those in metal 1 have gained energy since Fig. 67 represents what happens when the two sets of bands shift as indicated. Following Busch and Schade (1976), we write the energy balance equation as

$$(E_F - E_{F1}) - (E_F - E_{F2}) = -eV_{vta}, \tag{355}$$

where the difference between the energy gained and lost is the Volta potential or contact potential difference. Since both metals must always remain at constant potential, this difference shows up as a barrier at the gap or the contact. Combining the above two equations indicates that

$$W_1 - W_2 = -eV_{vta} \tag{356}$$

as the diagram shows. (We shall consider the effect of surface dipole layers on these results at the end of this section.)

Let us now apply (348) to the free electrons in the conduction band of metal 1 by writing it as

$$n_1 = (\pi/3)[8m/h^2)(E_{F1} - E_{C1})]^{3/2} \tag{357}$$

and similarly

$$n_2 = (\pi/3)[(8m/h^2)(E_{F2} - E_{C2})]^{3/2}, \tag{358}$$

where the quantities E_{C1} and E_{C2} represent a shift in the energy reference level to the bottom of the individual conduction bands. Making the assumption that a very small number of electrons are transferred from one metal to another during the establishment of equilibrium, Eqs. (357) and (358) indicate that the position of the Fermi level with respect to the bottom of the conduction band is unaltered in going from Fig. 65 to Fig. 66 since n_1 and n_2 are almost constant. Temporarily designating the energies of Fig. 67 by primes, we have

$$E'_{F1} - E'_{C1} = E_{F1} - E_{C1},$$
$$E'_{F2} - E'_{C2} = E_{F2} - E_{C2},$$

and subtracting, with $E'_{F1} = E'_{F2}$, gives

$$E'_{C2} - E'_{C1} = (E_{C2} - E_{C1}) - (E_{F2} - E_{F1}) \tag{357a}$$

or by (355)

$$E'_{C2} - E'_{C1} = -eV_{gal} = -(eV_{vta}) + (E_{C2} - E_{C1}), \tag{358a}$$

where $-eV_{gal}$, the Galvani potential difference, is stated by van der Ziel (1976) to be the difference in energy between electrons at rest in the two materials. It may also be expressed as

$$-eV_{gal} = -(\chi_2 - W_2) + (\chi_1 - W_1)$$
$$= (\chi_1 - \chi_2) + (-eV_{vta}). \tag{359}$$

The Volta potential can be measured by the Kelvin (or Zisman) method (van der Ziel, 1976), but the Galvani potential cannot be measured and—as we shall see—does not even have a universally accepted definition.

Next, we consider the metal–semiconductor contact or Schottky barrier. Again taking the vacuum level as common, the situation for an isolated metal and an N-type semiconductor, with $W_M > W_{SC}$, is shown in Fig. 68a, while Fig. 68b represents the bands when they are in contact. There is a transfer of charge from the material with the smaller work function (in this case, the semiconductor) and this side of the contact can sustain a space-charge, so that the bands bend upward from the Fermi level as the electrons move out. The band bending is controlled by Eq. (31), and the example shown in Fig. 68b even indicates an inversion layer. The Volta potential establishes a diffusion barrier of magnitude:

$$-eV_D = W_M - W_{SC}. \tag{360}$$

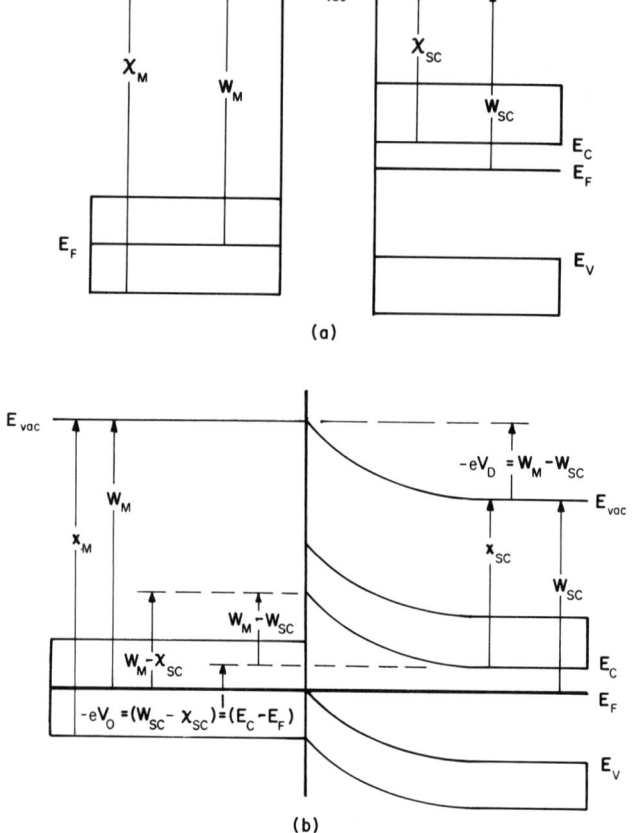

FIG. 68. A metal–semiconductor pair (a) before contact and (b) after contact.

2. THE THEORY OF SEMICONDUCTING JUNCTIONS

This is the energy required to transfer an electron from the semiconductor to the metal. To go the other way, an electron at the Fermi level in the metal must be given $(W_M - \chi_{SC})$ units of energy to just reach the bottom of the conduction band. The effective equilibrium barrier is then

$$-eV_0 = (W_M - \chi_S) - (W_M - W_{SC})$$
$$= W_{SC} - \chi_{SC} \qquad (361)$$

as shown; note that we differentiate between the diffusion barrier $-eV_D$ in the semiconductor and the net barrier $-eV_0$ of the Schottky diode. An electron that attains the bottom of the conduction band at the contact will be urged by the depletion-layer field into the neutral region, where it has a lower energy. The net energy gained is $(E_C - E_F)$, which is identical to $(W_{SC} - \chi_{SC})$.

An alternative interpretation of the effect of the depletion layer has been given by Kano (1972), who argues that the energy needed for an electron to go from the metal to the semiconductor is $(W_M - W_{SC})$, so that this quantity—rather than $(W_M - \chi_{SC})$—represents the barrier for this process. The barrier in the other direction is this quantity diminished by $(E_C - E_F)$ or

$$(W_M - W_{SC}) - (W_{SC} - \chi_{SC}) = W_M - 2W_{SC} + \chi_{SC}.$$

The flaw in the logic here is the inherent assumption that electrons are going from the Fermi level in the metal to the Fermi level in the semiconductor; they are going only as far as the bottom of the conduction band, so that the net energy involved is $(W_M - \chi_{SC})$, where

$$(W_M - \chi_{SC}) > (W_M - W_{SC}).$$

Three other possibilities should be considered. If we let $W_M < W_{SC}$, the bands will be found to bend down rather than up, and the contact is ohmic instead of blocking, since diffusion is enhanced. We can examine this situation by using a very intuitive analogy. The characteristic equation (105) for the holes injected into the N region of a P^+N junction can be rewritten as

$$J_{hN} = (eD_h p_{OP}/L_h)\exp(eV_D/kT)[\exp(-eV_A/kT) - 1]. \qquad (362)$$

A similar expression (van der Ziel, 1976) holds for Schottky diodes; if we regard the action of this device as thermionic emission from metal to semiconductor, then the associated barrier is (Fig. 68b) $(W_M - \chi_{SC})$, and the current will be

$$J = J_S[\exp(-eV_A/kT) - 1], \qquad (363)$$

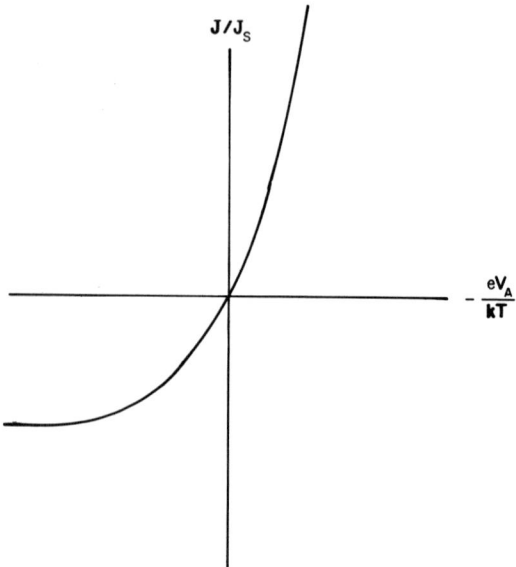

FIG. 69. The effect of an accumulation layer on a Schottky diode.

where
$$J_S = A^*T^2 \exp[-(W_M - \chi_{SC})/kT] \tag{364}$$

and where A^* is known as the modified Richardson constant. (The ordinary Richardson constant A, which comes out of thermionic emission theory, does not take into account the effective mass of the electron.) The significance of Eq. (362) is that it involves eV_D, a negative quantity, in the exponent. Hence, when the barrier in Fig. 68b slopes in the opposite way, the accumulation layer has associated with it an enormously enhanced reverse saturation current. We can indicate this schematically as shown in Fig. 69, from which we infer that the region near the origin can be regarded as "ohmic," that is, approximately linear. This corresponds to the band structure proposed in Fig. 62, and explains why it was assumed that a double layer with negative heights would produce the electrical properties of diode D in Fig. 59. The expression (363), which we make plausible because of its similarity to the PN-junction characteristic, assumes that thermionic emission is the dominant mechanism for carrier transfer. This theory is due to Bethe (1942); it is an alternate to the earlier Schottky and Spenke (1939) treatment, which assumed a diffusion-limited mechanism.

The two types of contacts just discussed have their counterparts when the semiconductor is P-type, except that $W_M > W_{SC}$ gives an ohmic contact and vice versa. All of this, however, ignores the realities of interface states, which we do not consider here. What is of importance is that the possibility of band bending at the contact changes the situation from the discontinuity in E_{vac} at B in Fig. 67 to the continuous behavior of Fig. 68. We visualize the transition from Fig. 68a to Fig. 68b by letting the electrons near the contact cross into the metal. The levels E_{vac}, E_C, and E_V right at the contact need not move since the combination of band bending and a bodily shift of E_F in the semiconductor by an amount equal to the voltage or diffusion potential accomplishes the necessary task of making the Fermi level uniform everywhere.

The same argument applies to the *PN* homojunction. Figure 70 shows the isolated components; we refer back to Fig. 3 for the actual junction, and add a vacuum level to obtain Fig. 71. Again, we imagine that this diagram has been generated by shifting electrons and holes in such a way as to make E_F uniform and, at the same time, keep E_{vac}, E_C, E_1, and E_V at the junction unaltered. This implies that the *P*-region bands bend upward, the *N*-region bands go the opposite way and—since the affinity is a constant of the material—the vacuum level bends with the bands.

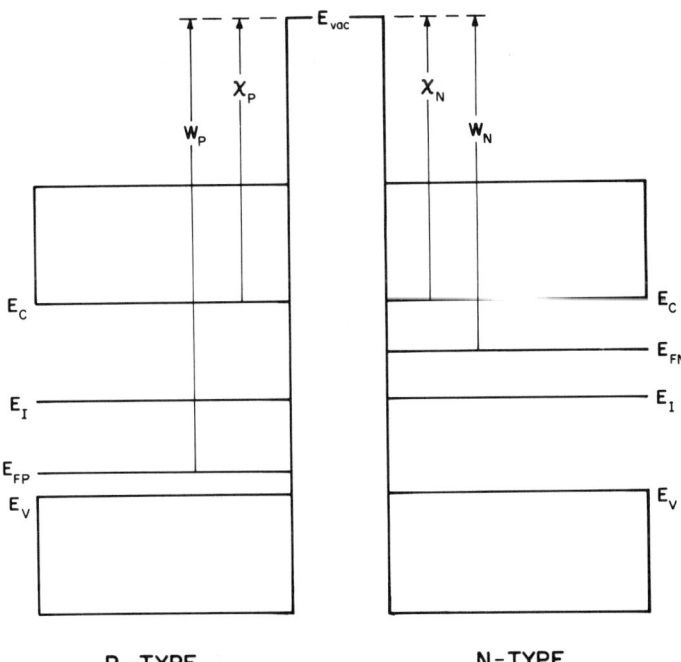

FIG. 70. Energy-band diagram for the components of a homojunction.

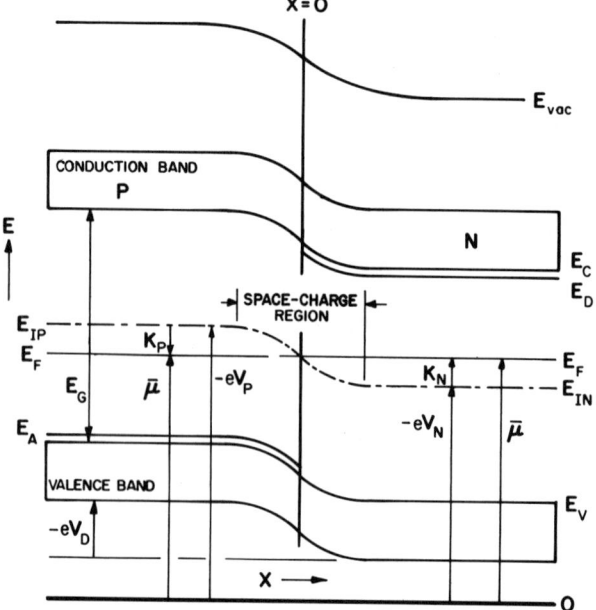

FIG. 71. A *PN* junction in equilibrium.

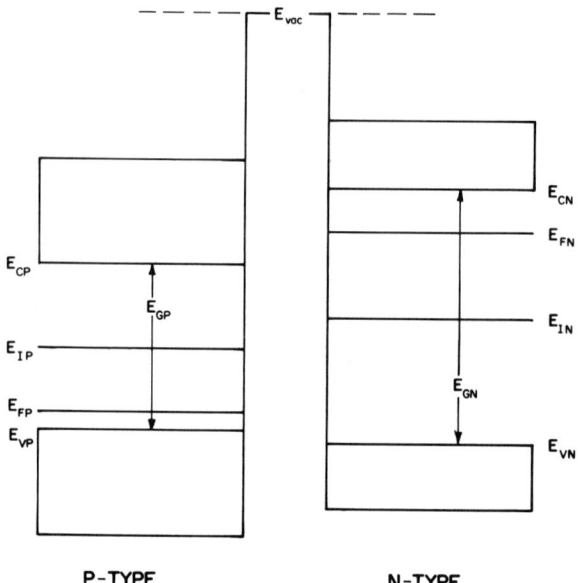

FIG. 72. Energy-band diagram for the components of a heterojunction.

The final example is the heterojunction; the isolated situation (Fig. 72) is like Fig. 70 but with discontinuities in E_C, E_I, and E_V, and Fig. 51 is the result of bringing the two semiconductors together. Now, only E_{vac} is continuous because—as the common reference level—it was that way in the separated materials. Hence, we see the basis for the Anderson model; it represents the natural extension to heterojunctions of concepts which were satisfactory for Figs. 67, 68, and 71. This procedure is logical and self-consistent, but it completely ignores the continuity conditions for the electric field and electrostatic potential which we have previously used for *PN* junctions, and uses the reasoning that if the vacuum level is common for the separated materials, then it must be the only continuous level in the heterojunction.

A completely different approach to these various devices is that of Spenke (1958), who acknowledges the use of unpublished work of Schottky. He indicates that the common theme in the literature in English is to start with the widely separated materials and base the equilibrium energy-band structure on the readjustments which occur as the distance between them is gradually reduced. This is the way the band structures were determined in all the examples taken from van der Ziel, and the method appears to go at least as far back as Torrey and Whitmer (1948).

The Spenke–Schottky approach, on the other hand, makes the point that two materials—no matter how widely separated—cannot be part of the same equilibrium system unless the Fermi levels have already reached a common value. In addition, they do not explicitly use the concept of vacuum level, but instead introduce what is known as the macropotential (this will be defined shortly), a quantity which plays a role somewhat like that of the vacuum level in the theories we have just considered. As we have mentioned a number of times in our treatment of heterojunctions, one of the controversial features associated with the vacuum level is the question of its continuity. This difficulty is avoided in the Schottky–Spenke approach by the use of a fixed energy reference level corresponding to an electron at rest and at a large distance away: basically, then, the difference between what we are about to describe and what has already been considered is that we no longer bring together two completely isolated materials and ask how their bands readjust as they influence one another and reach an equilibrium condition. Instead, we start with the system already in equilibrium (although the two materials need not be touching or even in electrical contact) and determine the resulting band shapes on the basis of the material properties. Before examining the consequences of this principle, we introduce the energy designation scheme of Spenke.

The calculations which lead to Eqs. (1) and (2) can be extended so that they apply to both metals and semiconductors. The density dn of electrons in a solid over the energy range $E, E + dE$, is the product of the density of states dG in this range and the probability $f(E)$ that a state with energy E is occupied, or

$$dn = f(E)\,dG,$$

where dG is given by (350) and $f(E)$ is the Fermi function, Eq. (24). Using the argument of (357) or (358), we may write

$$dG = 4\pi(2m/h^2)^{3/2}(E - E_{\text{pot}})^{1/2}\,dE,$$

where a factor of 2 has been included for spin degeneracy, and where E_{pot} is the minimum potential energy available to an electron (e.g., the bottom of a well or of a conduction band). This can be expressed as

$$dG = (2/\sqrt{\pi})N_S\eta^{1/2}\,d\eta,$$

where the effective density of states, defined as

$$N_S = 2(2\pi mkT/h^2)^{3/2},$$

is like (3) and (4) and where

$$\eta = (E - E_{\text{pot}})/kT. \tag{364a}$$

Then the total density of electrons is

$$n = \int_{E_{\text{pot}}}^{\infty} f(E)\,dG$$

$$= (2/\sqrt{\pi})N_S \int_0^{\infty} \frac{\eta^{1/2}\,\eta}{\exp(\eta - \zeta) + 1}, \tag{365}$$

where

$$\zeta = (E_F - E_{\text{pot}})/kT. \tag{366}$$

In dealing with semiconductors, it is customary to restrict the Fermi level to a region not less than $2kT$ from the bottom of the conduction band (or the top of the valence band). That is, $\zeta \leq -2$ so that the smallest possible value of $\exp(\eta - \zeta)$ is 7.4 and the integral in (365) has an approximate value given by

$$\int_0^{\infty} \exp(\zeta - \eta)\eta\,d\eta = (\sqrt{\pi}/2)e^{\zeta}$$

so that

$$n = N_S e^{\zeta}. \tag{367}$$

Similar arguments produce the corresponding expression for holes, so that we obtain

$$n = N_C \exp\zeta_e \tag{367a}$$

and
$$p = N_V \exp \zeta_h, \quad (367b)$$

where ζ_e is a relabeling of ζ in (366) and where

$$\zeta_h = (E_{pot} - E_F)/kT. \quad (367c)$$

These expressions are what we have already used as Eqs. (1) and (2), with E_{pot} being identified as E_C or E_V, respectively. The integral on the right of (365) is known as the Fermi–Dirac integral and has been studied extensively (Beer, 1963). On the other hand, if $\zeta \geq 2$, then we have to investigate the approximations numerically. Taking $\zeta = 2$, we obtain the following values:

η	0	1	1.0	2.1	2.5	5.0
$e^{\eta-2}$	0.14	0.37	0.90	1.11	1.65	20.09

Hence, for $0 \leq \eta \leq \zeta$, the exponential term is negligible and (365) becomes

$$n = (2/\sqrt{\pi})N_S \int_0^\zeta \sqrt{\eta}\, d\eta = \frac{2}{\sqrt{\pi}} N_S \frac{2}{3} \zeta^{3/2}$$

$$= 8(2m)^{3/2}\pi(E_F - E_{pot})^{3/2}/3h^3 \quad (368)$$

or

$$E_F - E_{pot} = (h^2/8m)(3n/\pi)^{2/3}, \quad (369)$$

which is a form of (348). That portion of the integrand from ζ to ∞ which is not included above is small, making very little contribution to the integral. This is especially true for very low values of T, which is why we re-obtain (348). Actually, the approximation in the range $1.9 < \eta < 2.0$ does not appear to be very good, but when we compare (367) and (368) with an exact numerical integration (Fig. 73), we see that the agreement is excellent for $|\zeta| \geq 2$. This graph expresses (367) as

$$\zeta = \log_e(n/N_S) \quad (370)$$

and (368) as

$$\zeta = [\tfrac{3}{4}\sqrt{\pi}(n/N_S)]^{2/3} = 1.21(n/N_S)^{2/3}. \quad (371)$$

It implies that one kind of approximation is valid when the Fermi level is at least $2kT$ below the bottom of the conduction band while the other one holds when E_F is at least $2kT$ into this band.

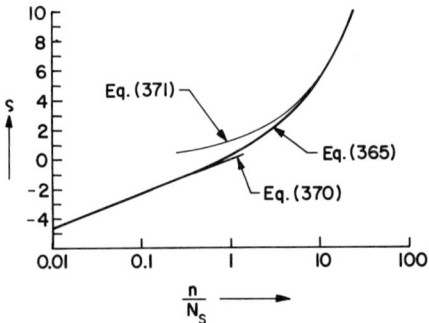

FIG. 73. The integral of Eq. (365) and its approximations.

The first example to consider is the contact between two different metals, as previously analyzed via Fig. 67. It is convenient, in fact, to again start with two widely separated metals, but this time, we require that the Fermi levels E_{F_1} and E_{F_2} have a common value with respect to whatever level is used as a reference. This is shown in Fig. 74, which is similar to the right-hand half of Fig. 67. The differences, however, lie in the absence of a vacuum level and in the use of Spenke's conventions (1958). He chooses the energy reference as some arbitrary level above the external surface energy. The electrostatic potential $-eV$, called the macropotential, is a measure of the energy possessed by an electron just at the surface. One assumption incorporated into the figure (which we shall shortly remove) is that there are no double layers. Hence, there is no external field and the macropotential is everywhere constant outside so that its value just at the surface is a convenient reference for

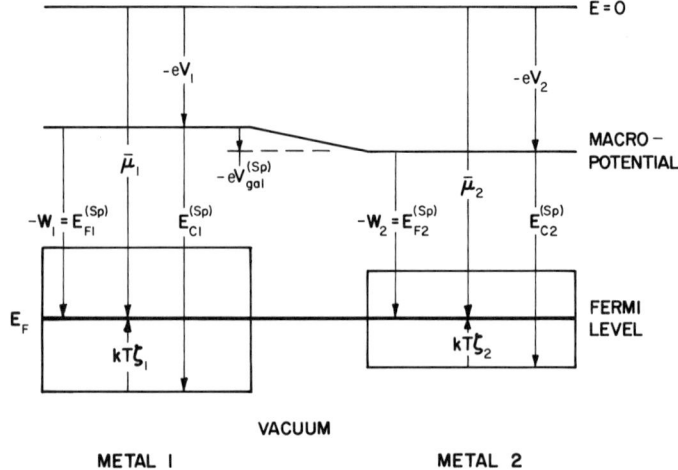

FIG. 74. The Spenke model for two separated metals, assuming no surface dipoles.

two other quantities: the positions $E_F^{(Sp)}$ of the Fermi level and $E_C^{(Sp)}$ for the bottom of the conduction band. The superscripts emphasize the fact that Spenke's reference point is not the same as we used earlier and in addition— as we shall now see—his Fermi energy has a different physical meaning. The quantity $(E_F^{(Sp)} - E_C^{(Sp)})/kT$ can, in fact, be identified as ζ of (366); thus (for either metal)

$$\bar{\mu} = -eV + E_C^{(Sp)} - kT\zeta \qquad (372)$$

and $E_C^{(Sp)} - kT\zeta$ is the chemical potential as we have previously used it. Another feature of Fig. 74 is that the work functions are simply $-E_{F1}^{(Sp)}$ and $-E_{F2}^{(Sp)}$, since there are no dipole layers. If the Fermi level is everywhere constant, then the macropotentials must be different, and the difference represents the contact potential between the two metals. This quantity is identified by Spenke as the Galvani potential $-eV_{gal}^{(Sp)}$, which differs from the definition already given in Fig. 67. (However, the algebraic sum of the two Galvani potentials in Fig. 67 would be zero; this is true for the present figure, and also true for an arbitrary combination of metals.) The conventions of Fig. 74 imply that the symbol $E_F^{(Sp)}$ actually denotes a chemical potential, as is obvious from the relation $\bar{\mu} = -eV + E_F^{(Sp)}$; the work functions are also chemical potentials, in accordance with the Herring and Nichols (1949) definition.

To justify the equality of $\bar{\mu}_1$ and $\bar{\mu}_2$, we use the argument that some elections are always leaving a metal at any finite temperature due to thermionic emission, and others are returning. For two metals that may be widely separated but which are in equilibrium with their surroundings (including each other), the detailed balance principle used in Section 5 requires that the emission and absorption currents be identical. Assuming one-dimensional motion, the number dn_1 of electrons crossing a unit surface of metal 1 with momentum between p_1 and $p_1 + dp_1$ is, in the notation of (365),

$$dn_1 = \frac{(2/h^3)\,dp_1}{\exp[(E_1 - E_{F1})/kT] + 1},$$

where the quantity $2/h^3$ is simply dG of (351) expressed in terms of density per unit volume of phase space, or

$$dn_1 = \frac{(2/h^3)\,dp_1}{\exp(\eta_1 - \zeta_1) + 1}.$$

We can identify η_1 in terms of the kinetic energy $p_1^2/2mkT$, so that

$$dn_1 = \frac{(2/h^3)\,dp_1}{\exp[(p_1^2/2m) + E_{pot\,1} - E_{F1}]/kT] + 1}. \qquad (373)$$

There will be a similar expression for the other metal; applying conservation of energy shows that

$$p_1^2/2m + E_{pot\,1} = p_2^2/2m + E_{pot\,2} \qquad (373a)$$

and differentiating

$$v_1 \, dv_1 = v_2 \, dv_2$$

or

$$v_1 \, dp_1 = v_2 \, dp_2. \tag{374}$$

But the product $v_1 \, dn_1$ measures the current density dJ_1 emitted by metal 1. There is a corresponding expression for dJ_2, and when these are combined with (373) and its analog for metal 2, plus (374), we find that E_{F1} has to be the same as E_{F2}. The Fermi levels therefore have identical positions. Of course, this equalization may be a very slow process at temperatures in the vicinity of 300°K, but, in principle, we can argue that the initial situation must be as depicted in Fig. 74.

This argument can be easily modified to incorporate the surface double layers (Fig. 75). Note that—in the absence of information to the contrary—we are assuming that the dipole layers can even have opposite polarities. The work functions now involve the double-layer potentials, in accordance with (353), and the hypothetical Galvani potential is replaced with the actual Volta difference, so that the macropotential resembles the vacuum level behavior shown in Fig. 67. Furthermore, Fig. 75 indicates that

$$-eV_{\text{vta}} = W_1 - W_2, \tag{375}$$

which we recognize as Eq. (356). We have thus reached our previous conclusion in a completely different manner. That is, we have not used the concept of vacuum level, but have instead expressed the work functions in terms of the macropotential modified by the dipole energy.

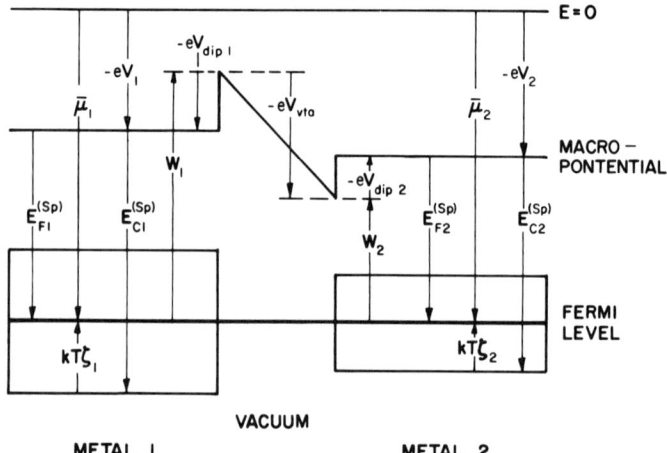

FIG. 75. The effect of including surface dipoles in Fig. 74.

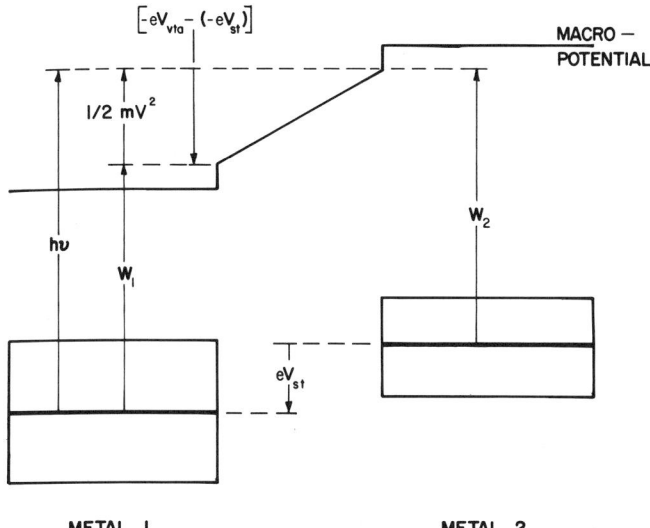

FIG. 76. An interpretation of the photoelectric equation.

An important corollary of this analysis deals with the photoelectric effect. Suppose that metal 1 receives a photon of energy $h\nu$, which causes the release of an electron lying at the Fermi level. We shall assume that it has enough kinetic energy to proceed to metal 2, and we can measure this energy by applying a stopping potential V_{st} between the two metals. That is, $-eV_{st}$ is the energy just necessary to cut off the photoelectric current. Energy conservation appears to require that

$$h\nu - W_1 = eV_{st}, \qquad (376)$$

which is a form of the well-known Einstein photoelectric equation (Nussbaum, 1962; Sproull, 1963). As pointed out by James (1973), this conclusion is incorrect. Figure 76 shows that if the electron just reaches metal 2 with zero energy, then

$$h\nu = -(-eV_{st}) + W_2, \qquad (377)$$

and this is the correct version of (376). Also,

$$h\nu = \tfrac{1}{2}mv^2 + W_1,$$

where $\tfrac{1}{2}mv^2$ is the kinetic energy needed for the transfer from metal 1 to metal 2.

For the case of a metal and an N-type semiconductor, we show in Fig. 77 a situation not considered in Fig. 68; the two materials have reached thermal equilibrium, but they are in contact only through electron emission and absorption. The difference in potential between metal and semiconductor is

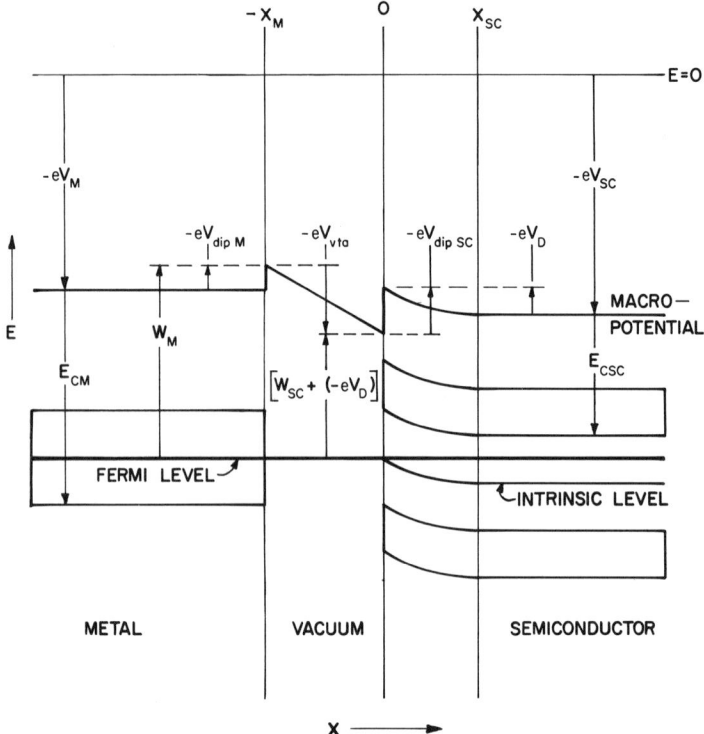

FIG. 77 The Spenke model for an isolated metal and an N-type semiconductor.

a combination of the Volta potential, the diffusion barrier, and the dipole terms. An effective work function of magnitude $[W_{SC} + (-eV_D)]$ is shown at the semiconductor surface $(x = 0)$; this represents the actual value augmented by an amount $-eV_D$ representing the bending of the bands in the semiconductor. Then

$$W_M = W_{SC} + (-eV_D) + (-eV_{vta}). \tag{378}$$

This can be used as a boundary condition at the metal–vacuum interface $(x = -x_M)$. At $x = 0$, the quantity D_n must be continuous, so that

$$\varepsilon E = \varepsilon_0 E_0, \tag{379}$$

where E is the field in the semiconductor and E_0 is the field in the vacuum, given by

$$-eV_{vta} = -ex_M E_0. \tag{380}$$

The third boundary condition is that $E = 0$ at the edge $(x = x_{SC})$ of the space-charge region in the semiconductor. Solutions to Poisson's equation using various approximations, and for other configurations also, have been given by Jordan (1965), Tobin (1968), Murgatroyd (1974), Skinner (1955), and Simmons (1970). The example of Fig. 77 shows that only E_F must be continuous. The boundary conditions given above do not impose this behavior on any other level, and the macropotential, in fact, is discontinuous at the dipole layers.

If the metal and the semiconductor are brought closer together, the bands in the latter bend even more, so that the diffusion potential increases in magnitude and the Volta potential in the vacuum decreases to maintain the validity of (378). This appears to violate the arguments given in connection with Fig. 75; it was stated that the two materials—even though not in contact—are in equilibrium, and the Volta potential is independent of the separation distance (which may even be zero). The situation is different for a semiconductor; the possibility of sustaining a space charge means that the boundary condition, a combination of Eqs. (379) and (380), is satisfied in a different way. This implies that the potentials in both semiconductor and vacuum are a function of spacing, with the final condition as shown in Fig. 78. Note that the Volta potential of Fig. 77 gets absorbed into the semiconductor and that at the junction the common value of the macropotential

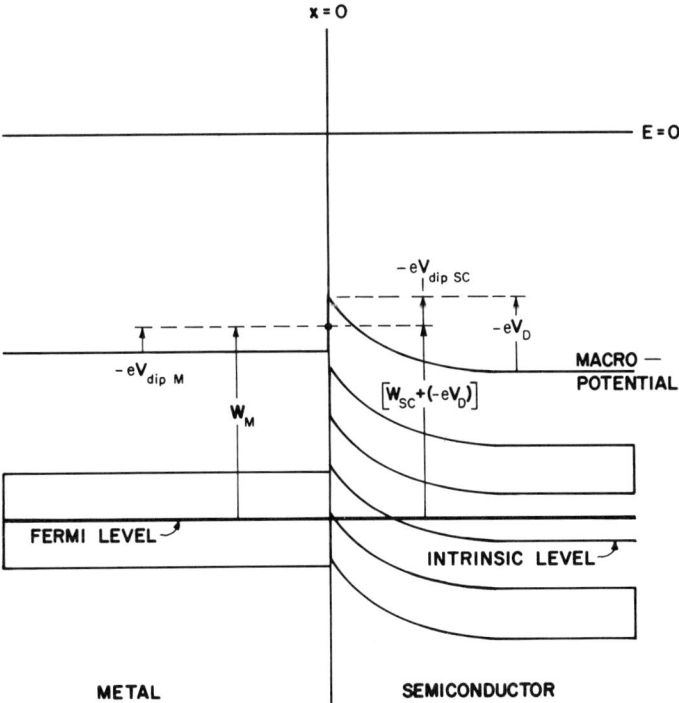

FIG. 78. A Schottky diode.

implies that

$$W_M = W_{SC} - eV_D,$$

agreeing with Eq. (360) and Fig. 68b. Thus, the two theories reach the same conclusion. Spenke remarks that there are some implicit assumptions as well; the dipole layers probably interact (see also Kroemer, 1975) and it is difficult to conceive of two dissimilar materials "touching" each other on an atomic scale.

It is not actually necessary to have an illustration for the semiconductor *PN* junction. The macropotentials at each surface and the dipole energies are identical for the two separated portions, so that when they are brought together, all levels join continuously, as shown for E_I in Figs. 3 or 5. The junction is crystallographically ideal, and the dipoles—being of identical orientation amd magnitude—cause the macropotential to be continuous at the junction (Fig. 79), since their contributions cancel. Hence, no assumptions are necessary to establish the relation

$$W_P - W_N = -eV_D. \tag{381}$$

These diagrams use the conventions of Section 2; to consider the effect of shifting the potential reference from the intrinsic level to the surface, we

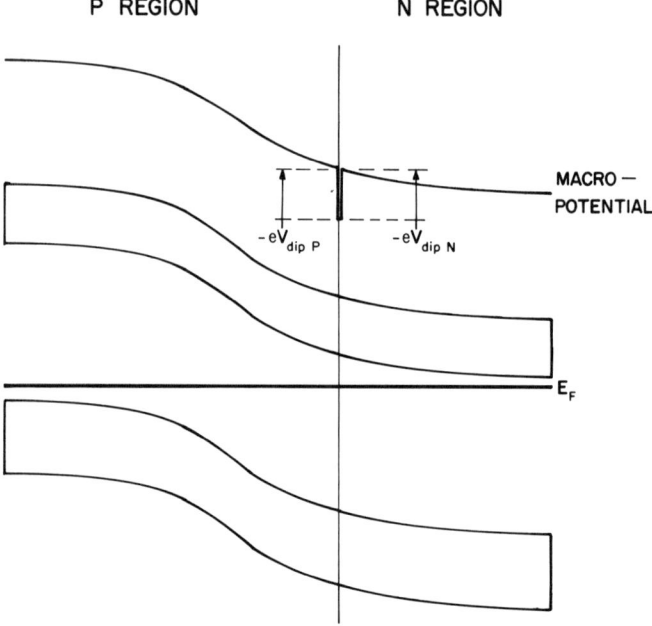

FIG. 79. Macropotential for a homojunction.

2. THE THEORY OF SEMICONDUCTING JUNCTIONS

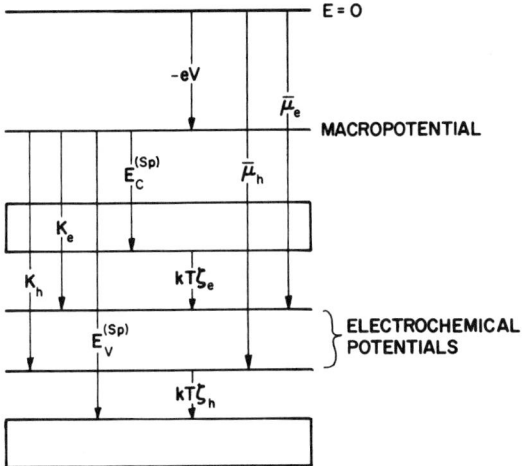

FIG. 80. Definitions of potentials as used by Spenke.

show an arbitrary portion of a space-charge region under low-level injection in Fig. 80. The chemical potentials, in the notation of (372), are

$$K_e = E_C^{(Sp)} + kT\xi_e = E_C^{(Sp)} + kT \log(n/N_C) \quad (382)$$

and

$$K_h = E_V^{(Sp)} + kT\zeta_h = E_V^{(Sp)} - kT \log(p/N_V). \quad (383)$$

If equilibrium is restored, the two electrochemical potentials merge to form a common Fermi level, giving

$$kT(\zeta_e + \zeta_h) = -E_C^{(Sp)} + E_V^{(Sp)} = E_G \quad (384)$$

and

$$\begin{aligned}\bar{\mu}_h &= -eV + E_V^{(Sp)} - kT\zeta_h \\ &= -eV + E_V^{(Sp)} - E_G + kT\zeta_e \\ &= \bar{\mu}_e = \bar{\mu}.\end{aligned} \quad (385)$$

Equations (384) and (385) are internally consistent, so that the definitions of Fig. 80 represent simply an alternative to Fig 2. We have thus shown that the approaches based on either the continuity of the vacuum level or of the behavior of the macropotential give similar results for Schottky diodes and for PN junctions and—in the latter case—the boundary conditions can be equally well applied to the intrinsic level.

Applying Spenke's method to heterojunctions raises some interesting problems. This is something which he did not apparently include in his own work; the second edition (Spenke, 1957) appeared prior to the original paper of Anderson. In order to construct a heterojunction energy-band diagram as Spenke might have viewed it, it is necessary to have numerical values for the surface dipole potentials and the macropotentials. We could use the results of the pseudopotential calculations referred to in Frensley and Kroemer (1976) and Harrison (1977), but it has already been noted that the inherent approximations in such computations make their use a bit risky. What we shall do, therefore, is to modify Fig. 54 by deleting the vacuum level and incorporating the macropotential. We have chosen to use Fig. 54, rather than Fig. 51, because we wish to show that the former is compatible

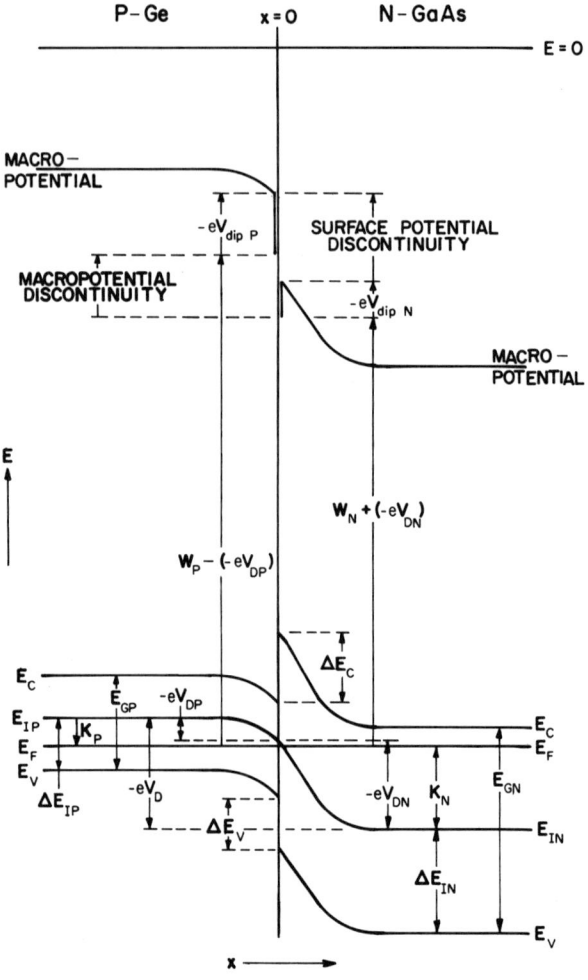

FIG. 81. Band structure for a heterojunction in accordance with the Schottky–Spenke approach.

with Spenke's treatment. Figure 81 shows such a diagram, using the constants given in Table III plus the calculations based on the assumed behavior of E_1. We can locate the two values of the macropotential—as well as the discontinuity between these values—at the junction, because their distance above the Fermi level must be determined by the magnitudes of the work functions in the Ge and the GaAs, respectively. We have assumed that the dipole potentials are similar to those for the homojunction of Fig. 79. The signs are taken to be the same, since this seems reasonable for two materials whose lattice structures are similar enough to form an ideal heterojunction. The magnitudes, on the other hand, could be quite different. Therefore, we do not know the location of the two macropotentials in the neutral regions of this heterojunction, and we are guessing that they could also be quite different. This point is of no particular significance, however; what is important is that there is a discontinuity in the macropotential. Although we have emphasized that the electrostatic potential through a junction is required to be continuous in our original exposition of heterojunction, we realize that the macropotential—which represents the combined contributions of the periodic lattice structure (i.e., the energy bands) and the various processes which produce the surface dipole layer (unbalanced forces, thermionic emission, etc)—can have a discontinuity. We have already considered the possibility of a discontinuity in the vacuum level along the lines of the Ashcroft and Mermin (1976) treatment. Figure 81 proposes similar behavior for both the macropotential and the surface potential of the heterojunction. These discontinuities then result in the creation of external fields; similar (but not identical) situations have been postulated for different faces of metallic and of insulating crystals (Harper, 1967).

It should be clear from the above discussion and the associated figure that an extension of Spenke's approach to heterojunctions is fully compatible with Fig. 54 and, in fact, represents only changes in that model which occur at the surface or externally. That is, the resulting band structures can be identical. On the other hand, the Spenke and the Anderson treatments would be contradictory, for requiring the vacuum level to be continuous at the junction would remove the gap in the macropotential. It should not be inferred from this conclusion that Spenke's point of view then serves as an argument for one model or the other; what we wish to bring out of this discussion is a recognition of the fact that heterojunctions present special difficulties which at this point have not been fully resolved.

It is worth noting that the questions raised above were to some extent anticipated in a remarkable review paper of Domenicali (1954), which appeared at about the time that junction theory was just beginning to be a separate, full-scale discipline. Although this paper was primarily devoted to irreversible thermodynamics and its applications to thermoelectricity, the concept of the equilibrium Volta potential is treated at length in one of the Appendices. The arguments are based on the arrangement shown in Fig. 66. Let us consider the situation before the two ends at B are brought into contact. The Schottky–Spenke picture has thermionic emission taking place (at a very slow rate) and electrons being transferred from one metal to another. Let us assume that this emission occurs at such a negligible rate that we can ignore it in comparison to any electron transfer by conduction. Then we initially have two metals which are in equilibrium with themselves and their surroundings, and the region around them is completely field-free. The electrostatic potential in the space (i.e., the vacuum) around the two separated metals is everywhere uniform, so for convenience we use it as the reference and set it equal to zero. It can be argued that this process is equivalent to Spenke's designation of a macropotential or Anderson's specification of a vacuum level; we shall comment on this point just below.

Now restore the situation of Fig. 66 by having metal 1 touch metal 2 at B, allowing electron transfer and Fermi level equalization to take place, and compute the change in electrochemical potential for a closed path involving the two metals. Starting at D, an arbitrary place on the external flat surface of metal 2, proceeding to a similar point C on metal 1, entering metal 1, passing through the interface at B, and leaving metal 1 via its flat face to return to D, we would expect the individual electrochemical potential changes for the various steps in this process to add up to zero. The reason is that we have not altered the temperature, pressure, or electronic composition of the system by executing this cyclic process. Furthermore, the interior electrochemical potential for the entire system is everywhere constant, so that whatever changes do occur must be associated with a difference between D and C and with differences between exterior and interior at these two points.

To express this behavior of the electrochemical potential quantitatively, we realize that there is a change in the external conditions after the two metals are joined. The charge on the flat faces produces a field in the vacuum, the potential is no longer constant, and we need a reference point which is a long way from the system. Thus, we have reintroduced a vacuum level, but with a significant difference. We shall simply use it as a convenient zero point; unlike Fig. 51, it has no structure, nor is it necessary to postulate any continuity conditions. Now the fact that all the electrochemical potential differences in the closed path described above add up to zero means that the difference from one flat face to another in the vacuum is equal to the contributions (including proper signs) obtained in crossing the two metal–vacuum interfaces. But there is no chemical potential in a vacuum, and the difference between the flat faces reduces to the difference in electrostatic potentials; this is the quantity we have already identified as the Volta potential or contact potential difference. Furthermore, the vanishing of the chemical potential in the vacuum means that each of the two terms across an interface represents the difference between an electrochemical potential inside a metal and an electrostatic potential just outside, which is one of the standard definitions of work function. Domenicali's thought experiment, therefore, represents a simple proof of one of the fundamental properties of contacts: the Volta potential is given by the difference in work functions. The same argument will, of course, apply to Schottky barriers, homojunctions, and heterojunctions. With reference to the last of these, accepting this relation between Volta potential and work functions and at the same time ascribing structure to the vacuum level represents a contradiction in definitions. This simple, almost intuitive argument then reinforces our contention that whatever sort of boundary conditions apply to a heterojunction, they must at least be expressible as internal—rather than external—restrictions.

VI. Summary

It has been the purpose of this article to review the fundamental physics of semiconductor junctions. Using classical statistics and the band theory of solids as a foundation, we have shown how to determine the equilibrium condition of homojunctions, heterojunctions, and devices involving metallic regions. Then we have considered the *PN* junction under low-level injection conditions, presenting the analysis which represents the remarkable achievement of Shockley. The approximations which he used to establish a soluble differential equation were considered, as were those associated with his boundary conditions. The depletion approximation of Schottky was also introduced and compared with exact, numerical calculations.

Next, it was pointed out that one of the most drastic of the approximations just cited is that of direct recombination; its use is satisfactory for germanium, but gives very misleading characteristic curves in the case of silicon. The theory of recombination via imperfections (traps or recombination centers) was developed and the relation between the trapping mechanisms and the electrical properties of the junction was examined. Another fairly drastic approximation in the original Shockley theory results in a boundary condition that assumes majority-carrier injection is essentially negligible. The difficulties associated with this statement were examined as part of a review of the whole topic of boundary conditions, especially those of Fletcher, Rittner, Hauser, and van der Ziel. The various inconsistencies and misconceptions which have been published (including some errors of the present author) were considered in detail with the hope that the dispute that keeps showing up in the literature will be finally settled.

The logical culmination of the process which starts with the low-level approximations is the exact, computer solution of the van Roosbroeck model for a series of silicon diodes, and this was reviewed in a fairly detailed manner in Section 8. One of the most interesting aspects of this work is that it confirms many of the approximations that have been used but also shows when these approximations might no longer be justified. It should be realized that none of the numerical treatments which were considered or cited in this portion of this article investigated in any detail the behavior of the electrochemical potentials as a function of injected current or the resulting effect on the various boundary conditions; this is a problem which is still unfinished business.

The next major topic to be considered was that of heterojunctions. It was shown that there are at least two different ways to extend the theory previously outlined to such devices; at the time of writing, it is believed that the experimental evidence in favor of one model or the other is not definite enough to make a firm statement about which one to choose, and this problem, also, comes under the heading of unfinished business. As part of an attempt to examine the foundations of semiconductor theory as it applies to heterojunctions, the final subject which was considered was a comparison of the two very different ways of looking at the equilibrium structure of combinations of metal and semiconductors. One school of thought starts with two widely separated and noninteracting materials which have a common vacuum level and examines how the bands and free carriers readjust under an interaction produced by bringing them together. The other point of view considers the two materials as always influencing each other, no matter how far apart, and examines the changes that take place when a contact or a junction is formed. As far as we could tell, it turned out that the end results (with the possible exception of the heterojunction) were indistinguishable.

Acknowledgments

The discussion of boundary conditions presented in this article owes much to a number of conversations with my colleague, Prof. R. M. Warner, Jr. Similarly, I have received a great deal of help through correspondence with a former colleague, Prof. K. M. van Vliet of the Université de Montreal. Finally, I would like to acknowledge the contribution of the coeditor, Dr. A. C. Beer, who corrected many errors and made numerous helpful suggestions.

References

Adams, M. J., and Nussbaum, A. (197). *Solid-State Electron.* **22**, 783.
Agakhangan, T. M. (1966). *Radio Eng. Electron. Phys.* (*USSR*) **11**, 240.
Anderson, R. L. (1962). *Solid-State Electron.* **5**, 341.

2. THE THEORY OF SEMICONDUCTING JUNCTIONS

Arandjelovic, V. (1970). *Solid-State Electron.* **13**, 865.
Armstrong, H. L., Metz, E. D., and Weiman, I. (1956). *IEEE Trans. Electron Devices* **3**, 86.
Ashcroft, N. W., and Mermin, N. D. (1976). "Solid State Physics." Holt, New York.
Baranov, L. I., Gamanyuk, V. B., and Usanov, D. A. (1968). *Radio Eng. Electron. Phys. (USSR)* **13**, 1245.
Barber, H. D. (1969). *Solid-State Electron.* **12**, 425.
Barber, H. D. (1968). Personal communication.
Beer, A. C. (1963). "Galvanomagnetic Effects in Semiconductor," Solid State Physics, Supplement 4. Academic Press, New York.
Bellman, R. E., and Kalaba, R. E. (1965). "Quasilinearization and Nonlinear Boundary Value Problems." Am. Elsevier, New York.
Bethe, H. A. (1942). *Mass. Inst. Technol., Radiat. Lab. Rep.* No. 43-12.
Bleaney, B. I, and Bleaney, B. (1976). "Electromagnetic Theory," 3rd ed. Oxford Univ, Press, London and New York.
Brancus, D., and Dolocan, V. (1972). *Int. J. Electron.* **32**, 137.
Buckingham, M. J., and Faulkner, E. A. (1969). *Radio Electron. Eng.* **38**, 33.
Bullis, W. M. (1967). *Solid-State Electron.* **9**, 145.
Busch, G., and Schade, H. (1976). "Lectures on Solid State Physics." Pergamon, Oxford.
Callen, E. (1957). *Am. J. Phys.* **25**, 138.
Callen, H. B. (1960). "Thermodynamics." Wiley, New York.
Calzolari, P. U., and Graffi, S. (1972). *Solid-State Electron.* **15**, 1003.
Caughey, D. M., and Thomas, R. E. (1967). *Proc. IEEE* **55**, 2192.
Chang, Y. F. (1965). *J. Appl. Phys.* **36**, 3350.
Chevychekelov, A. D. (1960). *Sov. Phys.—Solid State* **1**, 1102.
Choo, S. C. (1968). *Solid-State Electron.* **11**, 1069.
Choo, S. C. (1971a). *Solid-State Electron.* **14**, 1201.
Choo, S. C. (1971b). *IEEE Trans. Electron Devices* **18**, 574.
Choo, S. C. (1972). *IEEE Trans. Electron Devices* **19**, 954.
Chou, S. (1971). *Solid-State Electron.* **14**, 811.
Davies, L. W. (1957). *Proc. Phys. Soc. London, Sect. B* **70**, 885.
Davydov, B. (1938). *J. Tech. Phys. (USSR)* **5**, 87.
de Mari, A. (1968). *Solid-State Electron.* **11**, 33.
Domenicali, C. A. (1954). *Rev. Mod. Phys.* **26**, 237.
Faulkner, E. A., and Buckingham, M. J. (1968). *Electron. Lett.* **4**, 359.
Fletcher, N. H. (1957). *J. Electron.* **2**, 609.
Frensley, W. R., and Kroemer, H. (1976). *J. Vac. Sci. Technol.* **13**, 810.
Fulkerson, D. E. (1968). *IEEE Trans. Electron Devices* **15**, 404.
Fulkerson, D. E., and Nussbaum, A. (1966). *Solid-State Electron.* **9**, 709.
Gaertner, W. W. (1960). "Transistors: Principles, Design, and Applications." Van Nostrand-Rheinhold, Princeton, New Jersey.
Gandhi, S. K. (1968). "The Theory and Practice of Microelectronics." Wiley, New York.
Graham, Jr., E. D., and Hauser, J. R., (1972). *Solid-State Electron.* **15**, 303.
Graham, Jr., E. D., and Hauser, J. R. (1970). NSF Report on Grant GK-13752. North Carolina State Univ., Raleigh.
Grove, A. S. (1967). "Physics and Technology of Semiconductor Devices." Wiley, New York.
Guckel, H., Thomas, D. C., Iyengar, S. V., and Demirkol, A. (1977). *Solid-State Electron.* **20**, 647.
Gummel, H. K. (1964). *IEEE Trans. Electron Devices* **11**, 455.
Gummel, H. K. (1967). *Solid-State Electron.* **10**, 209.
Hagstrum, H. D. (1976). *Surf. Sci.* **54**, 197.

Hall, R. N. (1952). *Phys. Rev.* **87**, 387.
Hall, R. N. (1954). *Proc. IRE* **40**, 1512.
Harper, W. R. (1967). "Contact and Frictional Electrification." Oxford Univ. Press, London and New York.
Harrick, N. J. (1962). *J. Appl. Phys.* **29**, 764.
Harrison, W. A. (1977). *J. Vac. Technol.* **14**, 1016.
Hauser, J. R., (1965). *Proc. IEEE* **53**, 743.
Hauser, J. R., (1971). *Solid-State Electron.* **14**, 133.
Hauser, J. R., and Littlejohn, M. A. (1968). *Solid-State Electron.* **11**, 667.
Herlet, A. (1968). *Solid-State Electron.* **11**, 717.
Herring, C., and Nichols, M. H. (1949). *Rev. Mod. Phys.* **21**, 185.
Howes, M. J., and Read, T. G. (1972). *Proc. IEEE* **60**, 329.
Jain, G. C., and Al-Rifai, R. M. S. (1967). *J. Appl. Phys.* **38**, 2701.
Jain, R. K., and van Overstraeten, R. J. (1974). *IEEE Trans. Electron Devices* **21**, 155.
James, A. N. (1973). *Phys. Educ.* **8**, 382.
Jordan, A. G. (1965). *IEEE Trans. Educ.* **8**, 131.
Kano, K. (1972). "Physical and Solid State Electronics." Addison-Wesley, Reading, Massachusetts.
Klein, N. (1976). *IEEE Trans. Electron Devices* **17**, 1094.
Klopfenstein, R. W. (1975). *IEEE Trans. Electron Devices* **22**, 329.
Kroemer, H. (1957). *Proc. IRE* **46**, 1535.
Kroemer, H. (1975). *Crit. Rev. Solid State Sci.* **5**, 555.
Lade, R. W., and Poncelet, C. G. (1964). *Proc. IEEE* **52**, 629.
Lieb, D. P., Jackson, B. D., and Root, C. D. (1962). *IEEE Trans. Electron Devices* **9**, 143.
Lorrain, P., and Corson, D. R. (1970). "Electromagnetic Fields and Waves," 2nd ed. Freeman, San Francisco, California.
Low, G. G. E. (1955). *Proc. Phys. Soc. London, Sect. B* **68**, 310.
McKelvey, J. P. (1966). "Solid State and Semiconductor Physics." Harper, New York.
Matz, A. W. (1960). *In* "Solid State Physics in Electronics and Telecommunications" (M. Désirant and J. L. Michiels, eds.), Vol. 2, p. 1000. Academic Press, New York.
Milnes, A. G., and Feucht, D. L. (1972). "Heterojunctions and Metal-Semiconductor Junctions." Academic Press, New York.
Misawa, T. (1955). *J. Phys. Soc. Jpn.* **10**, 362.
Misawa, T. (1958). *J. Phys. Soc. Jpn.* **11**, 128.
Moll, J. L. (1964). "Physics of Semiconductors." McGraw-Hill, New York.
Moore, R. M. (1969). *IEEE Trans. Electron Devices* **16**, 186.
Moore, W. J. (1962). "Physical Chemistry," 3rd ed. Prentice-Hall, Englewood Cliffs, New Jersey.
Murgatroyd, P. N. (1974). *Am. J. Phys.* **42**, 677.
Nussbaum, A. (1962). "Semiconductor Device Physics." Prentice-Hall, Englewood Cliffs, New Jersey.
Nussbaum, A. (1966). "Electromagnetic and Quantum Properties of Materials." Prentice-Hall, Englewood Cliffs, New Jersey.
Nussbaum, A. (1969). *Solid-State Electron.* **12**, 177.
Nussbaum, A. (1973). *Phys. Status Solidi A* **19**, 441.
Nussbaum, A. (1975). *Solid-State Electron.* **18**, 107.
Nussbaum, A. (1978). *Solid-State Electron.* **21**, 1178.
Nussbaum, A., Cook, E. L., and Korn, D. M. (1977). *Solid-State Electron.* **20**, 583.
Oprea, D. M., and Mandt, T. (1974). Personal communication.
Padovani, F. A. (1971). *In* "Semiconductors and Semimetals" (R. K. Willardson and A. C. Beer, eds.), Vol. 7A, p. 75. Academic Press, New York.

Passau, P., and van Styvendael, M. (1960). In "Solid State Physics in Electronics and Telecommunications" (M. Désirant and J. L. Michiels, eds.), Vol. 1, p. 407. Academic Press, New York.
Rhoderick, E. H. (1970). J. Phys. D **3**, 1153.
Rhoderick, E. H. (1974). Inst. Phys. Conf. Ser. No. 22, p. 3.
Riben, A. R., and Feucht, D. L. (1966). Solid-State Electron. **9**, 1055.
Rittner, E. S. (1954). Phys. Rev. **94**, 1161.
Ross, B., and Madigan, J. R. (1961). Phys. Rev. **108**, 1428.
Sah, C. T. (1966). IEEE Trans. Electron. Devices **13**, 839.
Sah, C. T. (1967a). Proc. IEEE **55**, 654.
Sah, C. T. (1967b). Proc. IEEE **55**, 672, 681.
Sah, C. T., Noyce, R. N., and Shockley, W. (1957). Proc. IRE **45**, 1228.
Sai-Halasz, G. A., Tsu, R., and Esaki, L. (1977). Appl. Phys. Lett. **30**, 651.
Sakaki, H., Chang, L. L., Ludeke, R., Chang, C.-A., Sai-Halasz, G. A., and Esaki, L. (1977). Appl. Phys. Lett. **31**, 211.
Sanchez, M. (1967a). Electron. Lett. **3**, 117.
Sanchez, M. (1967b). Electron. Lett. **3**, 160.
Sanchez, M. (1967c). Electron. Lett. **3**, 176.
Sanchez, M. (1967d). Electron. Lett. **3**, 223.
Sanchez, M. (1968). Z. Natur Forsch., Teil A **23**, 1135.
Schottky, W. (1942). Z. Phys. **118**, 539.
Schottky, W., and Spenke, E. (1939). Wiss. Veroeff. Siemens-Werken **18**, 1.
Sharma, B. L., and Purohit, R. K. (1972). "Semiconductor Heterojunctions." Pergamon, Oxford.
Shay, J. L., Wagner, S., and Phillips, J. C. (1976). Appl. Phys. Lett. **28**, 31.
Shields, J. (1959). Proc. Inst. Electr. Eng., Part B: Suppl. **106**, No. 15, 342.
Shockley, W. (1949). Bell Syst. Tech. J. **28**, 435.
Shockley, W. (1950). "Electrons and Holes in Semiconductors." Van Nostrand-Rheinhold, Princeton, New Jersey.
Shockley, W. (1958). Proc. IRE **46**, 973.
Shockley, W., and Read, W. T., Jr. (1952). Phys. Rev. **87**, 835.
Simmons, J. G. (1970). In "Handbook of Thin Film Technology" (L. I. Maisel and R. Glang, eds.), Chap. 14. McGraw-Hill, New York.
Skinner, S. M. (1955). J. Appl. Phys. **26**, 498, 509.
Spenke, E. (1957). "Electronische Halbleiter." Springer-Verlag, Berlin and New York.
Spenke, E. (1958). "Electronic Semiconductors." McGraw-Hill, New York.
Sproull, R. L. (1963). "Modern Physics," 2nd ed. Wiley, New York.
Stöckmann, F. (1973). Phys. Status Solidi A **20**, 217.
Stone, P. S. (1971). Ph.D. Thesis, Univ. of Minnesota, Minneapolis.
Sze, S. M. (1969). "Physics of Semiconductor Devices." Wiley, New York.
Tansley, T. L. (1971). In "Semiconductors and Semimetals" (R. K. Willardson and A. C. Beer, eds.), Vol. 7A, p. 294. Academic Press, New York.
Tobin, M. C. (1968). Am. J. Phys. **36**, 944.
Torrey, H. C., and Whitmer, C. A. (1948). "Crystal Rectifiers." McGraw-Hill, New York.
van der Maesen, F., Greebe, C. A. A. J., and Nunnink, H. J. C. A. (1962). Philips Res. Rep. **17**, 479.
van der Ziel, A. (1973). Solid-State Electron. **16**, 1509.
van der Ziel, A. (1976). "Solid State Physical Electronics," 3rd ed. Prentice-Hall, Englewood Cliffs, New Jersey.
van de Wiele, F., and Demoulin, E. (1970). Solid-State Electron. **13**, 717.

Vandorpe, D., and Xuong, N. H. (1971). *Electron. Lett.* **7**, 47.
van Roosbroeck, W. (1950). *Bill Syst. Tech. J.* **29**, 560.
van Vliet, K. M. (1966). *Solid-State Electron.* **9**, 185.
van Vliet, K. M. (1979). *Solid-State Electron.* **22**, 443.
van Vliet, K. M., and van der Ziel, A. (1977). *Solid-State Electron.* **20**, 931.
Wagner, S., Shay, J. L., Bachmann, K. J., and Buehler, E. (1975). *Appl. Phys. Lett.* **26**, 229.
Wannier, G. H. (1960). "Elements of Solid State Theory." Cambridge Univ. Press, London and New York.
Webster, W. M. (1954). *Proc. IRE* **42**, 914.
Wigner, E., and Bardeen, J. (1935). *Phys. Rev.* **48**, 84.
Wilson, P. R. (1969). *Solid-State Electron.* **12**, 675.

CHAPTER 3

NEA Semiconductor Photoemitters

John S. Escher

	LIST OF SYMBOLS	196
I.	INTRODUCTION	197
	1. Scope	197
	2. Earlier NEA Review Articles	198
	3. Surface Research Tools	199
II.	PHOTOEMISSION FROM METALS	201
III.	PHOTOEMISSION FROM SEMICONDUCTORS	205
IV.	WORK FUNCTION LOWERING WITH CESIUM AND OXYGEN	211
V.	THE NEA SEMICONDUCTOR PHOTOEMITTER	213
VI.	BASIC NEA-RELATED SURFACE STUDIES	216
	4. Clean Semiconductor Surfaces—Atomic Structure	216
	5. Clean Semiconductor Surfaces—Electronic Structure	220
	6. Activation Related Studies—III–V's	224
	7. Activation Related Studies—Silicon	230
VII.	THE THREE-STEP MODEL OF PHOTOEMISSION	233
	8. Optical Absorption—Electron Generation	234
	9. Electron Transport to the Surface—Electron Diffusion	237
	10. Vacuum Emission—the Surface Escape Probability	242
VIII.	PHOTOCATHODE SENSITIVITY MEASUREMENTS	258
IX.	NEA DEVICES	260
	11. RM NEA Photomultiplier Tubes	260
	12. The Semitransparent (TM) GaAs Photocathode	263
	13. The Semitransparent (TM) Si Photocathode	265
	14. RM NEA Secondary Electron Emitters (RSE)	265
	15. TM Secondary Electron Emitters (TSE)	267
	16. Cold-Cathode Emitters	268
	17. The NEA GaAs Spin-Polarized Electron Source	269
X.	DARK CURRENT EMISSION FROM NEA CATHODES	270
XI.	SHELF LIFE AND OPERATING LIFE OF NEA DEVICES	274
XII.	BIAS-ASSISTED PHOTOEMITTERS	276
	18. TE Photoemission	276
	19. Direct Emitter Cathodes	282
	20. Heterojunction TE Cathodes	284
	21. Dark Current Emission	288
XIII.	SUMMARY	289
	REFERENCES	290

List of Symbols

a	Secondary electron range materials parameter	InGaP	Indium gallium phosphide
A	Ampere	InP	Indium phosphide
Å	Angstrom	IR	Infrared
AES	Auger electron spectroscopy	J_{dark}	Dark current density
Ag	Silver	K	Potassium
AlGaAs	Aluminum gallium arsenide	kcal	Kilocalorie
As	Arsenic	°K	Degrees kelvin
Au	Gold	k_B	Boltzmann's constant
b	Secondary electron range materials parameter	keV	Kilo-electron-volts
c	Speed of light	kV	Kilovolts
°C	Degrees centigrade	L	Electron diffusion length; conduction-band energy level
cm	Centimeter	l_e	Mean photoelectron escape depth
Cs	Cesium	LEED	Low-energy electron diffraction
E	Electric field	Li	Lithium
E_A	Electron affinity	lm	Lumen
E_b	Cs-O interfacial barrier height	mA	Milliampere
E_c	Conduction-band energy	MBE	Molecular beam epitaxy
EDC	Energy (electron) distribution curve	meV	Milli-electron-volts
E_F	Fermi energy	MTF	Modulation transfer function
E_g	Energy gap	n	Electron concentration
ELS	Electron loss spectroscopy	N	Impurity doping concentration
ENI	Equivalent noise input	Na	Sodium
E_p	Primary electron energy	NEA	Negative electron affinity
EPIP	Electrons per incident photon	N_i	Intrinsic carrier concentration
E_{ss}	Energy of surface states	P	Surface escape probability
eV	Electron volt	PE	Photoemission
E_V	Valence-band energy	PES	Photoelectron spectroscopy
EXAFS	Extended x-ray absorption fine structure	PMT	Photomultiplier tube
D	Electron diffusion constant	R	Primary electron range
f	Fraction of primary electrons reflected	Rb	Rubidium
		R_B	Back surface optical reflectance
G	Total dynode chain gain	R_F	Front surface optical reflectance
Ga	Gallium	RM	Reflection mode
GaAlP	Gallium aluminum phosphide	S	Relative surface recombination velocity
GaAs	Gallium arsenide	SA	Sputter–Anneal
GaAsP	Gallium arsenide phosphide	S_b	Back surface recombination velocity
GaP	Gallium phosphide	Si	Silicon
Ge	Germanium	S_R	RM lumen light integrated sensitivity
$G(x)$	Electron generation function	T	Temperature in °K
h	Planck's constant	t_c	Cathode thickness
HC	Heat cleaned	TE	Transferred Electron
Hz	Hertz	TM	Transmission Mode
I_{dark}	Dark current	t_s	Substrate thickness
I_0	Incident light intensity	TSE	Transmission secondary (electron) emission
InAs	Indium arsenide	UHV	Ultrahigh vacuum
InGaAs	Indium gallium arsenide	UPS	Ultraviolet photoelectron spectroscopy
InGaAsP	Indium gallium arsenide phosphide		

V	Volts	ε_p	Average energy to create a secondary electron
VC	Vacuum cleave		
V_d	Diffusion velocity	ε_s	Static relative dielectric constant
W	Depletion depth	λ	Wavelength of light
X	Conduction-band energy level	μ	Electron mobility
Y	Yield	μA	Microamperes
Zn	Zinc	ν	Photon frequency in hertz
$\alpha_{c,s}$	Optical absorption coefficient (cathode, substrate)	σ	Integrated cathode sensitivity (mA/W)
Γ	Conduction-band energy level	τ	Electron lifetime
$\delta_{n,p}$	Relative Fermi level	Φ	Work function
δ_R	Reflection secondary electron gain	$\Phi_{n,p}$	Schottky barrier height (n, p-type semiconductor)
ε_0	Permittivity of free space		

I. Introduction[†]

1. SCOPE

This chapter is an overview of the physics and technology of electron emission into vacuum from p-type semiconductors which have demonstrated, in many cases, outstanding performance in developmental and commercial devices including photoemitters, secondary electron emitters, and cold-cathode emitters. The emphasis here is on the negative electron affinity (NEA) semiconductor surface (III–V's and Si), which is characterized by a high probability for thermalized electron emission into vacuum. GaAs is the most important NEA surface from both an applications standpoint and in terms of basic surface physics studies to date. The chapter concludes with a review of externally biased photocathodes which have been designed for the 1–2 μm wavelength range.

Although the focus here is partly an outline of the surface physics relevant to an understanding of the potential and limitations of the NEA semiconductor cathode, it is clear that much of the work in this area overlaps broader areas. These include the subjects of Schottky barrier formation, chemisorption onto atomically clean surfaces, surface atomic and electronic structure, and heterojunctions between semiconductors. These areas of investigation are developing rapidly with new insights relating to the semiconductor photoemitter being published regularly.

From an applications standpoint, the NEA photocathode was initially put forward by many as being potentially superior to all conventional photoemitters and a wide range of devices and applications would soon follow. Such optimistic predictions have only partially been realized to date as the

---

[†] This work was sponsored by the U.S. Army Night Vision and Electro-Optics Laboratory, Ft. Belvoir, Virginia; and Varian Associates, Inc., Palo Alto, California.

challenging engineering difficulties of transferring a laboratory demonstration into a viable, usable device soon emerged. However, some of these difficulties are being overcome. A major goal of NEA development over the past ten years has been a semitransparent NEA GaAs cathode suitable for low-light-level applications. This development has succeeded with InGaP–GaAs and GaAlAs–GaAs cathodes. Other relatively less demanding applications such as the Cs/GaP dynode and the GaAs reflection-mode photomultiplier have been well-established commercial products for many years. These and other NEA devices are reviewed in Section IX. Whatever success III–V NEA devices have achieved is due to a large extent to the parallel development of III–V epitaxial growth technology. Many of the early growth techniques for GaAs–AlGaAs–GaP, GaAs–GaP, GaAs–GaAsP–GaP, GaAs–InGaP–GaP, GaAs–AlGaAs–GaAs, GaAsSb–GaAs, InGaAs–GaAs, InGaAs–InGaAs–InP, and InGaAsP–InP were first developed for NEA photocathodes. A review of these technologies, however, is beyond the scope of this chapter. A partial review is given by Fisher and Martinelli (1974) and van Oirschot et al. (1977).

2. Earlier NEA Review Articles

The earliest developments of the NEA semiconductor photoemitter, secondary emitter, and cold cathode covering the period from about 1965 to 1973 are reviewed in articles by Sonnenberg (1970a), Bell and Spicer (1970), Williams and Tietjen (1971), Petrov (1971), Spicer and Bell (1972), and van Laar (1973). The most extensive review of NEA cathode technology and physics of emission is the monograph by Bell (1973) covering the developments up to early 1973. The 1974 review articles by Martinelli and Fisher (1974) and Fisher and Martinelli (1974) cover most of the developmental NEA devices and the early materials work on image tube development. The influence of surface electronic structures, together with results on recent devices, are summarized in the article by Spicer (1977). A detailed discussion and the employment of the diffusion model applied to NEA reflection- and transmission-mode photocathodes is emphasized in the extensive articles by Rougeot and Baud (1976, 1979). More recent semitransparent materials and photoemission developments are discussed in the NEA review article by Piaget et al. (1977c).

Conventional and NEA photoemitters have been compared in articles by Sommer (1973a,b) and Thomas (1973). The extensive article by Zwicker (1977) covering both conventional and NEA photoemitters is recommended to the reader not already somewhat familiar with photoemissive devices and desiring a current review. Conventional, or so-called classical photoemitters have been reviewed in the article by Sommer and Spicer (1965) and in the book by Sommer (1968).

3. NEA SEMICONDUCTOR PHOTOEMITTERS 199

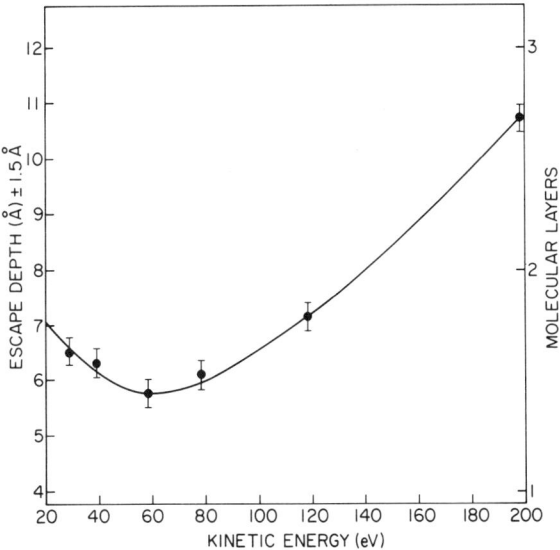

FIG. 1. Experimental escape depth for photoelectrons from GaAs (110). The extremely short escape depth for electrons in this energy range is the basis for many surface-sensitive research techniques. (From Pianetta *et al.*, 1978b.)

3. SURFACE RESEARCH TOOLS

The term "surface" should be clarified at this point. By surface is meant the outer few atomic layers. In this sense, electron emission from conventional and NEA semiconductor photocathodes originates from the bulk. Only for a range of relatively high photon energies is the escape depth for photoexcited electrons sufficiently short to approach surface photoemission (Lindau and Spicer, 1974). An understanding of the structure and electronic nature of the emitting surface of a semiconductor photocathode is essential, however, for a full description of the photoemission process. The successful escape of a photoelectron into vacuum is largely affected by 5–10 Å of the cathode's surface. Fortunately, sufficiently surface-sensitive techniques have been developed that, combined with theory, can give detailed surface information. These surface-sensitive tools can be classified into two general categories: (1) those that use particle beams to probe the surface, and (2) those that use electromagnetic radiation. The surface sensitivity of most of these techniques is due to the very short escape depth into vacuum of electrons in solids, which can be ~ 10 Å. See Fig. 1.

Among the most commonly used techniques in the first classification are low-energy electron diffraction, LEED, Auger electron spectroscopy (AES), and electron energy loss spectroscopy (ELS). These methods use electron beams to probe the surface looking for the spatial distribution of elastically scattered electrons (LEED) or inelastically scattered primary electrons from plasmons, valence band, or core level excitations (ELS), or for characteristic electron energies of emitted electrons from an atom whose core electrons have been excited by the incident primary beam (AES). LEED is especially helpful in analysis of surface crystallography, while AES is useful as a semiquantitative indication of the type and chemical nature of the surface. ELS is useful in studying changes in empty electronic surface states and the chemical nature of the surface. LEED has been reviewed by Jona and Marcus (1977) and Pendry (1974), AES by Riviere (1972) and Chang (1974), and ELS as used by Rowe et al. (1975) and Ludeke and Esaki (1974). A relatively new technique for semiconductor surface analysis is the diffraction of a highly collimated beam of He atoms from the surface which gives complementary structural information to LEED (Cardillo and Becker, 1978).

Photoemission is the basis for the second class of surface techniques. The most basic photoemission measurements are quantum yield and the electron energy distribution curve, EDC. Quantum yield is the number of photoemitted electrons per incident photon, $Y(EPIP)$, and is sometimes expressed as a percentage. If corrections are made to account for the surface reflectance of the photoemitter, then the quantum yield is expressed in electrons per absorbed photon. The EDC measures the spread of kinetic energy of the emitted electrons, usually for a fixed incident photon energy. The quantum yield is essentially the integral over all emitted electron energies of the EDC. If the angle of electron emission is also measured relative to that of the surface and incident photon beam, then the term angularly resolved or angular-dependent photoemission is used. The significance of the energy distribution curve (EDC) is that it contains a great deal of surface and bulk electronic band structure information (e.g., Berglund and Spicer, 1964; Wagner and Spicer, 1974; Petersen and Hagstrom, 1978).

Historically, most photoemission EDCs were divided into ultraviolet and x-ray ranges of photon energy due in part to availability of high-intensity sources. Terms such as UPS and XPS are generally used. More recently a number of research groups have been using synchrotron radiation associated with high-energy particle accelerators as a high-intensity, tunable, polarized source of photons. A number of new photoemission spectroscopy (PES) techniques are possible with the added flexibility of continuously varying the incident photon energy. Studies of both filled and empty surface states is now possible by PES. EDC techniques have been reviewed by Derbenwick et al. (1974) and Fischer (1969) while the use of synchrotron radiation in UPS studies is reviewed by Spicer (1978) and Eastman (1975). A promising new surface-sensitive photoemission technique for bond length determination is the use of synchrotron radiation for extended x-ray absorption fine-structure studies (EXAFS) (Stohr, 1979). Williams (1978) has reviewed many of the techniques mentioned above. A more detailed discussion of experiment and theory of PES applied to solids is given in the volumes edited by Cardona and Levy (1978) and Levy and Cardona (1979).

II. Photoemission from Metals

The earliest confirmation of Einstein's photoelectric theory (Einstein, 1905) came within ten years of 1905 with photoemission measurements on metals, although general acceptance of the theory came much later (e.g., Hughes and DuBridge, 1932). Figure 2 is an energy-band diagram for photoemission from a clean metal surface. E_F is the Fermi level separating the highest-energy filled electron states from a continuum of higher-lying unfilled states. The work function Φ is a measure of the minimum energy needed to remove an electron from E_F into vacuum. One of Einstein's contentions was that the photoemitted electron's energy E would be proportional to the radiation frequency of the incident light v. This can be written as

$$E = hv - \Phi, \tag{1}$$

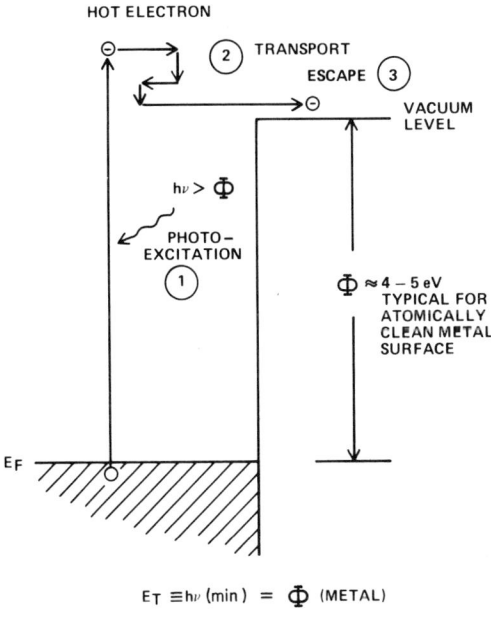

FIG. 2. Electron energy versus distance schematic for a clean metal surface. The nearly elastic scattering shown for the photogenerated electron within the metal is electron–phonon scattering. E_T is the minimum photon energy which will yield photoemission.

where h is Planck's constant, 6.625×10^{-34} J sec. Einstein also proposed that the emitted electron current into vacuum would be linear with the intensity of the incident light. A significant feature of Eq. (1) is the quantum nature of light in which the incident light "particles" or photons can be thought of as having energy $h\nu$ that can be imparted to an electron within the metal. The photoexcited electron can then potentially escape into vacuum if its energy is above the work function.

The quantum efficiency from a metal for photon energies near threshold (i.e., photon energies just above the work function) is generally quite low, $\sim 10^{-4}$. (See Fig. 3.) The reason for this is that the surface optical reflectance from metals is generally quite high and therefore the number of absorbed photons is low. Also, photoexcited electrons in metals, which are now hot electrons, have a high probability of suffering strong inelastic electron–electron collisions before reaching the vacuum interface. A single electron–electron collision is sufficient to reduce a photoexcited electron's initial energy by a large fraction, thereby reducing its probability of emission to essentially zero. The electron–electron scattering length in a metal is a strong function of electron energy varying from ~ 10 Å at 100 eV to several thousand angstroms for energies less than 1 eV. Generally, other scattering processes such as electron–phonon and impurity scattering must be included in calculations of the total attenuation length for hot electrons in metals (or semiconductors),

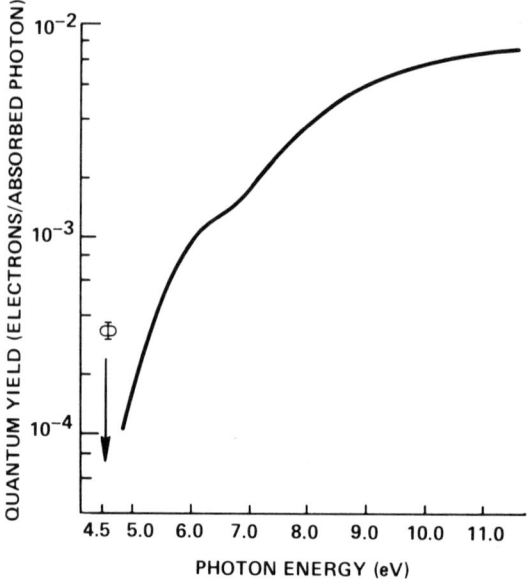

FIG. 3. Photoemission from a clean copper surface. The threshold of the yield from a metal is the work function, Φ. (From Krolikowski and Spicer, 1969.)

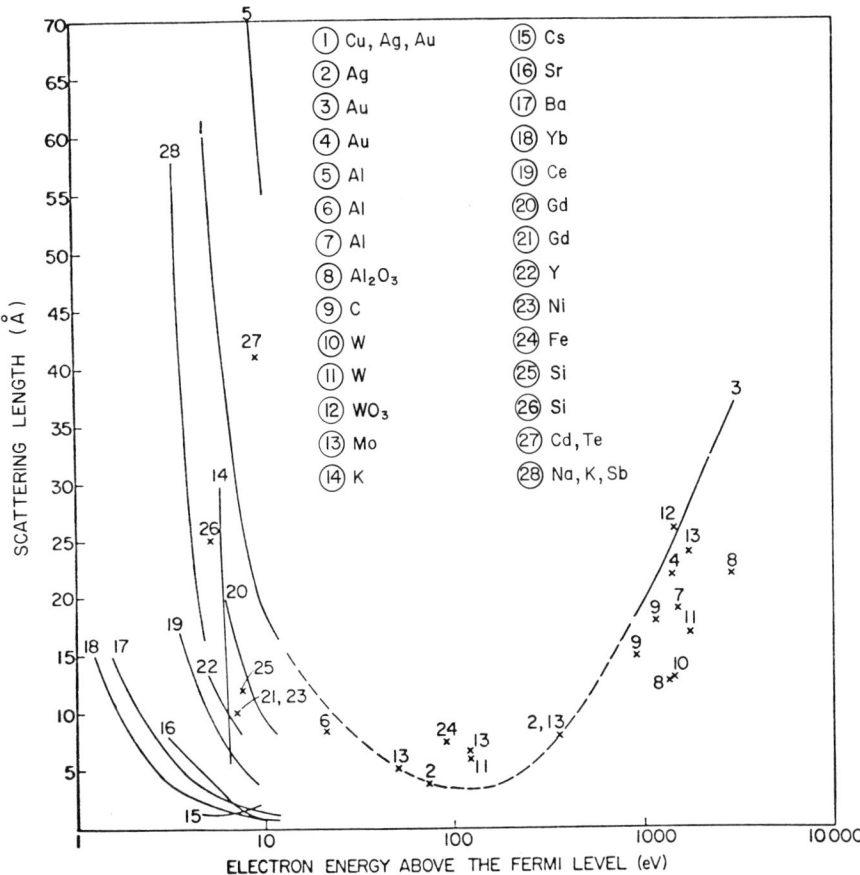

FIG. 4. Scattering length for hot electrons in metals from a compilation by Lindau and Spicer (1974).

especially at the lower energies (Crowell and Sze, 1967). A compilation of experimentally measured electron scattering lengths in metals is given by Lindau and Spicer (1974), and Powell (1974). A summary of results is shown in Fig. 4. Typical optical absorption coefficients $\alpha(h\nu)$ for metals are on the order of 10^5–10^6/cm in the visible and UV range (e.g., Ehrenreich and Phillip, 1962). which implies a mean optical penetration of $1/\alpha \simeq 10^2$–10^3 Å. Hence the majority of the photoelectrons are generated much deeper than the electron escape depth for all but the lowest-energy electrons. Most photoelectrons either never reach the vacuum interface or arrive with insufficient energy to escape.

The work function of a metal can be determined photoelectrically by measuring the minimum photon energy, $h\nu(\min)$, at which photoemission is

TABLE I
Periodic System of The Elements and Work Function Values in eV[a]

IA	IIA	IIIB	IVB	VB	VIB	VIIB	IB	IB	IB	IIB	IIIA	IVA	VA	VIA	VIIA
3 Li 2.9	4 Be 4.98										5 B 4.45	6 C 5.0			
11 Na 2.75	12 Mg 3.66										13 Al 4.28	14 Si 4.85	15 P ⋯	16 S ⋯	
19 K 2.30	20 Ca 2.87	21 Sc 3.5	22 Ti 4.33	23 V 4.3	24 Cr 4.5	25 Mn 4.1	26 Fe 4.5	27 Co 5.0	28 Ni 5.15	29 Cu 4.65	30 Zn 4.33	31 Ga 4.2	32 Ge 5.0	33 As 3.75	34 Se 5.9	Br
37 Rb 2.16	38 Sr 2.59	39 Y 3.1	40 Zr 4.05	41 Nb 4.3	42 Mo 4.6	43 Tc ⋯	44 Ru 4.71	45 Rh 4.98	46 Pd 5.12	47 Ag 4.26	48 Cd 4.22	49 In 4.12	50 Sn 4.42	51 Sb 4.55	52 Te 4.95	I
55 Cs 2.14	56 Ba 2.7	57 La 3.5	72 Hf 3.9	73 Ta 4.25	74 W 4.55	75 Re 4.96	76 Os 4.83	77 Ir 5.27	78 Pt 5.65	79 Au 5.1	80 Hg 4.49	81 Tl 3.84	82 Pb 4.25	83 Bi 4.22	84 Po ⋯	At
87 Fr ⋯	88 Ra ⋯	89 Ac ⋯	90 Th 3.4	91 Pa ⋯	92 U 3.63											
			58 Ce 2.9	59 Pr ⋯	60 Nd 3.2	61 Pm ⋯	62 Sm 2.7	63 Eu 2.5	64 Gd 3.1	65 Tb 3.0	66 Dy ⋯	67 Ho ⋯	68 Er ⋯	69 Tm ⋯	70 Yb ⋯	71 Lu 3.3
			90 Th 3.4	91 Pa ⋯	92 U 3.63	93 Np ⋯										

[a] Reprinted from compilation by Michaelson (1977).

observed. A theory of the temperature dependence of the photoelectric threshold from metals has been given by Fowler (1931). The low-temperature result of Fowler's theory is often used for metal work function measurements (e.g., Eastman, 1970) and analysis of internal photoemission experiments on Schottky barriers (e.g., Sze, 1969, p. 404). According to the Fowler theory the quantum yield Y near the photoelectric threshold can be written as

$$Y \sim (h\nu - \Phi)^2 \quad \text{for} \quad (h\nu - \Phi) \ll \Phi < h\nu \quad \text{and} \quad Y = 0 \quad \text{for} \quad h\nu < \Phi. \quad (2)$$

It is customary to plot $Y^{1/2}$ versus photon energy which should yield a straight line. The work function Φ is then extrapolated from the intercept at $Y = 0$. Work functions of clean metal surfaces range from a low of 2.0 eV for cesium to approximately 5.6 eV for platinum. A compilation of work function values for metals is given by Sommer (1968, p. 21), Eastman (1970), and most recently by Michaelson (1977). [See Table I.] Other important experimental work function techniques are contact potential (Kelvin probe), EDC threshold, and thermionic emission. These and other techniques are discussed by Sommer (1968, p. 13), Derbenwick et al. (1974), and Cardona and Levy (1978, p. 17). The relation between work function and electronegativity of the elements is discussed by Michaelson (1978) and a phenomenological relationship between Φ and sublimation entropies of the elements is discussed by Schade (1979).

III. Photoemission from Semiconductors

Photoemission from p-type semiconductors is the basis for most conventional (polycrystalline) and the newer (single-crystal) photoemissive devices. Semiconductor photoemission (n-type or p-type) is conceptually a three-step process consisting of photoexcitation of an electron from a filled electron energy state to an excited state above the vacuum level, transport to the surface, and escape over a surface barrier into vacuum (Fan, 1945; Apker et al., 1948; Spicer, 1958a). [See Figs. 5–7.] If the initial excited state is not so high as to result in pair production (impact ionization) (Sze, 1969, p. 59), i.e., $h\nu < 2–3E_g$, the dominant hot-electron energy loss mechanisms in a semiconductor become impurity or defect scattering and electron–phonon collisions (Powell et al., 1973). Electron scattering from optical phonons is usually most important in the high-purity single-crystal semiconductor case. Typical optical–phonon mean free paths are ~ 50 Å at 300°K. Since the average energy loss per phonon scattering is on the order of 50 meV, a large number of scattering events may occur before a photoexcited electron thermalizes to the bottom of a conduction-band minimum and eventually recombines. Hence the mean escape depth for $\sim 1–2$ eV hot electrons in a semiconductor is $\sim 100–200$ Å.

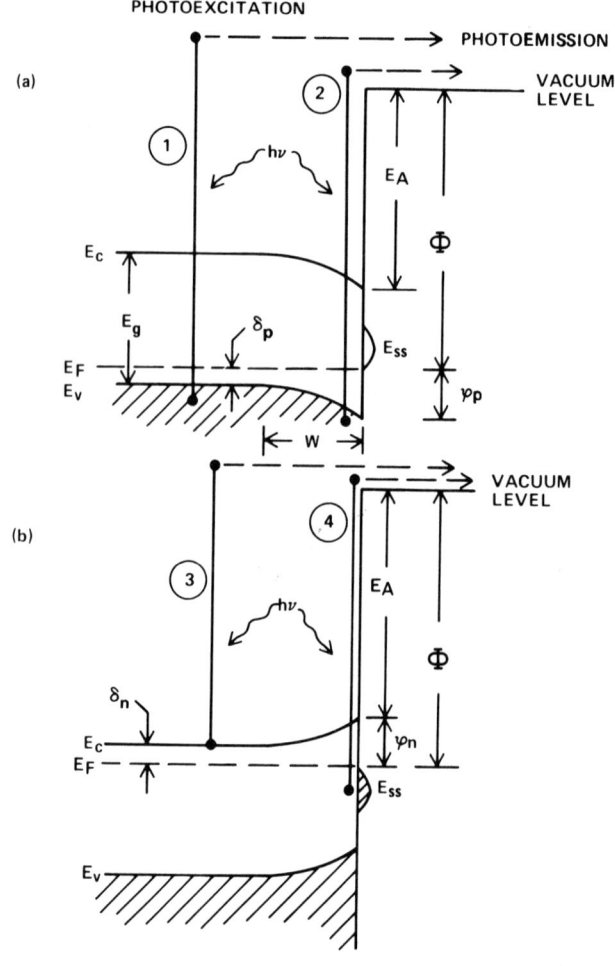

FIG. 5. Schematic electron energy-band diagram for n- and p-type semiconductors, E_c is the conduction-band minimum, E_F the Fermi level, E_V the valence-band minimum, E_{ss} the surface states, W the width of the band-bending region, E_A the electron affinity, $\Phi_{n,p}$ the Schottky barrier height for n- and p-type surfaces, and $h\nu$ the incident photon energy. The distribution of E_{ss} in this figure is schematic only (see Section VI). Photoemission originating from filled valence-band states within the flat band bulk and space-charge surface region are shown as processes (1) and (2). Photoemission from conduction-band states and surface states are shown as processes (3) and (4), respectively.

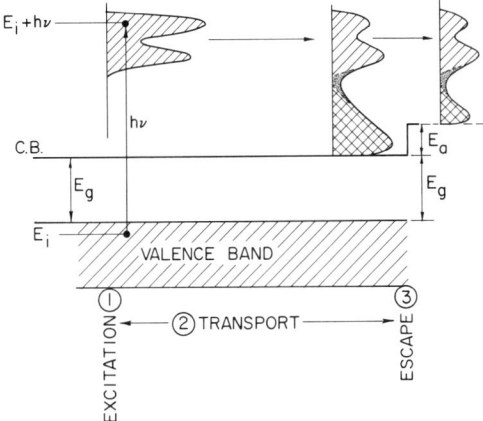

FIG. 6. Schematic energy-band diagram for semiconductor photoemission illustrating the three-step process. An initially optically excited electron distribution is shown on the left. Changes in this distribution occur as it approaches the emitting surface due to scattering. The emitted distribution is shown on the right after the lower-energy electrons have been blocked by the $E_A + E_g$ surface potential. (From Spicer, 1978.)

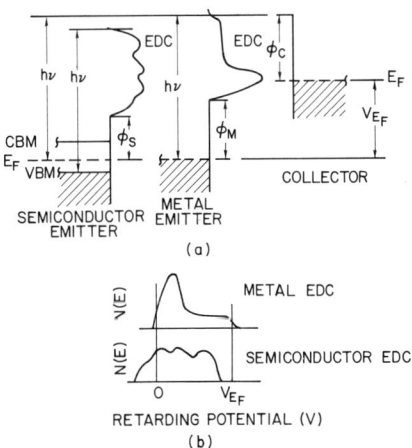

FIG. 7. (a) Schematic electron energy versus distance diagram for electron energy distribution curve (EDC) experiments from metal and semiconductor emitters. Photons of energy $h\nu$ are incident from the right and photoemitted electrons are collected by the metal collector. E_F is the Fermi level, Φ_s, Φ_M, and Φ_C are the semiconductor, metal, and collector work functions, respectively. $V_{E_F} = h\nu - \Phi_C$. (b) Schematic EDCs for metal and semiconductor surfaces. (From Gregory and Spicer, 1976.)

In general, photoemission from a semiconductor can originate from filled valence-band states within the flat band bulk or the space-charge surface region, processes (1) and (2) in Fig. 5, or from conduction-band states (bulk or near the surface) and surface states E_{ss}, processes (3) and (4) in Fig. 5. In almost all cases, however, excitation from the valence band dominates. Even in the case of an n^+ semiconductor, the density of electron states in the conduction band is still small relative to the valence band. Photoemission from surface states has been observed (e.g., Eastman and Grobman, 1972; Wagner and Spicer, 1972; Sebenne et al., 1975), but is also very small due to their local distribution in space (~10–20 Å) and small density of these states ($<10^{15}/cm^2$). Similar arguments hold for emission from defects or deep levels (Sommer and Spicer, 1965). Surface states and trapping levels may, however, play an important role in dark current emission from semiconductor photoemitters (Bell, 1969, 1970). (See Section X.)

Although the position of the Fermi level in the bulk is largely determined by bulk doping and the temperature, the Fermi level at the surface is most often dominated by E_{ss} in the case of III–V's and Si. For p-type semiconductors, the surface states effectively act as a localized positive surface charge which is compensated in the bulk by a negative space-charge region extend-

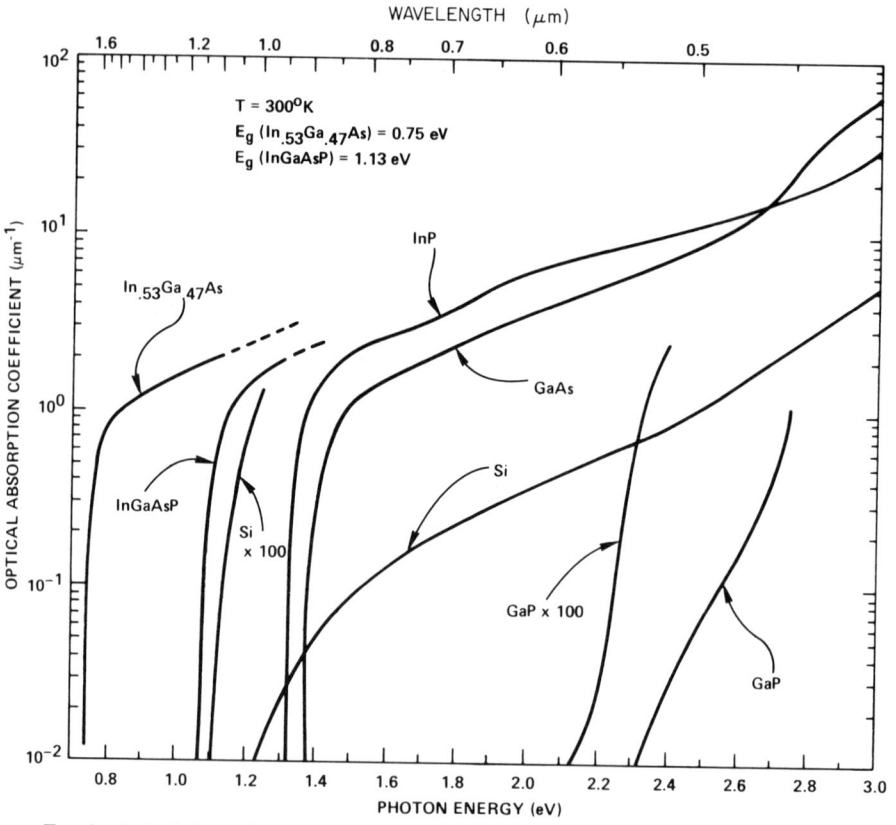

FIG. 8. Optical absorption spectra for some NEA semiconductors. InP, GaAs, and GaP data from Seraphin and Bennett (1967). Si data from Dash and Newman (1955). $In_{0.87}Ga_{0.13}As_{0.30}P_{0.70}$ and $In_{0.53}Ga_{0.47}As$ data from Escher (1979).

ing back into the bulk—analogous to a p–n^+ abrupt junction. Just as with bulk electronic states, surface states can be full or empty depending on their position relative to the Fermi level. An understanding of the nature and role of surface states is clearly important not only in the semiconductor photoemitter case, but also in many semiconductor–metal and semiconductor–insulator devices. (See Section VI.)

In the three-step model of photoemission it is assumed that optical excitation of electrons to higher-energy states is the same as that in bulk optical absorption. Bulk optical absorption coefficients for a number of III–V semiconductors and Si are shown in Fig. 8. GaAs, InP, InGaAsP, and InGaAs have relatively high absorption out to their threshold ($\sim E_g$) characteristic of direct-bandgap semiconductors. GaP and Si are indirect semiconductors and their optical absorption near threshold is relatively weak. The terms direct and indirect refer to whether the initial and final electron momenta are essentially the same (direct) or different because of phonon-assisted transitions (indirect) (Harrison, 1970). Figure 9 indicates several near-bandgap optical transitions possible for GaAs. Except for reflection losses and possible free-carrier absorption (Fan, 1967), semiconductors are essentially optically transparent for photon energies below E_g.

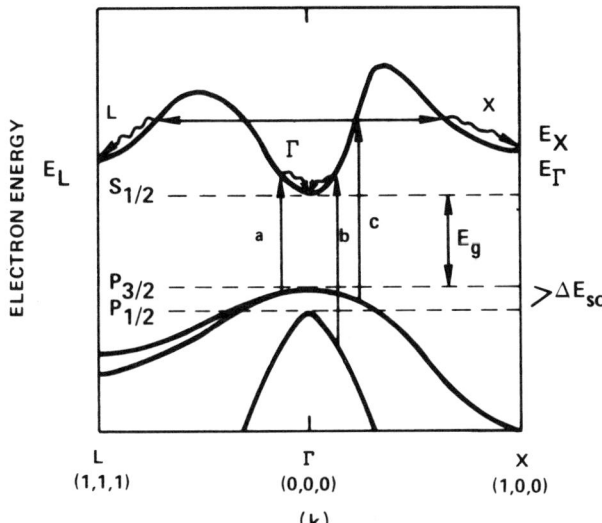

FIG. 9. Electron energy versus momentum diagram for GaAs near the fundamental bandgap. E_g (300°K) = 1.425 eV (Sturge, 1962); $E_L - E_\Gamma$ = 0.35 eV (James and Moll, 1969); $E_X - E_L$ = 0.17 eV (Aspnes et al., 1976); and E_{so} = 0.34 eV (Aspnes and Studna, 1973). a, b, and c are possible optical excitation transitions for incident photons of energy slightly greater than E_g.

In the case of many single-crystal semiconductors, one can conveniently control the width of the space-charge region W and bulk Fermi level position E_F by suitable choice of bulk impurity doping N. The standard Schottky barrier or abrupt one-sided p–n junction solution to the Poisson equation (no applied bias) can be used to relate W to N and Φ_p. Φ_p is defined as the Schottky barrier height on a p-type semiconductor surface. From Sze (1969, p. 370) and Fig. 5,

$$W = [(2\varepsilon_s\varepsilon_0/qN)(\Phi_p - \delta_p - k_B T/q)]^{1/2}, \qquad (3)$$

where q is the electronic charge, 1.602×10^{-19} coul, T the temperature in degrees Kelvin, k_B Boltzmann's constant, 1.38×10^{-23} J/°K, ε_s the semiconductor relative dielectric constant (static), and ε_0 the permittivity of free space, 8.854×10^{-12} coul^2/newton m^2. The relative Fermi level position $\delta_p \equiv E_F - E_v$ can be calculated from (Sze, 1969, p. 38)

$$\delta_p = \tfrac{1}{2}E_g - k_B T \ln(N/N_i), \qquad (4)$$

where N_i is the intrinsic carrier concentration and E_g is the energy bandgap. For GaAs (Si), $E_g = 1.425$ (1.12) eV, $N_i = 1.1 \times 10^7$ (1.06×10^{10})/cm^3 at 300°K (Sze, 1969, p. 58), and $\Phi_p \approx \tfrac{1}{3}E_g$ (Scheer and van Laar, 1967). For high doping levels, $N > 10^{16}$/cm^3, W is quite narrow, <100 Å, and $\delta_p < 0$. However, for $N < 10^{16}$/cm^3, $W > 1000$ Å and $\delta_p > 0.1$ eV. The ability to control the depletion depth is very important in the single-crystal semiconductor photoemitters and also in the interpretation of many PES experiments.

The work function (thermionic) is defined as the energy difference between the vacuum level and the Fermi level. For semiconductors, however, it is convenient to define another surface parameter, the electron affinity E_A, as the energy difference between the vacuum level and the conduction-band minimum E_c *at the surface.*

Note from Fig. 5 that the work function Φ is $E_g + E_A - \Phi_p$ for a p-type semiconductor while it is $E_A + \Phi_n$ for an n-type semiconductor. Therefore Φ can be dependent on the bulk Fermi level position if the surface-state density is so low that E_F is not "pinned" at the surface. The electron affinity, however, is independent of E_F and in that sense is a more fundamental parameter of the semiconductor–vacuum interface. Also note that the threshold for photoemission E_T from a semiconductor is not the work function but rather $E_g + E_A$, assuming $\Phi > E_g$ and W greater than the mean photoelectron escape depth l_e. For rather high doping levels, $N > 10^{18}$/cm^3, the width of the band-bending region W becomes comparable to the mean photoelectron escape depth, and photoemission from the flat valence band region is now possible. For p-type semiconductors, E_T moves toward lower photon energies as $W \to 0$ and approaches the work function assuming $\Phi > E_g$. For n-type semiconductors, E_T remains $E_g + E_A$ but the magnitude of the

TABLE II

SEMICONDUCTOR THRESHOLDS, E_T

	E_T	
	p-type	n-type
$W \gg l_e$ $\Phi > E_g$	$E_A + E_g = \Phi + \Phi_p$	$E'_A + E_g = \Phi + E_g - \Phi_n$
$W \ll l_e$ $\Phi > E_g$	$E_A + E_g - \Phi_p + \delta_p = \Phi + \delta_p \approx \Phi$	$E_A + E_g$
$W \gg l_e$ $\Phi < E_g$	$E_A + E_g$	$E_A + E_g$
$W \ll l_e$ $\Phi < E_g$	E_g NEA surface	$E_A + E_g$

near-threshold yield will likely decrease due to weaker optical absorption within the band-bending region as $W \to 0$. (See Table II.)

The internal electric field within the band-bending region is clearly more favorable for electron emission for the p-type case since the field tends to accelerate photoexcited electrons into vacuum rather than back into the bulk. Therefore from this simple energy-band model of semiconductor photoemission p-type semiconductors will have a lower photon energy (longer wavelength) threshold response and will likely have a higher magnitude of yield than n-type semiconductors (Spicer, 1958b; Sommer, 1958; Scheer, 1960).

A detailed theory for photoelectric emission from semiconductors, including volume and surface-state yields, is given by Kane (1962). Nethercot (1974) gives a simple formula for predicting photoelectric thresholds of binary semiconductor compounds, based on individual electronegativities, that is in good agreement with most experiments. A more recent theoretical discussion of photoemission from solids is contained in the volume edited by Cardona and Levy (1978).

IV. Work Function Lowering with Cesium and Oxygen

It is clear from the work function values of metals in Table I and those of semiconductors, which are on the same order, that simple solids in themselves are incapable of photoemission in the visible or infrared region of the spectrum, i.e., $hv = 1.0–3.0$ eV. (Photon energy and wavelength are related by $\lambda = hc/hv$, where $hc = 12395$ Å/eV.) However, adsorption of approximately a monolayer thickness of cesium onto the atomically clean surface of

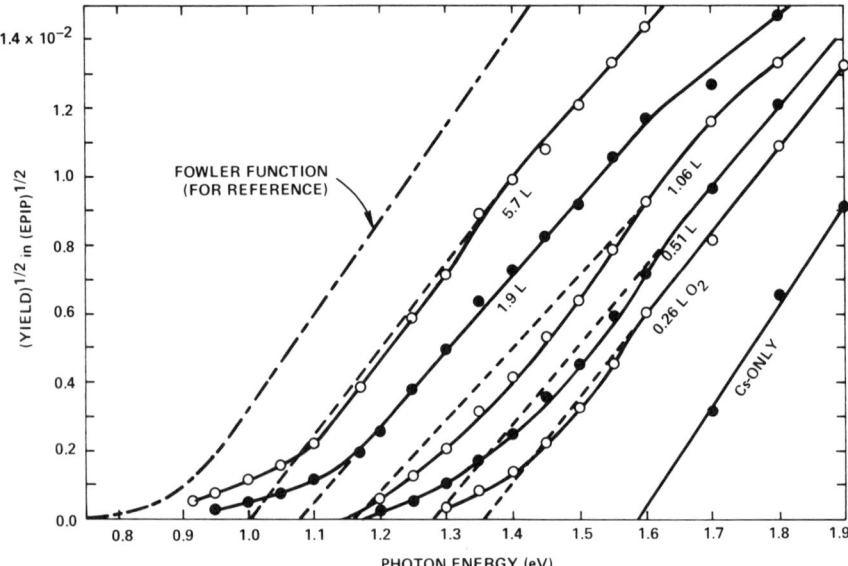

FIG. 10. Fowler plot of the quantum yield from Cs- and Cs + O-activated silver. The "0.26-LO_2" notation means 0.26-L exposure of the surface to oxygen. 1.0 L = 1 × 10^{-6} Torr sec. The linear extrapolation of the (yield)$^{1/2}$ to zero gives the photoemission threshold $E_T = \Phi$. (From Uebbing and James, 1970.)

most metals (and semiconductors) dramatically lowers the work function to ~1.6 eV. In the case of metals, the effect of cesium has been known for over fifty years (e.g., Kingdon and Langmuir, 1923; Taylor and Langmuir, 1933). Although cesium is the most effective element in lowering the function, the other alkali metals, lithium, sodium, potassium, and rubidium, are also effective (e.g., Lang, 1971). It has also been known for a long time that the adsorption of small amounts of oxygen onto a cesiated surface can reduce the metal work function slightly further (Kingdon, 1924). Several treatments of cesium and oxygen of the metal surface are sufficient to lower the photoelectric threshold to approximately 1.0 eV. The Cs + O coating process is often referred to as "activating" the surface. Figure 10 shows the threshold change for Cs + O activation of a clean silver surface. Fluorine in conjunction with cesium and oxygen is often as effective as cesium Cs + O in reducing the work function (e.g., Evans *et al.*, 1968; Garbe, 1970). In some cases Cs will chemically react with the metal or semiconductor giving rise to significantly different photoelectric behavior, e.g., Cs–Au (Sommer, 1943; Spicer, 1962) and Cs–CdTe (Spicer, 1979). The work function lowering capability of the alkali metals (especially cesium) is the basis for all photoemissive devices sensitive in the visible and near-infrared region.

Probably the first theoretical explanation of the changes in work function due to alkali adsorption is that by Langmuir (1932). In this model the initially adsorbed alkali metal atoms, being highly electropositive elements, readily give up their valence electrons to the substrate. There is an image charge induced in the substrate from the positive ions on the surface producing a dipole layer thereby lowering the work function. As the coverage increases, individual dipole strengths weaken from the combined electric field of all other dipoles. Therefore there will be a minimum in the work function versus coverage curve. Theories for alkali adsorption on metals include those of Muscat and Newns (1974), Lang (1971), and references therein. These theories are consistent with most experimental data which show an initial rapid decrease in work function with submonolayer coverage, followed by a less rapid decrease, a minimum, and then an increase up to a full monolayer coverage of the adsorbate ($\sim 10^{15}/\text{cm}^2$). Also consistent with experiments (e.g., Sommer and Spicer, 1965; Weber and Peria, 1969), the theory of Lang shows that the series Cs, Rb, K, Na, and Li give increasingly higher work function minima and density at which the minimum occurs. Factors such as polycrystalline versus single-crystal surfaces or crystal face orientation are generally second-order effects.

V. The NEA Semiconductor Photoemitter

The NEA semiconductor photoemitter is a p^+ semiconductor with $E_g > 1.0$ eV whose atomically clean surface has been suitably activated with cesium and oxygen to reduce the work function below the level of the conduction band *in the bulk*. See Fig. 11. The nearly degenerate p-type doping concentration, $N \sim 10^{18}$–$10^{19}/\text{cm}^3$, is such that W is on the order of the optical–phonon mean free path, $\gtrsim 100$ Å. As a result, photoexcited electrons which thermalized into quasicquilibrium in the lowest conduction-band minimum (e.g., Γ in GaAs) near the surface can escape. These electrons dominate the photoemission threshold yield. The photoelectron escape depth is no longer limited by the hot-electron scattering length l_e, but rather the minority-carrier electron diffusion length L. Since L can be on the order of 1 μm in single-crystal p^+ semiconductors, the increase in photoelectron escape depth and resulting quantum yield near threshold is significant. The term negative electron affinity refers to an effective electron affinity between the vacuum level and the conduction band in the bulk being less than zero. Strictly speaking, E_A is still positive as shown in the band diagram of Fig. 11. During activation with cesium and oxygen the photoemission threshold decreases from $E_g + E_A$ to simply E_g for the NEA surface. The activation of p^+ GaAs to NEA is shown in the experimental reflection-mode quantum yield curves in Fig. 12. Reflection mode, RM, refers to light incident on the electron

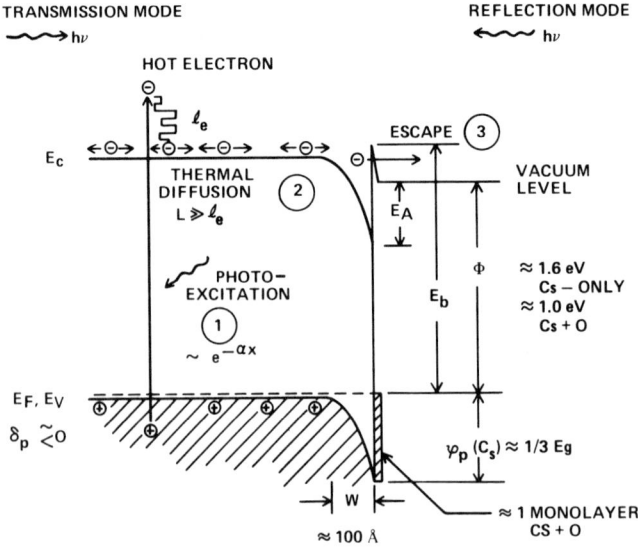

FIG. 11. Schematic energy-band diagram of a Cs + O/p-GaAs negative electron affinity (NEA) surface. L is the thermalized electron diffusion length, l_e the hot-electron scattering length, and E_b an interfacial Cs + O barrier height.

emitting surface of the cathode. Transmission mode, TM, refers to light incident on the back or substrate side of the cathode.

It is apparent from the above outline of photoemission and work function lowering that the NEA surface derives from a straightforward extension of basic semiconductor surface studies on band bending, Fermi level position, etc. (Spicer, 1958a,b; Scheer, 1960; van Laar and Scheer, 1962; Gobeli and Allen, 1965; Allen and Gobeli, 1966). This is, in fact, largely how the NEA surface was proposed and first demonstrated with vacuum-cleaved cesiated p-GaAs (110) surfaces (Scheer and van Laar, 1965). Following previous experience with metals, it was soon experimentally demonstrated by Turnbull and Evans (1968) that Cs + O on GaAs gave superior results to Cs-only. Simultaneously and independently, Bell and Uebbing (1968) discovered Cs + O activation of InP. Cs/p-GaP was recognized very early as a NEA surface (Williams and Simon, 1967) and NEA was demonstrated on many other III–V surfaces soon afterward (see Fisher and Martinelli, 1974). On silicon, NEA was first achieved by Martinelli (1970a). More recently, the clean diamond (111) surface has been identified as exhibiting NEA by Eastman et al. (1979). (Diamond has a bandgap of ~5.5 eV.)

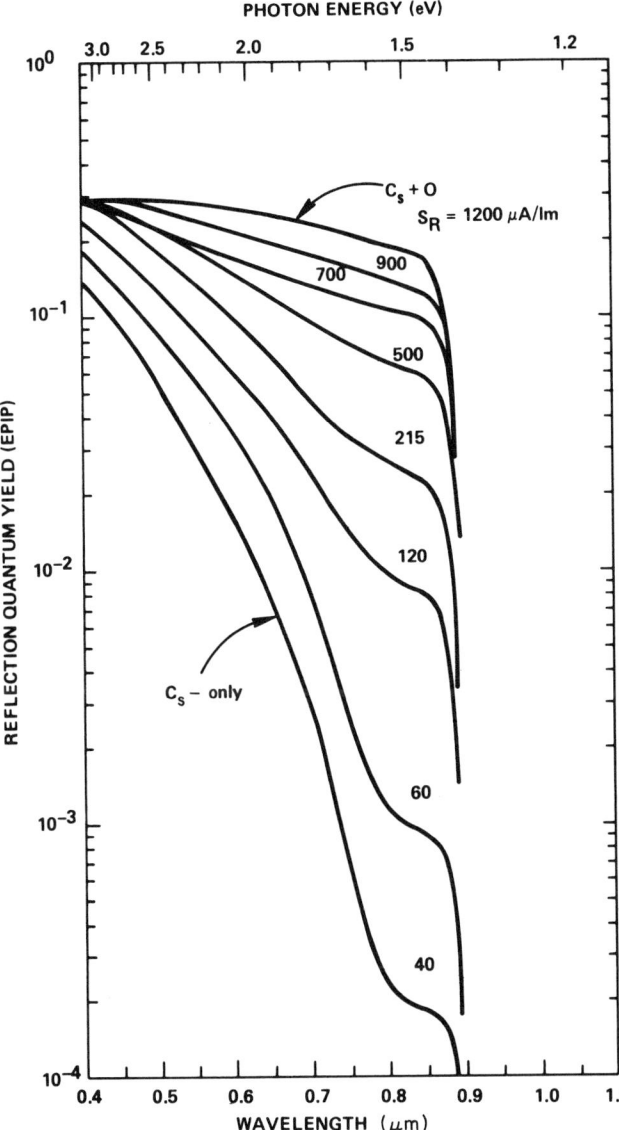

FIG. 12. GaAs experimental reflection-mode quantum yield curves measured during the activation of p-GaAs to NEA. The S_R values are a measure of the integrated sensitivity of the cathode to a standard light source (see Section VIII). The Cs-only yield curve has a noticeable inflection point around 1.6 eV. NEA has not quite been achieved on Cs-only and the inflection point is the approximate work function. The yield (Cs-only) below 1.6 eV is from electrons which are thermally excited over the work function.

A rush of basic and applied NEA research began in a number of laboratories during the late 1960s and early 1970s. These studies have yielded a much deeper understanding of clean and activated semiconductor surfaces. Although most studies have been on vacuum-cleaved surfaces, many features of the practical NEA surface have been extrapolated.

VI. Basic NEA-Related Surface Studies

This section is a brief overview of some of the more basic studies relating to NEA surfaces. The focus of these studies is the atomic and electronic structure of the clean semiconductor surface and what changes, if any, they undergo during the earliest stages of Schottky barrier formation and oxidation. The nature of the intrinsic (clean) surface is the starting point and has been the subject of extensive experimental studies ever since the widespread availability of ultrahigh vacuum (UHV) systems ($\approx$ 1960).

The earliest NEA-related surface studies were concerned, in part, with LEED studies of the clean and activated surface, the Fermi level position at the surface, how the work function changes with cesium coverage, and how much cesium is on the activated surface. More recent work, aided by improved surface sensitive techniques (Section I.3), has focused on the details of Cs and oxygen chemisorption, the nature of intrinsic and extrinsic electronic surface states, and reconstruction models of the clean surface. The basic studies have used, for the most part, bulk-grown samples cleaved in ultrahigh vacuum, i.e., the (110) orientation in the case of III–V's and the (111) in the case of silicon. Most NEA devices, however, use epitaxially grown (100) or (111) B-oriented surfaces which are in situ vacuum cleaned of (most) residual contaminants. Furthermore, in some of the activation-related studies, the exposure of the clean surface to cesium or oxygen is continued many orders of magnitude past the exposure for optimal NEA photoemission in order to study the limiting cases of full Schottky barrier formation, surface oxidation, or thick cesium oxide. Further differences between "ideal" and "real" NEA surfaces and activation methods are included in the discussions below.

4. CLEAN SEMICONDUCTOR SURFACE—ATOMIC STRUCTURE

A nearly atomically clean semiconductor surface (as judged by AES) is essential for successful work function lowering with Cs and oxygen and achievement of NEA (Uebbing, 1969, 1970). This is especially true in regard to the NEA Si (100) surface (Goldstein, 1973; Kapitsa et al., 1976; Holtom and Gundry, 1977) There are only four generally employed in situ vacuum techniques, however, to achieve the required cleanliness. These are cleaving of bulk samples, (VC) (Derbenwick et al., 1974); epitaxial growth by vacuum deposition, e.g., molecular beam epitaxy, (MBE) (Cho and Arthur, 1975; Joyce and Foxon, 1977; Luscher, 1977); ion-bombardment annealing, e.g.,

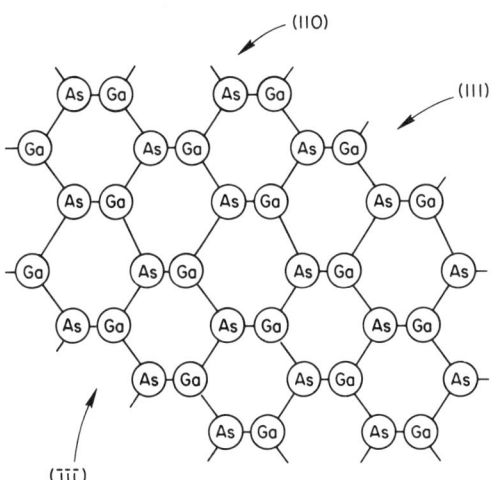

FIG. 13. Schematic termination of the GaAs lattice along three low index planes. The view is along the (110) direction. The (110) is the natural cleavage face for GaAs and many other III–V's. The (111) Ga face is often called the A face and the (1̄1̄1̄) As the B face. Real semiconductor surfaces show slight rearrangement or "reconstruction" of the positions of the top few atomic layers of the surface from their natural bulk positions. (From Pianetta et al., 1978a.)

argon sputter–annealing, (SA) (Wehner, 1975); and simple vacuum heat cleaning to near the congruent evaporation temperature, (HC) (Jona, 1965a; Liu et al. 1969; James et al. 1971a; Miyao et al. 1978). The first two procedures are preferred for more fundamental surface studies, while the latter two have been employed for actual device studies and fabrication. Although NEA can be achieved on any of the low index faces of GaAs, e.g., (100), (111) A and B, and (110) (James et al., 1971a), only the (100) face of Si can be activated to NEA (Goldstein, 1973; Martinelli, 1973a). [See Fig. 13.] The natural cleavage face for Si, however, is the (111) and therefore some caution should be given to relating Si (111) surface studies to the NEA Si (100) surface.

It is well established that the atomic position of the topmost layers of most semiconductor surfaces are not in their bulk positions. Furthermore, it is apparent that even small "relaxations" or even localized defects in the surface can have a significant effect on surface electronic states, chemisorption, etc. (Spicer et al., 1979; Chye et al., 1979). Surface reconstruction occurs on VC surfaces as well as surfaces cleaned by other surface preparation techniques (Mark et al., 1977). Experimental LEED intensity spectra (I–V) measurements combined with structural model calculations have been particularly powerful techniques in determining surface crystallography (Pendry, 1974; Jona and Marcus, 1977; Tong, 1978). In the case of GaAs (110), which has equal numbers of surface Ga and As atoms, the topmost As atoms are believed to be displaced outward (approximately 0.3 Å), while the topmost Ga atoms are

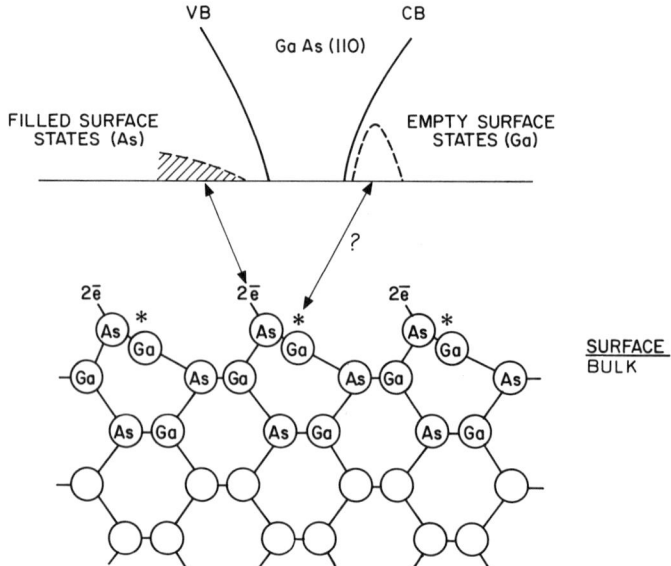

FIG. 14. Schematic diagram of the surface state distribution near the fundamental bandgap of GaAs (110) and the atomic surface reconstruction for this surface. VB and CB refer to the valence- and conduction-band edges, respectively. Other faces of GaAs likely have a somewhat different electronic and atomic structure than the (110). (From Pianetta et al., 1978a.)

displaced inward (approximately 0.5 Å) relative to the ideal surface (Lubinsky et al., 1976; Khan et al., 1978a,b). Slight subsurface displacements are also consistent with LEED data (Kahn et al., 1978a). [See Fig. 14.]

There is also LEED evidence for reconstruction on the polar faces of GaAs, i.e., the (111) Ga, (111) As, and the (100) (MacRae, 1966; Cho and Arthur, 1975; van Bommel et al., 1978; Jacobi et al., 1979). These surfaces must be prepared by SA, HC, or MBE techniques. It is not unusual to observe several different LEED patterns on these surfaces, depending on particular vacuum heat treatment cycles, etc. (e.g., Jona, 1965b; Arthur, 1974; Ranke Jacobi, 1977; Drathen et al., 1978). Structural models for the polar faces of GaAs have not been established to date. Although a nearly atomically clean surface is a prerequisite for subsequent achievement of NEA on GaAs, a particular LEED pattern or any LEED pattern at all on the clean surface is not (Goldstein, 1975; Stocker, 1975).

In the case of the NEA Si (100) surface, SA surface preparation techniques are generally used and a reconstructed (2 × 1) LEED pattern observed (Jona, 1965b; Goldstein, 1973; Gundry et al., 1974; Holtom and Gundry, 1977; Jona et al., 1977; White and Woodruff, 1977). A (4 × 2) reconstruction was also observed in the earliest LEED studies on Si (100) (Lander and Morrison, 1962) and confirmed by thermal He beam diffraction studies (Cardillo and

Becker, 1978). However, Gundry et al., (1974) have attributed this structure to (111) facets. They could not activate this surface to NEA. Only the Si (100) (2 × 1) surface has been successfully activated to NEA.

Over a half dozen structural models for the Si (100) (2 × 1) surface have been proposed. These can be divided into two broad classifications: those which involve vacancies in the top atomic layer and those with no vacancies but which include relatively large top surface atomic displacements (including rotations) from bulk positions (Tong and Maldonado, 1978). The recent LEED data and surface models suggest that the (2 × 1) reconstruction may involve as many as four or five surface layers (Appelbaum and Hamann, 1978). A modification of the early surface model of Schlier and Farnsworth (1957), in which adjacent rows of Si surface atoms rotate to form double rows and parallel caves, has been successfully used in a detailed model of the Si NEA surface by Levine (1973). [See Section VI.7.] There is no general agreement on a structural model for the Si (100) (2 × 1) surface; however, most evidence is against the vacancy-type reconstruction and favors some form of pairing. In the case of VC Si (111), the as-cleaved surface generally exhibits a (2 × 1) LEED pattern while a vacuum-annealing cycle yields a (7 × 7) pattern (Schlier and Farnsworth, 1959; Mönch, 1977; Mönch et al., 1979).

In contrast to nearly "ideal" (reconstructed) VC or MBE surfaces, SA and HC surfaces are likely much more disordered. SA GaAs (110) surfaces can show nearly identical LEED patterns with VC surfaces, but exhibit a much different surface electronic structure—presumably due to residual sputter damage within the unit cell itself that is not fully annealed (Skeath et al., 1978). In the case of III–V's, the HC surface and to a lesser extent the SA surface will likely suffer preferential column V vacancies, perhaps extending several atomic layers below the surface (Liu et al., 1969; Cho, 1971; Foxon et al., 1973; Trueba et al., 1974; Bayliss and Kirk, 1975, 1976). If the vacuum HC temperature-time schedule is increased much beyond that just sufficient to remove the residual oxygen on the surface, a visually obvious surface decomposition occurs in the case of III–V's (Foxon et al., 1973; Fisher and Martinelli, 1974). [See Fig. 15.] The largest features on the surface consist of "balls" or droplets of the group III metal on the order of a micron or more in diameter. A much finer (200–1000 Å) thermal etch structure has been noted by Bradley et al. (1977) on vacuum HC GaAs (111) B surfaces. Rectangular pits with (111) facets were observed by Foxon et al. (1973) on HC (100) surfaces. Thermally etched Si (111) surfaces (850°C, 1 h) were noted by Jona (1965a). Farrow (1975) has pointed out, however, that a reasonably high partial pressure of P during vacuum HC of InP can yield a nearly stoichiometric atomically clean surface. Vacuum HC of GaAs substrates in an As partial pressure of $\sim 10^{-6}$ Torr is often employed with MBE growth studies (e.g., Foxon et al., 1973; Cho and Dernier, 1978) but has not been reported for NEA III–V device experiments.

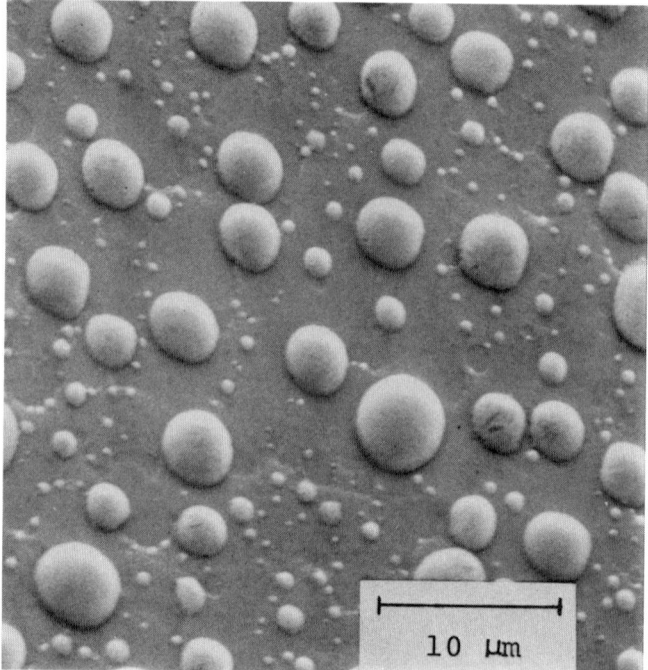

FIG. 15. Scanning electron microscope picture of an InP surface which has been heat cleaned in vacuum at a temperature considerably above that necessary to remove residual surface contaminants. The droplets are indium.

LEED studies of SA and HC surfaces show that (111) and (100) surfaces of GaAs often exhibit (110) facets (MacRae, 1966; Chen, 1971; James *et al.*, 1971a; Arthur, 1974). Residual carbon on the surface enhances this tendency according to Arthur (1974). A practical consequence of a nonplanar cathode surface is a potentially higher transverse electron velocity distribution than from a perfectly planar emitting surface. The transverse velocity is directly related to the spatial resolution of proximity focused image tubes (Martinelli, 1973b; Bradley *et al.*, 1977). In summary, it is likely that most "real" NEA III–V surfaces employed in device fabrication are to some degree nonstoichiometric and faceted.

5. CLEAN SEMICONDUCTOR SURFACES—ELECTRONIC STRUCTURE

It has long been recognized that abrupt termination of the bulk lattice, i.e., breaking of bonds at the surface, should lead to a new set of localized electronic states different from that of the bulk (Tamm, 1933; Shockley, 1939). If there is a sufficiently high density of these localized surface states and they happen to fall within the bandgap (e.g., E_{ss} in Fig. 5), the bands will bend at the surface. Experimental evidence for the effect of surface states (band bending) was later recognized in a number of semiconductor–metal and semiconductor–insulator studies (e.g., Shockley and Pearson, 1948), including semiconductor photoemission (Spicer, 1958b).

The earliest UHV studies ($\lesssim$ 1970) were concerned, in part, with measuring work function, electron affinity, and Fermi level position at the surface of silicon (Scheer, 1960; van Laar and Scheer, 1962; Gobeli and Allen, 1962; Allen and Gobeli, 1964, 1966; Henzler, 1967; Callcott, 1967; Fischer, 1968; van Laar and Scheer, 1968; Mönch, 1970) and a number of III–V's (Gobeli and Allen, 1965; Fischer, 1966a,b; van Laar and Scheer, 1967; Fischer *et al.*, 1967; Scheer and van Laar, 1967; van Laar and Scheer, 1968). Many of these included studies with cesium. More recent experiments, aided by newer surface-sensitive techniques (Section I.3), have focused on measuring the distribution and density of filled and empty surface states directly (e.g., Ludeke and Esaki, 1974; Lapeyre and Anderson, 1975; Spicer *et al.*, 1977).

In the case of VC GaAs (110) surfaces, it is generally agreed that for "good quality" cleaves, the Fermi level at the surface is not "pinned", but is very close to its bulk position—i.e., there is little, if any, band bending (van Laar and Scheer, 1967; Scheer and van Laar, 1969; Huijser and van Laar, 1975; Lindau *et al.*, 1977). Similar conclusions have been reached in regard to VC InP, GaSb, InAs, and *p*-type GaP (Chye *et al.*, 1975; van Laar and Huijser, 1976; Huijser *et al.*, 1977; Williams *et al.*, 1977; Mönch and Clemens, 1979). Only *n*-type GaP consistently shows band bending with the Fermi level at 0.55–0.62 eV below the conduction-band edge (Huijser *et al.*, 1977; Norman *et al.*, 1977; Guichar *et al.*, 1979).

A surface state distribution for the "ideal" VC GaAs (110) surface is outlined in Fig. 14. [This model was originally proposed by Gregory *et al.* (1974) and slightly modified later by Spicer *et al.* (1976).] There are no intrinsic filled or empty surface states within the bandgap itself in agreement with the no Fermi level pinning experiments and theoretical calculations that take reconstruction into account (e.g., Chadi, 1978; Chelikowsky and Cohen, 1979). The filled surface states near the valence-band maximum are associated with surface As atoms (Ranke and Jacobi, 1977). The atomic surface reconstruction and resulting charge rearrangement is thought to result in the surface As atoms having two "dangling bond" electrons not involved with back bonding to the bulk crystal. The surface As atoms can contribute these electrons to the filled surface state band and may account for the much higher chemical activity of the GaAs (111) As (B face) relative to the (111) Ga (A face) (Gatos, 1961; Pianetta *et al.*, 1978a). It is interesting to note that earlier theoretical calculations of the distribution of surface states for the *un*reconstructed GaAs (110) surface indicated that there should be intrinsic surface states within the bandgap. Atomic reconstruction, however, moves these states outside the bandgap. GaP (110) seems to be an exception (Bertoni *et al.*, 1978).

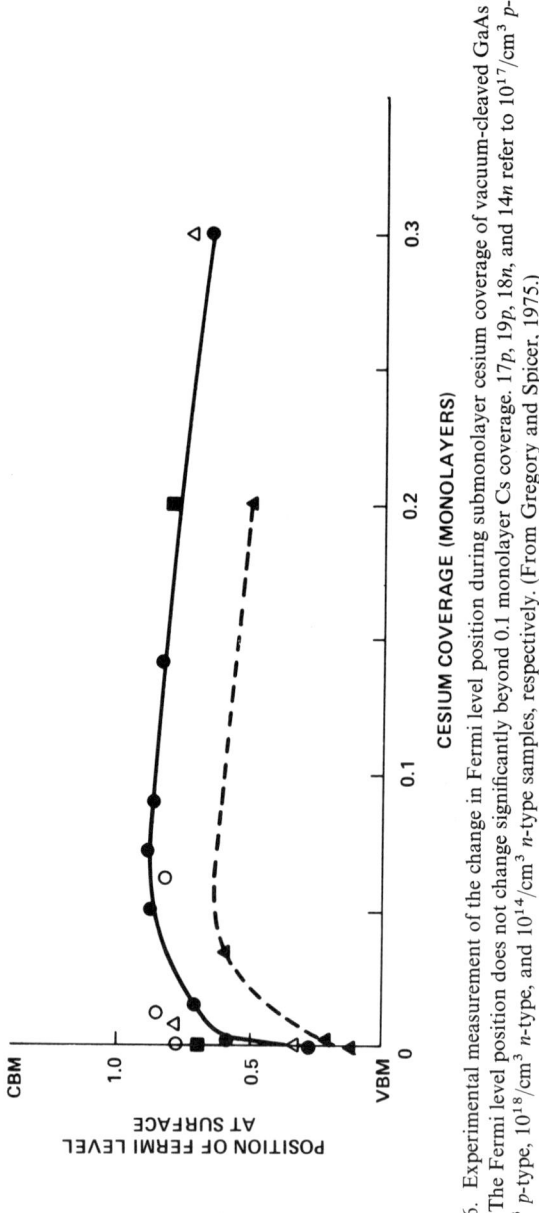

FIG. 16. Experimental measurement of the change in Fermi level position during submonolayer cesium coverage of vacuum-cleaved GaAs (110) surfaces. The Fermi level position does not change significantly beyond 0.1 monolayer Cs coverage. $17p$, $19p$, $18n$, and $14n$ refer to $10^{17}/cm^3$ p-type, $10^{19}/cm^3$ p-type, $10^{18}/cm^3$ n-type, and $10^{14}/cm^3$ n-type samples, respectively. (From Gregory and Spicer, 1975.)

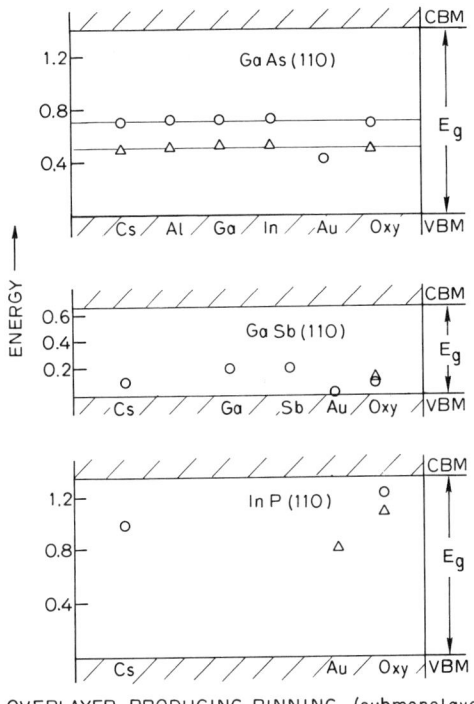

OVERLAYER PRODUCING PINNING (submonolayer)

FIG. 17. Approximate Fermi level pinning positions on *n*- and *p*-type vacuum-cleaved III–V surfaces with submonolayer coverage of adatoms. (From Spicer *et al.*, 1979.)

The "unpinned," clean (110) III–V surface, however, is *extremely* sensitive to the slightest perturbation. Gregory and Spicer (1975) found that only a tenth of a monolayer of Cs on GaAs (110) moved the Fermi level to near the middle of the bandgap at the surface. [See Fig. 16.] Lindau *et al.* (1978a) found that other metals and oxygen had a similar effect at ~ 0.1 monolayer coverage [Fig. 17]. The pinning position at about 0.1 monolayer coverage is very close to that observed with much heavier exposures—e.g., Schottky barriers heights (Sze, 1969, p. 397). Furthermore, MBE, HC, and SA cleaned III–V surfaces are usually "pinned" as are some strained or poorly VC surfaces (Massies *et al.*, 1979; Skeath *et al.*, 1978; Lindau *et al.*, 1977; Pianetta *et al.*, 1978c). At this time it is not clear exactly what the pinning mechanism is in the case of III–V's. A relaxing of the reconstructed surface back toward bulklike positions could be involved. Careful experiments have been made to see if the intrinsic surface states do move back into the bandgap during initial stages of oxidation or Schottky barrier formation. The evidence to date indicates that they do not move but rather new extrinsic states are created within the bandgap (e.g., Skeath *et al.*, 1979a). Skeath *et al.* (1978) found no evidence of intrinsic filled surface states within the bandgap on GaAs (111) B surfaces cleaned by HC or SA techniques, even though Fermi level pinning was consistently seen on these surfaces.

An interfacial vacancy model has been proposed by Spicer *et al.* (1979) and supported by Skeath *et al.* (1979a) and Lindau *et al.* (1978a) in which the heat of absorption of the metal on the semiconductor provides the driving force to create a low density of surface "defects." Only a few percent of the surface atoms are needed to cause Fermi level pinning (Gregory and Spicer, 1975; Spicer *et al.*, 1979). In the proposed vacancy model, GaAs, InP, and GaSb Fermi level pinning is due to a deficiency of the column V (anion) element at the surface. There is some support for this model from PES and AES experiments and the empirical observations of McCaldin *et al.* (1976). McCaldin *et al.* showed that the Schottky barrier height for holes Φ_p on III–V's contacted by Au depends only on the anion. The vacancy model is still speculative at the present even for the ideal VC (110) surface and further experiments will likely introduce new complexities.

Studies of the VC Si (111) surface show that the as-cleaved surface is pinned initially near midgap but that slight oxidation of the surface can remove the intrinsic state distribution from the bandgap (Wagner and Spicer, 1972, 1974). ELS studies of submonolayer metal coverages on SA Si (111) surfaces also indicate that the intrinsic surface states are removed by the metallic overlayer and suggest that metal-induced surface states (extrinsic states) are responsible for Fermi level pinning (Rowe *et al.*, 1975, 1977). As with III–V's, the exact nature of the extrinsic surface states is not known, but it is clear that the thermodynamics of chemisorption of essentially individual atoms or a single layer of atoms on the surface must be further investigated for a more complete understanding of surface states and Fermi level pinning (Margaritondo *et al.*, 1976). A microscopic model of Schottky barrier formation will likely follow along these lines as well (e.g., Chelikowsky, 1977).

6. ACTIVATION RELATED STUDIES—III–V'S

Essentially all of the III–V activation studies have been on GaAs surfaces. From the previous discussion, it is apparent that the Fermi level on all but ideal VC surfaces is likely pinned after in situ vacuum cleaning. Furthermore, from the results of Fig. 16, even the ideal VC surface is pinned after a small fraction of a monolayer Cs coverage. For *p*-GaAs, the Fermi level stabilizes ≈ 0.5 eV above the valence band maximum at the surface (e.g., Fig. 17; Uebbing and Bell, 1967). Therefore the primary effect of Cs is to lower the work function on the clean surface. There have been a number of experiments which have measured the work function change versus Cs coverage (Garbe, 1969a; Madey and Yates, 1971; Smith and Huchital, 1972; Clemens and

Mönch, 1975; Derrien and Arnaud d'Avitaya, 1977; Clemens et al., 1978). [See Fig. 18.] The change in work function is generally interpreted as Cs-induced electric dipoles, assuming that the chemisorbed Cs is at least partially ionized positively and that a compensating negative charge resides in surface states or within the band-bending region (e.g., Weber and Peria, 1969; Bell, 1973, p. 33; Clemens and Mönch, 1975). AES measurements versus Cs coverage have been reported by van Bommel and Crombeen (1974, 1976), Goldstein (1975), Stocker (1975), Derrien et al. (1976), Derrien and Arnaud d'Avitaya (1976, 1977), and van Bommel et al. (1978). [See Fig. 19.] Assuming a Cs monolayer coverage of 8.85×10^{14} atoms/cm^2 for the GaAs (110) surface (i.e., one Cs atom for each Ga and As surface atom) and a unity sticking

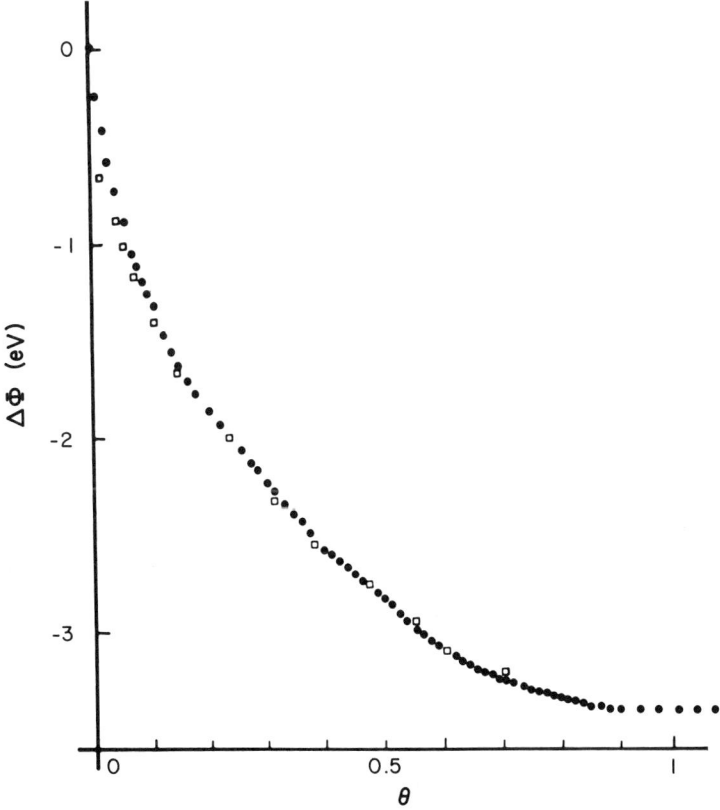

FIG. 18. Experimental work function change versus Cs coverage. $\Theta = 1.0$ corresponds to 4.6×10^{14} atoms/cm^2. (□) results from Clemens and Mönch (1975) and (●) from Derrien and Arnaud d'Avitaya (1977). The total work function change for Cs-only is about 3.4 eV and the minimum work function of the Cs/GaAs surface is about 1.6 eV. (From Derrien and Arnaud d'Avitaya, 1977.)

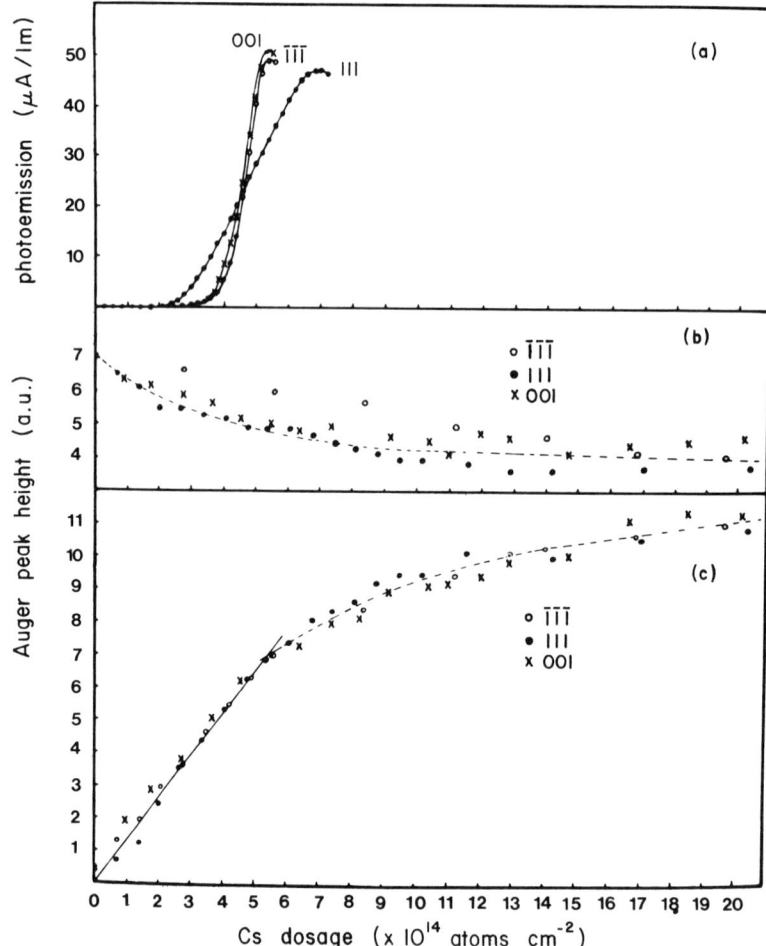

FIG. 19. (a) Reflection-mode photoemission sensitivity in μA/lm (see Section VIII) versus initial Cs coverage. The peak sensitivities of about 50 μA/lm are typical of NEA GaAs cathodes for Cs-only. (b) As (31-eV) Auger peak height versis Cs coverage. The surfaces are epitaxially grown GaAs (100), (111) A and B, and are sputter–anneal (SA) cleaned in vacuum. (From van Bommel et al., 1978.)

coefficient (Smith and Huchital, 1972), the maximum photoemission response for Cs-only occurs at about $\frac{1}{2}$ monolayer. This is in good agreement with the minimum work function coverage data. Similar Cs coverage estimates were deduced from ESCA measurements (Smith and Huchital, 1972).

LEED studies have also been done on Cs/GaAs surfaces to deduce possible ordering of the Cs on the clean surface (Weber and Peria, 1969; Mityagin *et al.*, 1972; van Bommel and Crombeen, 1974, 1976; Goldstein, 1975; Stocker, 1975; Derrien *et al.*, 1976 Derrien and Arnaud d'Avitaya, 1977; van Bommel *et al.*, 1978). Experiments on the (110) indicate some ordering of the Cs up to a $\frac{1}{2}$ monolayer coverage while most experiments on the other low index faces see no evidence of ordering. Crystallographic models for Cs/GaAs have been discussed by Mityagin *et al.* (1972), van Bommel and Crombeen (1974, 1976), van Bommel *et al.* (1978), and Clemens *et al.* (1978). Detailed assessment of these models must, however, await more definitive reconstruction models of the clean surface.

Thermal desorption studies of Cs/GaAs (Goldstein and Szostak, 1975a; Derrien and Arnaud d'Avitaya, 1976, 1977) indicate that the first $\frac{1}{2}$ monolayer of Cs is much more tightly bound to the surface than a complete monolayer. [See Fig. 20.] This behavior is consistent with increasing depolarization at the higher coverages, and may also be associated with new bonding sites for

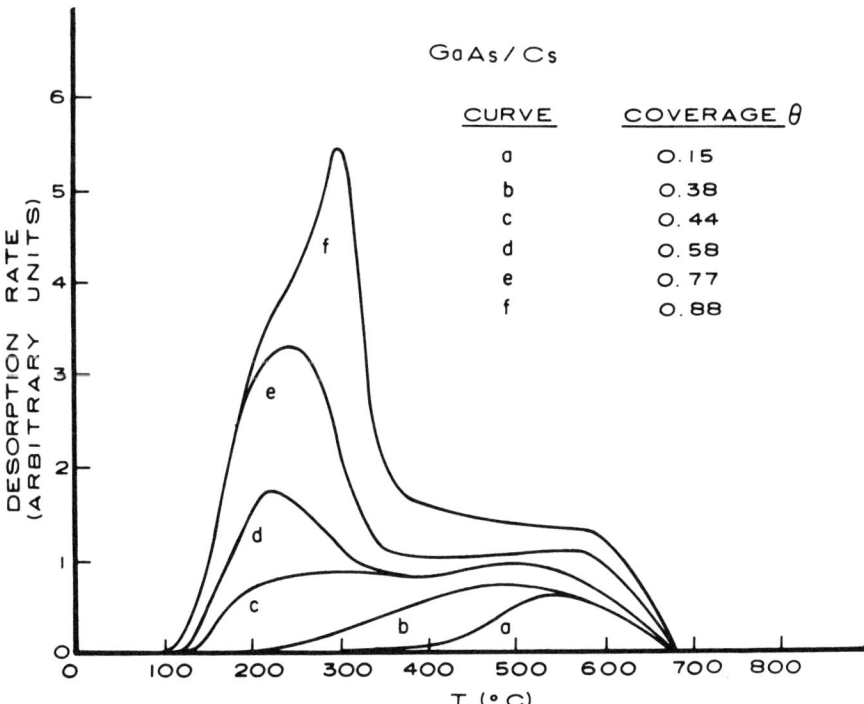

FIG. 20. Flash desorption curves of Cs from SA GaAs (100) surfaces for different initial Cs coverages, Θ. $\Theta = 1.0$ is defined as 7.2×10^{14} atoms/cm^2 here. The first $\frac{1}{3}$–$\frac{1}{2}$ monolayer of Cs is much more tightly bound to the surface than higher coverages. (From Goldstein and Szostak, 1975b.)

the second ½ monolayer of Cs. A Cs–GaAs binding energy of 40–60 kcal/mole for the first ½ monolayer is deduced from a thermodynamic analysis of experimental AES data by Derrien and Arnaud d'Avitaya (1977). ELS measurements by these same authors (Derrien and Arnaud d'Avitaya, 1976) suggest that Ga atoms are the initial binding sites for Cs on the (110) surface.

Following initial cesiation of the clean surface, activation continues with cesium and oxygen to a final photoemission peak. [Recall Fig. 12.] Approximately 1×10^{-6} Torr sec ($=1.0$ L) oxygen exposure is needed for optimal photoresponse from GaAs corresponding to a thickness of about 2–3 monolayers, i.e., about 10 Å (Sommer et al., 1970; Uebbing and James, 1970; Fisher et al., 1972; Gregory and Spicer, 1976). [See Fig. 21.] Photoelectric threshold and Kelvin work function measurements by Uebbing and James (1970) on silver and GaSb activated with Cs + O clearly indicate the presence of an interfacial electron barrier E_b between the substrate and the activation layer as is seen in Figs. 11 and 22. The origin of this barrier is not established to date. A heterojunction model between the cathode and a ∼2-eV bandgap Cs_2O semiconductor activation layer is discussed by Sonnenberg (1969), James and Uebbing (1970), Milton and Baer (1971), and Bell (1973, p. 39). PES measurements of Cs + O on VC GaAs (110) surfaces by Spicer et al. (1978) suggest an As–O dipole may be the origin of E_b. The practical effect

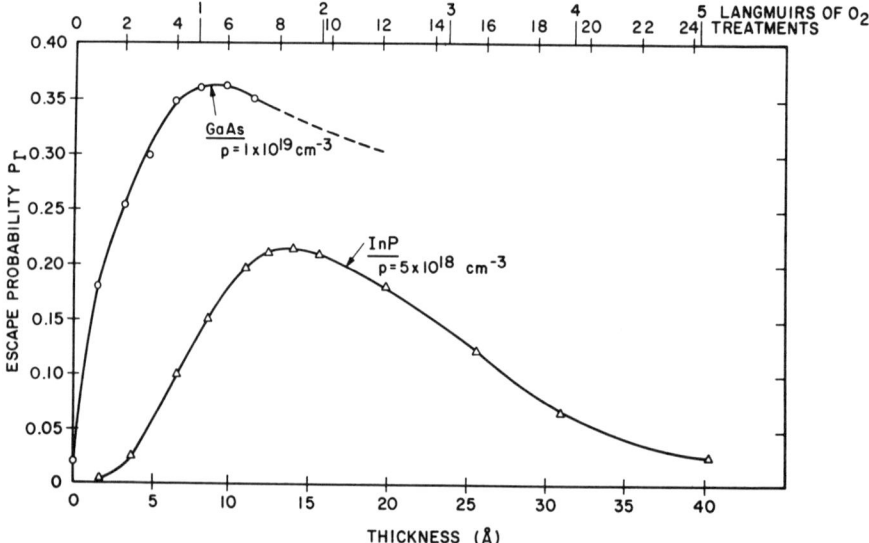

FIG. 21. Approximate Cs + O activator thickness versus surface escape probability (see Section VII.10) for NEA GaAs and InP surfaces. The total oxygen exposure for optimal photoemission is only about 1 L. (From Uebbing and James, 1970.)

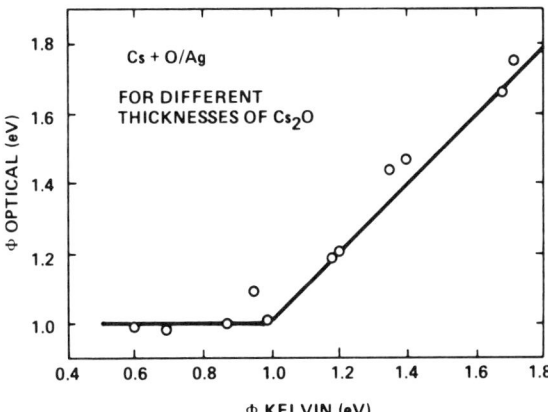

FIG. 22. Optical threshold (Fowler plot) work function versus Kelvin work function for different thickness of Cs + O activation layers on silver. The photoelectric threshold is limited by a ~1.0-eV interfacial electron barrier E_b at the Cs + O/Ag interface while the work function on the Cs + O/vacuum interface is as low as 0.6 eV. A ~1.2-eV semitransparent interfacial barrier is deduced from Cs + O activations on NEA III–V semiconductors. (From Uebbing and James, 1970.)

of an interfacial barrier is to significantly limit the photoelectron escape probability (and hence quantum efficiency) from III–V semiconductor photocathodes of bandgap less than E_b (Bell et al., 1971; Fisher et al., 1972) (Section VII.10).

The atomic arrangement and stoichiometry of the Cs + O NEA activation layer is still not established. Based on the idea of an enhanced dipole moment from a W–O–Cs arrangement (Langmuir and Kingdon, 1924; Langmuir and Taylor, 1944), Chen (1971) has proposed a Cs_2O–Cs–O–GaAs model. Fisher et al. (1972) have suggested a Cs–O–Cs–Cs GaAs structure which has some support from AES studies of Stocker (1975). Unfortunately, there are many possible cesium oxides ranging from essentially metallic Cs_7O to semiconducting CsO_3 (Gregory et al., 1975). PES experiments on the oxidation of bulk Cs (at 100–140°K) by Gregory et al. (1975) suggest that the oxygen ions initially dissolve below the Cs surface. Furthermore, a sharp minimum occurs in the work function of bulk Cs during the oxidation process. The role of the oxygen in the activation process of III–V NEA surfaces is likely quite complex, based on the many suboxides of cesium which are possible and the interaction between oxygen and the clean GaAs surface itself (e.g., Spicer et al., 1976, 1978; Lindau et al., 1978b; Mele and Joannopoulos, 1978; Ranke and Jacobi, 1979; Thauault et al., 1979; Chye et al., 1979). A theoretical model for the lower work function of monolayer cesium oxides compared with that of bulk cesium is given by Burt and Heine (1978) and a semiquantitative model of the electronic structure of metallic and nonmetallic cesium suboxides and their relation to NEA III–V surfaces is given by Clark (1975).

The Cs + O activation layer thickness required for optimal thermalized electron emission from III–V surfaces is somewhat dependent on the cathode bandgap. The lower the bandgap, the thicker the activation layer needed, e.g., Fig. 21. Lower bandgap cathodes require a lower work function which implies a thicker activation layer despite increased electron losses in transport through the Cs + O (James et al., 1968; Sonnenberg, 1972).

A relatively few NEA-related basic studies have been made of other III–V surfaces. Experimental studies of VC GaP (110) surfaces indicate a similar reconstruction to the GaAs (110), i.e., P atoms displaced outward and Ga inward (e.g., Miller and Haneman, 1979). Van Bommel and Crombeen (1978) have made LEED, AES, thermal desorption, and photoemission yield measurements on SA Cs/GaP (100) surfaces. The Cs adsorption studies on GaP show behavior quite similar to that of GaAs, i.e., Fig. 19. LEED indicates that the Cs is adsorbed amorphously but that heating a Cs/GaP surface can induce several ordered structures. In contrast to GaAs, however, the thermal desorption studies show three distinct desorption peaks at 160, 360, and 550°C. Jacobi (1975), using photoemission and ELS on MBE GaP (111) B surfaces, has identified empty and filled surface states in general agreement with the surface-state model outlined in Fig. 14 for GaAs. Piaget et al. (1977b) have deduced from their analysis of EDC measurements, $\Phi_p = 1.65$ and 1.4 eV for Cs and Cs + O, respectively, on HC GaP (100) surfaces. This is much higher than typical Schottky barrier heights on p-type GaP of 0.70–0.75 eV (e.g., Scheer and van Laar, 1967; Sze, 1969, p. 398) and is unusual in that Φ_p changes after the initial cesiation. This is not seen with VC GaAs (110) studies (Skeath et al. 1979b). Fischer (1966b) measured $\Phi_p = 0.85$ eV from EDC experiments with Cs-only on VC GaP (110) surfaces. PES experiments on clean and cesiated VC InP (110) show behavior quite similar to that of GaAs (110) (e.g., Fisher, 1966a; Uebbing and Bell, 1968; Chye et al., 1976).

7. ACTIVATION RELATED STUDIES—SILICON

There have been a number of adsorption studies of Cs on Si (111) surfaces (Scheer, 1960; Allen and Gobeli, 1966; Fischer, 1968; Weber and Peria, 1969; Mönch, 1970; Wagner et al., 1977). As in the case with GaAs, submonolayer coverage of Cs rapidly pins the Fermi level at the surface (Mönch, 1970). From PES studies of VC, p^+-Si (111), Wagner and Spicer (1974) measured the Fermi level about 0.4 eV above the valence-band maximum at the surface for both clean and oxidized (10^6 L) silicon which is consistent with the work function and photoemission threshold measurements on Cs + O activated

Si (100) of Martinelli (1974a). Earlier studies with Cs indicated that the Fermi level was even higher above the valence band, i.e., 0.8–1.0 eV, but this is hard to justify with Schottky barrier results (e.g., Scheer and van Laar, 1967; Sze, 1969, p. 399). In contrast to GaAs, however, the intrinsic distribution of surface states is moved out of the bandgap and new Cs-induced (extrinsic) states appear (e.g., Rowe *et al.*, 1977). Oxygen adsorption studies on VC Si (111) also indicate a complex surface chemistry in the initial stages (e.g., Garner *et al.*, 1978; Carriere and Deville, 1979).

Levine (1973) has proposed a detailed structural and electronic model for the NEA Si (100) surface that is based, in part, on a (2 × 1) reconstruction outlined in Fig. 23. The surface atoms are assumed to rotate into adjacent rows of surface "pedestal" and "cave" sites. Cs adsorption is confined to pedestal sites. Subsequent oxygen adsorption is primarily into "cave" sites. The surface dielectric constant, deduced from AES experiments (Goldstein, 1973), is then used by Levine to compute a 2.8-Å Cs–O–Si dipole length which is in good agreement with the structural model. Additional support to Levine's model comes from the initial LEED, AES, and thermal desorption experiments of Goldstein (1975) and the work of Gundry *et al.* (1974), Holtom and Gundry (1977), and Koval' *et al.* (1978). Figure 24 shows a schematic of the initial cesiation of Si (100) indicating photoresponse, work function, and three distinct Cs LEED patterns observed during cesiation (Holtom and Gundry, 1977). For successful activation to NEA (assuming a clean, well-ordered silicon surface), the cesiation must be continued somewhat past the photoemission peak (Gundry *et al.*, 1974). [See Fig. 25.] Holtom and Gundry (1977) have interpreted this observation and the Cs LEED data in terms of

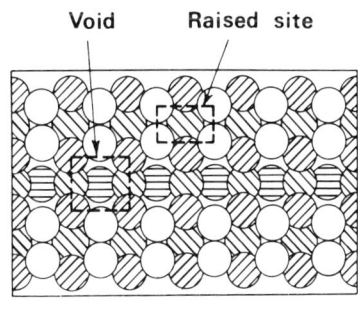

FIG. 23. Reconstruction model of the (2 × 1) Si (100) surface used by Levine (1973) in a detailed model of the NEA Si surface. The raised or "pedestal" sites are sites for Cs adsorption and the void or "cave" for oxygen, thereby forming a Cs + O dipole configuration. (From Holtom and Gundry, 1977.)

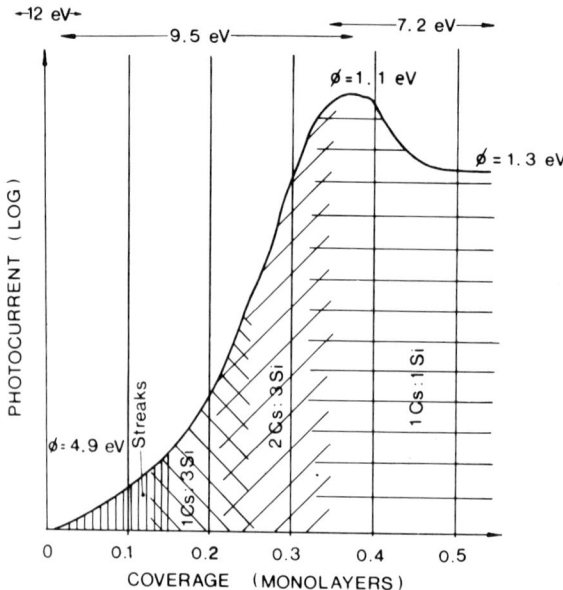

FIG. 24. Schematic diagram of the initial cesiation of the clean (2 × 1) reconstructed Si (100) surface. Three Cs-related LEED patterns are observed with the final pattern and full cesiation essential for successful achievement of NEA. (From Holtom and Gundry, 1977.)

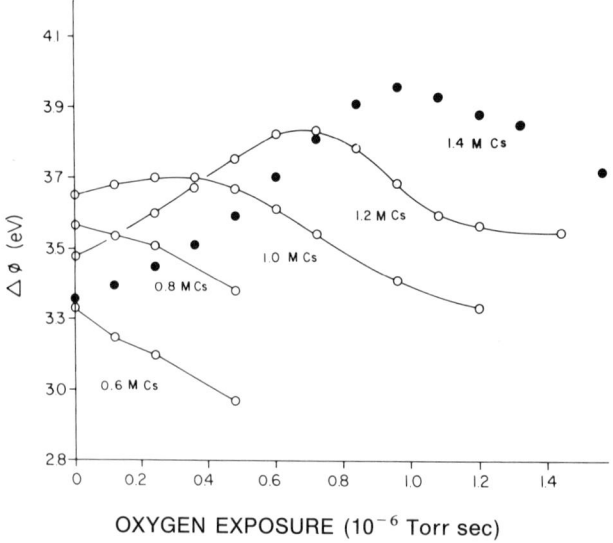

FIG. 25. Change in work function versus oxygen exposure beginning with different initial Cs coverages of the Si (100) surface. If the Si surface is not fully cesiated, subsequent adsorption of oxygen increases the work function. Cs coverage is in terms of $M = 6.25 \times 10^{14}/cm^2$ and the oxygen exposure is measured in Langmuirs. (From Edwards and Peria, 1978.)

specific Cs adsorption sites on the reconstructed Si surface discussed by Levine. The full cesiation condition is characterized by complete filling of all pedestal sites and a change in Cs–Si bonding compared with low Cs coverage due to the development of Cs–Cs metallic bonds along the (110) directions. Koval' *et al.* (1978) found that Cs/Si (100) significantly increased the rate of oxide formation compared with the clean surface and that silicon oxide could be seen by ELS measurements after only 1.0-L exposure of the cesiated surface. The measurements of Koval' *et al.* also suggest that the oxygen penetrates below the cesiated silicon surface consistent with the Levine model and earlier AES measurements of Goldstein.

The work function of the optimally activated surface is reported to be 0.85–0.93 eV from Kelvin probe measurements (Howorth *et al.*, 1975b; Edwards and Peria, 1978). From the temperature dependence of the thermionic emission (dark current), Martinelli (1974a) deduced the temperature-independent portion of the work function to be 1.06 eV. There is no experimental evidence of an interfacial barrier analogous to E_b observed on the III–V's.

It is interesting to note that activation with Rb + O/Si (100) is essentially the same as Cs + O/Si (100) (Martinelli, 1973a) while Rb + O/InAsP (111) B yields inferior photoemission results to Cs + O (Bell *et al.*, 1971). Garbe (1970) found that Cs + F on VC GaAs (110) yielded excellent NEA photoemission while Martinelli (1973a) reported that Si (100) could not be activated to NEA with Cs + F. Cs + O activation of the Ge (100) surface is quite similar to the Si (100) except that NEA is not achieved due to the low bandgap of Ge (0.67 eV) (e.g., Goldstein and Martinelli, 1973; Edwards and Peria, 1978).

VII. The Three-Step Model of Photoemission

In this section a more detailed discussion is presented of the "three-step model" of photoemission (e.g., Fan, 1945; Berglund and Spicer, 1964) as it is generally used in the analysis of NEA surfaces (James *et al.*, 1968; James and Moll, 1969; Antypas *et al.*, 1970; Liu *et al.*, 1970; Allen, 1971, 1973; Spicer and Bell, 1972; Bell, 1973; van Laar, 1973; Liu *et al.*, 1973; Frank and Garbe, 1973, 1974; James, 1974; Martinelli and Fisher, 1974; Clark, 1976; Rougeot and Baud, 1976, 1979). The model has been successfully used for a variety of NEA devices to deduce minority carrier electron diffusion lengths and important properties of the cathode-substrate and cathode-vacuum interface. The three-step model is also commonly used in the analysis of a wide variety of photoemission experiments (e.g., Cardona and Levy, 1978, p. 84; Spicer, 1978).

8. Optical Absorption—Electron Generation

It is assumed that for photon energies larger than the bandgap of the semiconductor the absorption of a photon creates a single electron–hole pair and that the photoexcitation of electrons from the valence band is governed by the bulk optical absorption coefficient $\alpha(h\nu)$. Consider the semitransparent photocathode shown in Fig. 26. Light of intensity I_0 is incident normally either from the front (electron emitting) surface of the

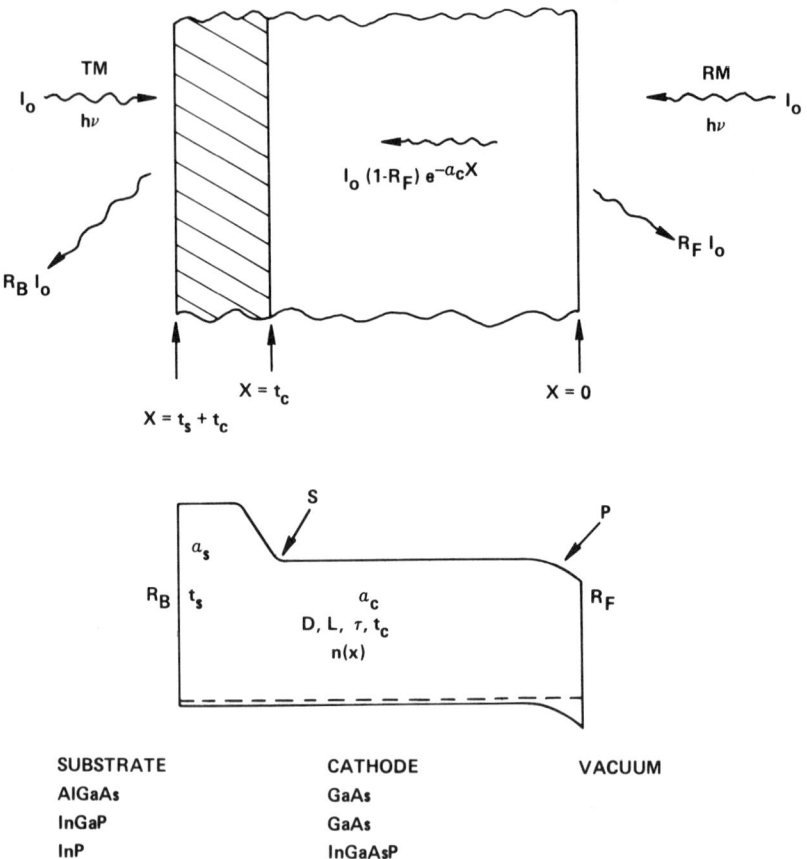

FIG. 26. Schematic cross section of a semitransparent semiconductor photocathode. RM and TM refer to reflection-mode and transmission-mode operation, t_c the thickness of the active cathode, t_s the substrate thickness, R_F and R_B the optical reflectances of the cathode and substrate surfaces, S and P the back surface recombination velocity (relative) and cathode surface escape probability, α_c and α_s the cathode and substrate optical absorption coefficients, D and L the minority-carrier electron diffusion constant and diffusion length, and τ and $n(x)$ the electron lifetime and density, respectively.

cathode, RM (reflection mode), or the back (substrate) surface, TM (transmission mode). The substrate is assumed to be a higher bandgap semiconductor whose lattice constant matches the cathode, and therefore the optical reflectance at the interface between the substrate and cathode is negligible due to the similar indices of refraction. The front and back reflectances of the cathode substrate are R_F and R_B, respectively. It is assumed that the thickness of the cathode t_c is such that for photon energies near the bandgap of the cathode, multiple internal reflections will occur. For these assumptions, the generation rate $G(x)\,dx$ of electron–hole pairs at a plane x within the cathode is given by

$$G(x, hv, t) = G_1 \exp(-\alpha_c x) + G_2 \exp(+\alpha_c x), \tag{5}$$

where

$$G_1 = \frac{\alpha_c I_0 (1 - R_F)}{1 - R_F R_B \exp[-2(\alpha_c t_c + \alpha_s t_s)]}, \tag{6}$$

$$G_2 = G_1 R_B \exp[-2(\alpha_c t_c + \alpha_s t_s)]$$

for RM and

$$G_1 = \frac{\alpha_c I_0 R_F (1 - R_B) \exp(-\alpha_c t_c - \alpha_s t_s)}{1 - R_F R_B \exp[-2(\alpha_c t_c + \alpha_s t_s)]}, \tag{7}$$

$$G_2 = G_1 / R_F$$

for TM.

In most cases, the variation in R_F and R_B with photon energy can be ignored and an average value over the relatively small range of photon energies used for analysis can be employed. If the substrate is coated with an antireflection layer, however, R_B versus hv should best be measured. For Si and III–V's, $R_F \approx 25$–35%.

For the case of a simple bar pattern suitable for calculating the spatial resolution characteristics (MTF) of a cathode, a two-dimensional generation function can be used of the form

$$G(x, y) = G(x)[A + B \cos(\omega_s y)], \tag{8}$$

where the cathode is assumed to be the y-z plane, the bars of frequency ω_s run parallel to the z axis, and $G(x)$ is the generation function normal to the cathode surface (Bell, 1973, p. 115; Fisher and Martinelli, 1974).

Secondary electron emission can also be analyzed by the three-step model. High-energy electrons (up to 20 kV) are incident upon the cathode (either RM or TM). A large percentage of the incident electrons penetrate into the cathode losing energy mostly by impact ionization. The generated secondary electrons rapidly thermalize into conduction-band minima and can thereby diffuse to the emitting surface. If the energy of the incident primary electron beam E_p is below 1 kV, the depth of secondary electron generation is shallow—a few hundred angstroms. In this case a good fraction of the emitted secondary electrons may not have fully thermalized into a conduction band before vacuum emission (Piaget *et al.*, 1975, 1977b). For the higher-energy primary electron beams, it has been found empirically that the depth variation (range) of secondary electron generation may be approximated by the simple relation

$$G(x) = [E_p(1-f)/\varepsilon_p R], \quad 0 \leqslant x \leqslant R$$

and (9)

$$G(x) = 0, \quad x > R$$

where ε_p is the average energy to create a secondary (ionization energy), R the range of penetration of the primary electron beam, and f the fraction of primary electrons reflected from the surface (backscattering coefficient). Experimental measurements show $f \approx 80\%$ for $E_p \sim 1.0$ eV, Cs + O/GaAs (Musatov *et al.*, 1976), and $f = 30$–40% for $E_p > 1$ kV (Sternglass 1954; Palmberg, 1967; Klein, 1972). It has been found empirically that the range over which secondary electrons are generated can be approximated by $R = aE_p^b$, where a and b are material parameters (e.g., Feldman, 1960). An empirical relation between ε_p and E_g,

$$\varepsilon_p \text{ (eV)} = 2.59 E_g \text{ (eV)} + 0.71, \tag{10}$$

closely fits a wide range of semiconductors (Kobayashi, 1972). Table III gives a, b, and ε_p parameters for Si, GaAs, and GaP. It is apparent that the range of secondary electron generation is no more than a few hundred angstroms for $E_p < 1$ kV and only about a micron at 10 kV.

TABLE III

PRIMARY ELECTRON BEAM RANGE PARAMETERS a AND b,[†] AND ε_p (eV)[‡]

	a (μm)	b	ε_p (eV)
Si	0.02	1.65	3.60
GaAs	0.027	1.46	4.40
GaP	0.009	2.00	6.54

[†] From Martinelli and Fisher (1974).
[‡] Calculated from Kobayashi (1972).

9. Electron Transport to the Surface—Electron Diffusion

Following photoexcitation, the electrons rapidly thermalize into various conduction-band minima depending on the incident photon energy and band structure of the cathode. Thermalized electrons are essentially free and diffuse randomly within the cathode until they recombine within the bulk, at the heterojunction interface, or fall into the bent-band region at the emitting surface. Although the density of minority carriers is small, the "diffusion equation," originally derived for a high density of particles, can be shown to give a suitably time-averaged description of the motion of thermalized electrons (Bell, 1973, p. 111). If there is an internal electric field, then electron motion can become a combination of both diffusion and drift. The time-independent diffusion equation is given by

$$-D\left(\frac{\partial^2 n}{\partial x^2} + \frac{\partial^2 n}{\partial y^2}\right) + \frac{n}{\tau} - \mu E\left(\frac{\partial n}{\partial x} + \frac{\partial n}{\partial y}\right) = G(x, y), \quad (11)$$

where $n(x, y)$ is the density of minority carriers, D the electron diffusion constant, τ the electron lifetime, μ the electron mobility, $G(x, y)$ the appropriate generation function, and E a constant electric field. The zero-field electron diffusion length L is $(D\tau)^{1/2}$ and the (zero-field) diffusion velocity v_d is $L/\tau = D/L$.

The simplest case to analyze is that of a thick, uniformly illuminated RM photocathode, i.e., $\alpha_c t_c \gg 1$. $G(x)$, from Fig. 26, becomes $(1 - R_F)\exp(-\alpha_c x)$. The two necessary boundary conditions in this case are $n(x) = 0$ for $x \to \infty$ and $n(x) = 0$ at $x = 0$. The emitting surface at $x = 0$ is essentially a perfect sink for electrons which fall into the narrow bent-band region. Electrons within this narrow region either recombine or are emitted into vacuum. The solution for $n(x)$ is

$$n(x) = A[\exp(-\alpha_c x) - \exp(-x/L)], \quad (12)$$

where

$$A = \frac{\alpha_c I_0 (1 - R_F)\tau}{1 - \alpha_c^2 L^2}$$

and the electron current flowing into the emitting surface region is given by $qD(dn/dx)|_{x=0}$. The quantum yield for collection of electrons into the bent-band region is the ratio of this current to the maximum possible photogenerated current, qI_0, i.e.,

$$Y(\text{Int.}) = \frac{\alpha_c L(1 - R_F)}{1 + \alpha_c L}, \quad (13)$$

Equation (13) describes the internal collection efficiency of the emitting surface for photogenerated electrons. It is convenient at this point to introduce an "escape probability" factor P defined as the fraction of the photocurrent which, having successfully reached the emitting surface, is actually emitted into vacuum. The photoemission quantum yield $Y(\text{PE})$ is then $PY(\text{Int.})$.

For analysis with experimental yield data, $Y(\text{PE})$ can be rewritten in the form

$$\frac{(1-R_F)}{Y(\text{PE})} = \frac{1}{P} + \frac{1}{PL\alpha_c}, \quad (14)$$

where a plot of the left-hand side of Eq. (14) versus $1/\alpha_c$ ideally should give a straight line. P and L can be obtained from the intercept and slope (e.g., Garbe, 1969a; Richard, 1973).

Generally the diffusion length is a bulk material property dependent on growth techniques, doping concentration, etc. (e.g., Garbe and Frank, 1969, 1970; Schade et al., 1971; James et al., 1971a; Allenson and Bass, 1976; Ettenberg et al., 1976). The surface escape probability is critically dependent on surface preparation techniques and to a lesser degree on bulk material properties such as doping concentration, surface orientation, etc. [See Section VII.3.] Note that for weak absorption, $\alpha_c L \ll 1$, $Y(\text{PE}) \simeq \alpha_c LP(1-R_F)$. Hence near photoemission threshold, the yield is especially sensitive to both the bulk diffusion length and optical absorption coefficient. Meaningful analysis of experimental yield data using Eq. (14) depends on three factors: (1) $\alpha_c t_c \gg 1$, (2) good α_c data, and (3) $Y(\text{PE})$ predominantly thermalized electron emission from a single conduction-band valley. The importance of factor 2 has been emphasized by Bell (1973, p. 66). Unfortunately subtle changes in the optical absorption near threshold due to a change in doping concentration (e.g., Burstein shift, Burstein, 1954) or dopant (e.g., Zn to Ge) can have a dominant effect on the analysis. [See Fig. 27.]

For the NEA GaAs cathode, the lowest conduction-band minimum E_Γ is 1.425 eV above the valence-band maximum and the next lowest conduction-band minimum E_L is about 0.35 eV above E_Γ (Aspnes et al., 1976; Eden et al., 1967). [Recall Fig. 9.] Therefore for photon energies between approximately 1.4 and 1.7 eV, P_Γ and L_Γ can be deduced from experimental $Y(\text{PE})$ data if the above three factors are satisfied. For somewhat higher photon energies some electrons will scatter and thermalize into the L minima, transport to the surface, and be emitted. The experimental yield can be similarly analyzed for P_L and L_L (e.g., James and Moll, 1969; Garbe and Frank, 1969; Garbe, 1969a). Note that prior to the work of Aspnes et al. (1976), it was generally assumed that the X minimum was the next higher

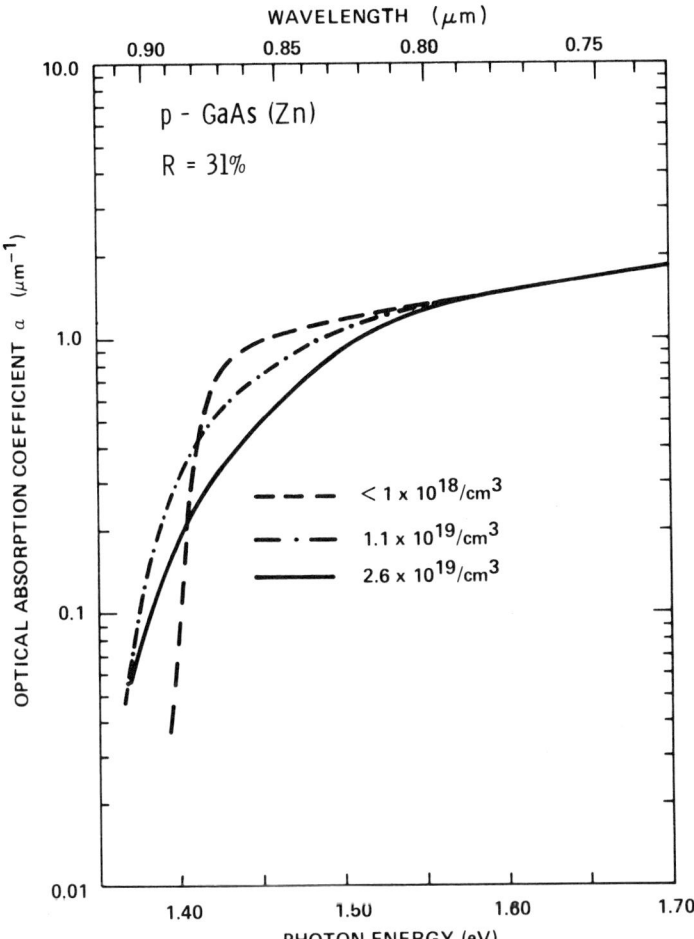

FIG. 27. Optical absorption coefficient for p-type, Zn-doped GaAs near threshold. R is the optical reflectance which was assumed to be constant over this photon energy range. (Data from Kudman and Seidel, 1962.)

conduction band above the Γ. The pre-1976 photoemission data for P_X and L_X is in fact P_L and L_L. Typical L_Γ values for high-quality epitaxially grown NEA GaAs RM cathodes are 2–5 μm (e.g., Olsen et al., 1977).

Solution of the diffusion equation for a semitransparent cathode, Fig. 26, is straightforward with trial solutions of the form $\pm\exp(\pm x/L)$ and $\pm\exp(\pm\alpha_c x)$. The boundary condition at the emitting surface is still $n(x = 0) = 0$. At the cathode-substrate interface, the boundary condition is conveniently described by the electron particle current which recombines at this interface, i.e.,

$$-D\frac{dn}{dx}\bigg|_{x=T} = S_b n(x)\big|_{x=T}, \tag{15}$$

where $S_b n(x)$ is an equivalent conduction current defined to characterize the interface (McKelvey, 1966, p. 350). S_b is the so-called surface recombination velocity which, for convenience of analysis, is measured relative to the bulk diffusion velocity v_d. The dimensionless parameter S_b/v_d is sometimes called the relative surface recombination velocity and will simply be called S. For almost all semiconductor–metal, semiconductor–vacuum, or semiconductor–semiconductor interfaces, S is $\gg 1.0$ for minority carriers. This is equivalent to having the cathode-substrate boundary condition of $n(x = t_c) = 0$. Devices requiring heterojunctions with low minority-carrier recombination, such as the TM photocathode, must achieve $S \gtrsim 1.0$ for optimal performance. This requirement places strict limits on the selection of cathode-substrate materials (e.g., James, 1974; Fisher and Martinelli, 1974; Antypas and Edgecumbe, 1975; van Oirschot et al., 1977). The solution for the uniformly illuminated semitransparent cathode is given by

$$Y(\text{PE}) = \frac{PG_2 L}{I_0(1 - \alpha_c^2 L^2)}\left[2K + \alpha_c L + 1 + \left(\frac{G_1}{G_2}\right)(1 - \alpha_c L)\right], \tag{16}$$

where

$$K \equiv [(S + \alpha_c L)\exp(\alpha_c t_c + t_c/L) - S + 1]/z$$
$$+ (G_1/G_2)[(S - \alpha_c L)\exp(-\alpha_c t_c + t_c/L) - S + 1]/z$$

and

$$z \equiv S - 1 - (1 + S)\exp(2t_c/L). \tag{17}$$

G_1 and G_2 are given by Eqs. (6) and (7) for RM and TM, respectively, and P is again the surface escape probability.

Analysis of semitransparent photoemission data can, in principle, give reasonable estimates for P, L, and S. It is almost essential, however, to have experimental yield data for both RM and TM. Even with this information, determination of unique values for L and S is difficult since there is often a trade-off between L and S values which give nearly equal "good fit" to the data. A relatively narrow range of L and S values can, however, be estimated after analyzing a number of different sets of yield curves from similarly prepared cathodes (e.g., Fisher et al., 1974; Fisher and Olsen, 1979). A simplified formula for TM yield suitable for graphical analysis is given by Liu et al. (1973).

There have been a number of semitransparent GaAs cathode model calculations indicating potential performance, yield curve shapes, optimal thickness requirements, etc. (e.g., Antypas et al., 1970; Allen, 1971, 1973;

Frank and Garbe, 1973; Bell, 1973; Fisher and Martinelli, 1974). These model calculations show that for near optimal TM response, (1) $t_c \approx 1.0$–2.0 μm, (2) $S \gtrsim 1.0$, and (3) $L \gtrsim t_c$. These requirements have been met with AlGaAs–GaAs, InGaP–GaAs, InGaAsP–InP, and passivated Si structures (e.g., Antypas and Edgecumbe, 1975; Fisher and Olsen, 1979; James, 1974; Howorth et al., 1976a). [See Fig. 28.] Figure 29 shows experimental and "best-fit" calculated yield curves for RM and TM from an AlGaAs–GaAs semitransparent cathode (Antypas and Edgecumbe, 1975). Diffusion model calculations for RM–NEA silicon have been made by Richard (1973), and TM/RM response versus t_c and S for NEA silicon is given by Howorth et al. (1976a).

Spatial frequency response calculations (MTF) using the diffusion equation are given by Bell (1973, p. 115) and Fisher and Martinelli (1974). Frequency response calculations are also given by Bell (1973, p. 115). A similar one-dimensional diffusion model analysis suitable for $\gtrsim 4$-keV x-ray TM detection with NEA GaAs is given by Van Speybroeck et al. (1974).

The diffusion model solution for RM secondary electron gain δ_R for $t_c \gg R$ has been calculated by Simon and Williams (1968) and is given by

$$\delta_R = (PE_p L/\varepsilon_p R)[1 - \exp(-R/L)], \qquad (18)$$

where P is the surface escape probability which was also used in Eq. (14), L the minority-carrier electron diffusion length, and the other parameters are

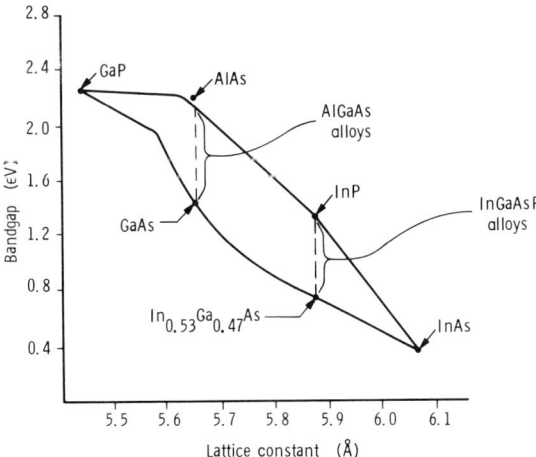

FIG. 28. Bandgap versus lattice constant (300°K) for a number of III–V semiconductors. The AlGaAs alloys are a close lattice match to GaAs. InGaAsP quaternary alloys can also be fabricated to lattice-match InP substrates. Lattice matching is essential for both high-quality epitaxial growth and low minority-carrier recombination at the substrate-epitaxial layer interface.

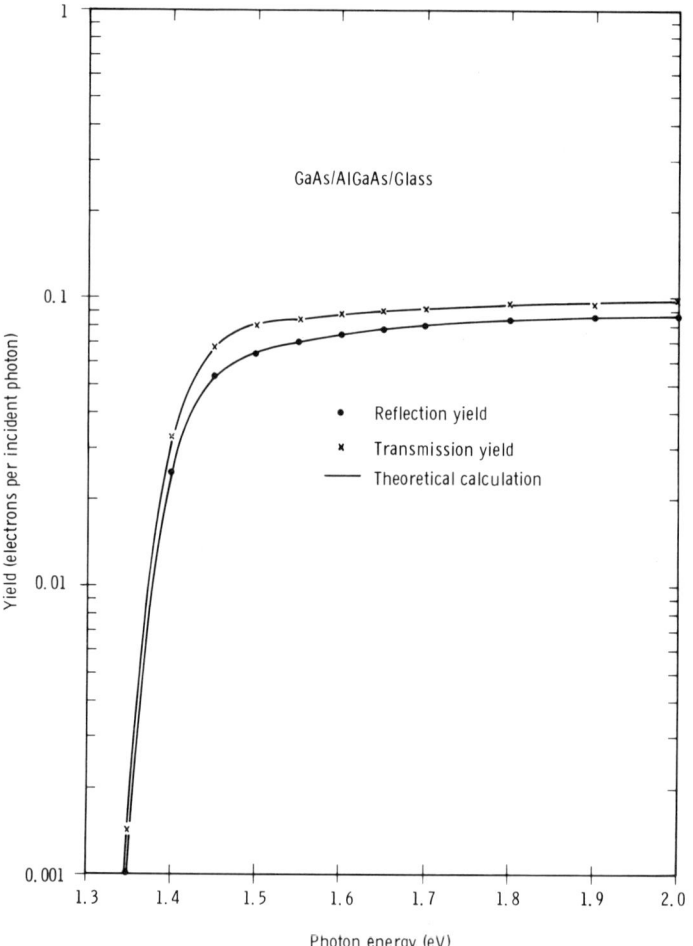

FIG. 29. Experimental and diffusion model calculation of RM and TM quantum yield from a AlGaAs–GaAs cathode. Best-fit parameters deduced from curve fitting are $L = 3.0\,\mu m$, $S = 1.0$, and $P = 13\%$. (From Antypas and Edgecumbe, 1975.)

the same as those defined in Eq. (9). Semitransparent secondary electron calculations of gain, MTF, and frequency response are given by Martinelli *et al.* (1972a), Bell (1973, p. 118), and Martinelli and Fisher (1974).

10. VACUUM EMISSION—THE SURFACE ESCAPE PROBABILITY

The fraction of photogenerated electrons incident upon the surface bent-band region which are successfully emitted into vacuum is called the surface escape probability P. Seven factors are known to have an influence on P. These are the following: (1) surface cleaning techniques, (2) activation with

Cs and oxygen, (3) bulk doping of the cathode, (4) the bulk band structure and crystal face of emission, (5) electric field, (6) temperature, and (7) strain. Each of these factors is discussed below.

a. Surface Preparation and Cleaning

Uebbing (1969, 1970) showed that surface carbon could not be effectively removed by vacuum HC procedures alone and that the surface escape probability from NEA GaAs was strongly dependent on residual carbon contamination. This fact places a premium on careful handling and storage of cathode surfaces prior to vacuum processing. A number of chemical cleaning procedures, surface passivation schemes, as well as UHV interlock systems, have been developed to minimize surface contamination (e.g., James *et al.*, 1971a).

The proper vacuum HC temperature, $T(HC)$, for a given cathode is determined empirically for a particular vacuum system. Subtle changes in sample size, cathode holder, etc., can affect $T(HC)$. The proper $T(HC)$ for GaAs is close to the highest congruent evaporation temperature T_c under Langmuir conditions (i.e., "free" evaporation from the surface). Goldstein *et al.* (1976), from Langmuir evaporation studies of GaAs, found $T_c = 663°C, 630°C,$ and $675°C$ for (100), (111) A and B faces of GaAs, respectively. Evaporation of GaAs under both Langmuir and Knudsen (i.e., equilibrium) conditions has been studied by Foxon *et al.* (1973). Similar evaporation studies of InP (100) surfaces have been done by Farrow (1974). Under Langmuir conditions, Farrow finds $T_c(\text{InP}) = 346°C$ in good agreement with preliminary data of Goldstein *et al.* (1976). Experimentally, however, successful HC of InP is difficult (e.g., Chang and Meijer, 1977). Above T_c there is preferential column V loss (e.g., As_2 and P_2 from GaAs and InP, respectively). Below T_c the evaporation is congruent. According to Goldstein *et al.* (1976), the relative evaporation rates for Ga and As on the surface appear to be the rate-limiting factor in the decomposition of GaAs and likely many other III–V semiconductors. $T_c(\text{GaP (100)}) = 680°C$ (Thurmond, 1965) and $T(\text{H-C})$ for Si is $\sim 1200°C$ (Koval' *et al.*, 1977).

There have been several reports of improved HC techniques for GaP surfaces. Piaget *et al.* (1977a) indicated that carbon-free, nonfaceted (100) surfaces could be obtained by an "insitu reaction with oxygen." Stupp *et al.* (1977) found that vacuum heating GaP surfaces at about 10^{-6} Torr H_2O at $600°C$ for a few minutes removed carbon contamination. By intentionally forming an oxide film on the GaP surface outside the vacuum system, Miyao *et al.* (1978) found that residual carbon contamination was subsequently removed by vacuum HC. The improvement in surface escape probability was about a factor of 4 in the case of Piaget *et al.* (1977a). The SA cleaning procedure can effectively remove all carbon contamination as measured by AES; however, the surface escape probabilities from SA NEA III–V surfaces have generally been somewhat lower than carefully prepared HC surfaces, presumably due to residual sputter damage to the cathode surface (Stocker, 1975; Skeath *et al.*, 1978). SA cleaning is often used, however, with Si due to the tenacious surface oxide. The use of AES has been particularly helpful in developing effective pre-activation handling and cleaning procedures. [See Figs. 30 and 31.]

b. Cs + O Activation Techniques

The Cs + O activation of the surface follows in situ UHV cleaning. Using photoemission as a monitor (e.g., a 10^{-2}–10^{-3} lm source, Section VIII), Cs is deposited onto the clean surface until a photoemission peak is observed. In the case of NEA Si (100) surfaces, it is particularly important that the cesiation not to be halted prematurely as subsequent oxidation of the surface may not yield NEA (Richard, 1973; Gundry *et al.*, 1974; Holtom and Gun-

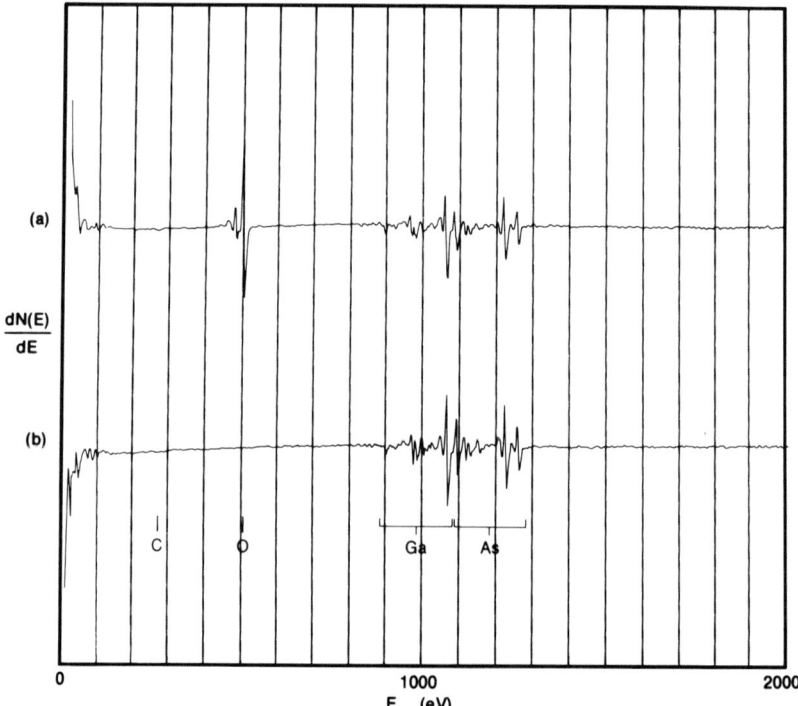

FIG. 30. Auger electron spectroscopy (AES) of a GaAs surface before vacuum heat cleaning (a) and after (b). No sputtering was used. Carbon is the most difficult surface contaminant to remove. Simple heating of the cathode in vacuum, in most cases, will not significantly remove carbon. The GaAs surface shown here before heat cleaning was chemically cleaned and etched prior to AES measurements.

3. NEA SEMICONDUCTOR PHOTOEMITTERS

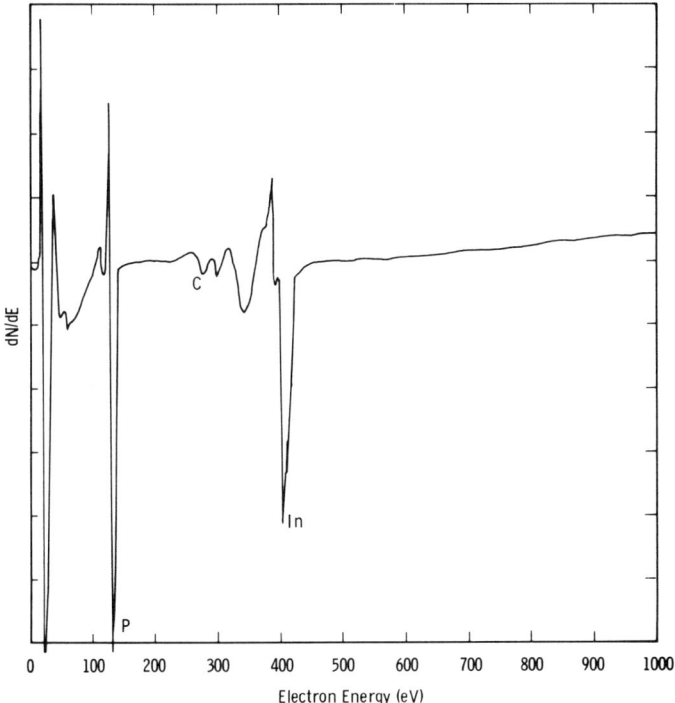

FIG. 31. AES spectrum of a vacuum heat-cleaned InP (100) surface.

dry, 1977; Edwards and Peria, 1978). [Recall Fig. 25.] With an NEA Si activation, the cesiation is usually stopped and a single oxygen treatment then yields maximum photoresponse. For III–V surfaces, a single oxidation is not sufficient and further Cs and oxygen are necessary. Simultaneous activation with Cs + O (after the initial Cs) is often found to give somewhat better near-threshold yields than "yo-yo" techniques (alternating Cs and O_2 exposures) (Sonnenberg, 1971; Piaget et al., 1976; Olsen et al., 1977). For optimal yield, the activation procedure must be performed near room temperature (e.g., Fisher, 1974). Thermal desorption studies have shown that Cs desorbs rapidly above 200°C from GaAs surfaces (Goldstein and Szostak, 1975a; Derrien and Arnaud d'Avitaya, 1977) and 200–300°C from Si (Goldstein, 1973; Edwards and Peria, 1978). Low-temperature activations on GaAs give poor results (Goldstein, 1975). The low-temperature results suggest that the Cs or Cs + O must be allowed some mobility on the surface in order to achieve maximum work function lowering. It has also been found that Cs must be deposited first. This may be due in part to the fact that oxygen has a 10^6 times lower sticking probability on a clean GaAs surface relative to a Cs/GaAs surface (Su et al., 1979; Skeath el al., 1979b) or to disruption of the "proper" Cs adsorption sites (e.g., Lindau et al., 1978b; Ranke and Jacobi, 1979). Similar considerations may hold for Si (Goldstein, 1973; Levine, 1973).

The escape probability from NEA GaAs and InGaAs surfaces can be enhanced significantly, in some cases, when a two-step activation procedure is used—the so-called "Hi-Lo" technique (Fisher and Fowler, 1973; Fisher, 1974; Stocker, 1975; Piaget et al., 1976; Olsen et al., 1977). In this process the cathode is vacuum heat cleaned and activated in a normal fashion discussed above. The activated cathode is then heat cleaned once again at a somewhat lower temperature (about 100°C lower), leaving residual Cs + O on the surface, and reactivated with Cs and Cs + O in the same manner as before. The yield near threshold on the second activation can be 10-500% higher than achieved on the initial activation. The reason for the enhancement in surface escape probability using the Hi-Lo technique is not known, however, subtle changes in the Cs + O stoichiometry or slight rearrangement of the surface atoms may be involved (e.g., Stocker, 1975).

c. Bulk Doping

For successful emission of thermalized electrons, the width of the bent-band region W must be on the order of 100 Å, i.e., the optical–phonon electron scattering mean free path. This implies a bulk cathode doping of $\sim 10^{19}/cm^3$. The reason for a narrow depletion region is clear from Fig. 11. The true electron affinity is positive; and, therefore, vacuum emission is only possible if thermalized electrons are accelerated within the bent-band region and emitted as hot electrons with energy $> E_A$ relative to the bottom of the conduction band at the surface. The relation between doping and the depletion depth was given in Eq. (3). It is clear that for a given work function both W and the amount of band bending at the surface, i.e., Φ_p in Fig. 11, should play a role in the electron escape process (e.g., James et al., 1971a). In general there is a trade-off between maximum surface escape probability and maximum quantum yield. For both Si and III–V surfaces, P generally increases with increasing p-type doping at least up to $10^{19}/cm^3$ (e.g., Garbe, 1969a; Howorth et al., 1976a). However, the bulk diffusion length of the cathode L decreases significantly at the higher p-type concentrations (Martinelli and Fisher, 1974; Howorth et al., 1976a). In the case of thick RM–NEA cathodes, $Y(PE) \sim PL$ near threshold [from Eq. (14)] and the trade-off is one for one.

d. Band Structure and Emission Face

There has been some theoretical work, notably by M. G. Burt and J. C. Inkson, to relate the surface escape probability to the bulk band structure and emission face of an ideal single-crystal NEA surface. The reason for these calculations is in part to account for some of the experimentally observed, but yet unexplained differences in surface escape probability from NEA surfaces and to aid in a deeper understanding of the electron escape process itself. Requirements of conservation of electron energy and momentum parallel to the surface dictate that only those electrons with momenta primarily normal to the surface will be emitted (Bell, 1973, p. 127, 1975; Burt and Inkson, 1975). Using this as a basis and taking the semiconductor wave functions (Bloch functions) into account, Burt and Inkson have calculated surface escape probabilities (transmission coefficients) for Γ electrons in GaAs for emission from (100), (110), (111) A, and (111) B faces (Burt and Inkson, 1976a,b) 1977). Emission of X electrons from GaAs (110) is also considered (Burt and Inkson, 1975, 1976c). These calculations show that the escape probability is a sensitive function of the amount of band bending independent of phonon scattering considerations. (Scattering in the band-bending region is not specifically included in these calculations.) The strong band-bending dependence is due to the fact that with more band bending the electron moves up higher in energy relative to the bottom of the conduction-band minimum in crossing the band-bending region. As a result, some electrons may gain enough energy to transfer to other minima, e.g., Γ to L in GaAs. The other minima may have a higher or lower probability of emission, depending on their momenta relative to the emitting surface. In general agreement with the experiments of James *et al.* (1971a), the (111) B face of GaAs is predicted to have a higher escape probability than the (111) A face due to the heteropolar nature of GaAs. The escape probability for Γ or X electrons from the (110) face, however, is calculated to be significantly lower than that for the other low index faces. Experimentally, this is not the case and Burt and Inkson (1976c) have suggested that multiple electron reflections within the band-bending region may be important. See for example the work of James and Moll (1969). Some of their data is shown later in Fig. 36. Most HC NEA III–V surfaces are faceted and therefore detailed comparison with these calculations, in most cases, is uncertain. James *et al.* (1971a) were careful, however, to avoid faceting in their experiments.

Calculations for NEA GaP are more difficult because of the complications from the larger band bending relative to Si or GaAs. There is a probability of large-momentum intervalley phonon scattering, e.g., X electrons into L and Γ minima (Piaget *et al.*, 1977a,b; Burt, 1978). [See Fig. 32.] Burt (1978) has calculated escape probabilities for GaP (111) B surfaces. Assuming thermalized electrons (20 meV) enter the band-bending region in the X_1 minimum, Burt allows them to transfer without energy loss via optical phonon scattering to the Γ or L minimum and to arrive at the surface at normal incidence. The escape probability for X_1 electrons is assumed to be zero from conservation of energy and the component of the electron's wave vector parallel to the surface. As in the case of GaAs, the escape probabilities for Γ and

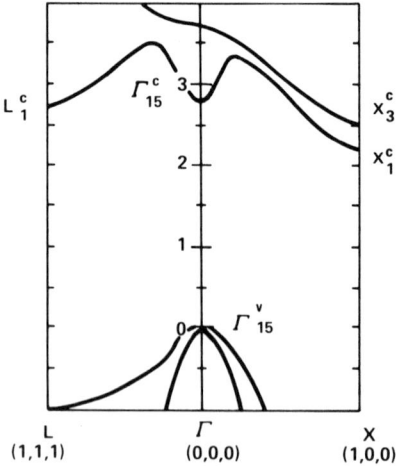

Fig. 32. Near-bandgap band structure of GaP (300°K). The lowest bandgap is the indirect gap $\Gamma^c_{15} - X^c_1 = 2.22$ eV. $X^c_3 - X^c_1 = 0.3$ eV and the lowest direct bandgap is $\Gamma^c_{15} - \Gamma^v_{15} = 2.78$ eV.

TABLE IV

Typical Experimental NEA Surface Escape Probabilities, P

Surface	P (%)		
	Cs-only	Cs + O	Valley
GaP (100)	N.D.	>50	Γ
	20–40	60–80	L_1
	5–15	30–40	X_1
GaAsP (111) B ($E_g \sim 1.9$ eV)	~5	50–70	Γ
GaAs (100), (111) A or B	~0.01	~50	L
	~0.01	20–40	Γ
Si (100)	≪0.01	10–25	X
InGaAsP (111) B ($E_g \sim 1.15$ eV)	≪0.01	1–10	Γ

L electrons are a function of the band bending. For ≥ 0.4 eV band bending, however, the calculated escape probabilities are essentially constant at 83% for Γ electrons and 22% for L electrons. Direct comparison of these GaP calculations with experiment is difficult since analysis of P_X, P_L, and P_Γ from the photoemission data requires assumptions about the fraction of electrons within each minimum, and these fractions are not well known. Miyao et al. (1979) have made a careful photoemission study of HC GaP (111) B surfaces, analyzing their data in terms of X_1 and L_1 electron emission. Their measured escape probabilities for X_1 electrons were quite low, $P_x \sim 1\%$. The P_L escape probability could not be determined uniquely but was $\sim 10X$ higher. P_Γ was not measured. Piaget et al. (1977b) experimentally found the escape probability for L electrons about a factor of 2 higher than for X electrons from nonfaceted HC GaP (100). Note that direct emission of L electrons is essentially forbidden by conservation of parallel momentum. Therefore, these electrons must have scattered into other regions of the Brillouin zone (Piaget et al., 1977b). See Table IV.

Bell (1973, p. 127, 1975) and Burt and Inkson (1975) have shown that simple application of conservation of energy and momentum for the case of NEA Si implies effective blockage of thermalized X electron emission from (111) and (110) surfaces consistent with the failure to achieve, experimentally, NEA on these surfaces. In the Levine (1973) model for NEA Si (Section VI.4), however, atomic reconstruction of the Si (100) surface is central and this too may be a factor. Since band bending is small relative to the higher conduction-band minima in Si, intervalley scattering would likely not be important within the band-bending region. For the NEA Si (100) surface Burt (1977) finds that the conduction-band states near the X point are dominated by plane waves with high momentum parallel to the surface and hence have a relatively low, but finite, probability of propagating into vacuum. Experimental escape probabilities from NEA Si, however, are somewhat better than from a ~ 1.1-eV bandgap NEA III–V cathode (Martinelli, 1970a; Howorth et al., 1975a, 1976a) and close to the $P = 33\%$ expected from a simple conservation of energy and momemtum model (Bell, 1973, p. 131).

The surface escape probability is also a function of the bandgap of the semiconductor. To a first approximation, $P = 1.0$ for $E_g > \Phi$ and $P = 0$ for $E_g < \Phi$ ignoring losses in the band-bending region, Cs + O, and energy-momentum considerations discussed above. Since Φ can be lowered to 0.6 0.8 eV with Cs + O on metal and semiconductor surfaces (Uebbing and James, 1970; James and Uebbing, 1970; Gregory et al., 1975), it was speculated (e.g., Sonnenberg, 1969; Williams, 1969; Klein, 1969) that the lowest bandgap semiconductor which would exhibit efficient thermalized electron emission would be 0.6–0.8 eV, i.e., 1.6–2.0 μm photoemission threshold. Experimentally, however, reasonably high quantum efficiencies ($>0.1\%$) can not be achieved from III–V cathodes with $E_g < 1.0$ eV (Antypas and James, 1970; James et al., 1971b; Sonnenberg, 1970, 1971; Jackson and Yee, 1971; Fisher et al., 1971, 1972; James et al., 1973). Bell et al. (1971) showed that the long wavelength experimental yield curve data of Fisher et al. (1971) for $E_g < 1.0$ eV InGaAs alloy cathodes is consistent with thermally assisted hot-electron emission over an interfacial barrier of 1.1–1.2 eV. The interfacial barrier, therefore, sets the practical long wavelength limit to efficient NEA photoemission (James and Uebbing, 1970). [See Fig. 33.]

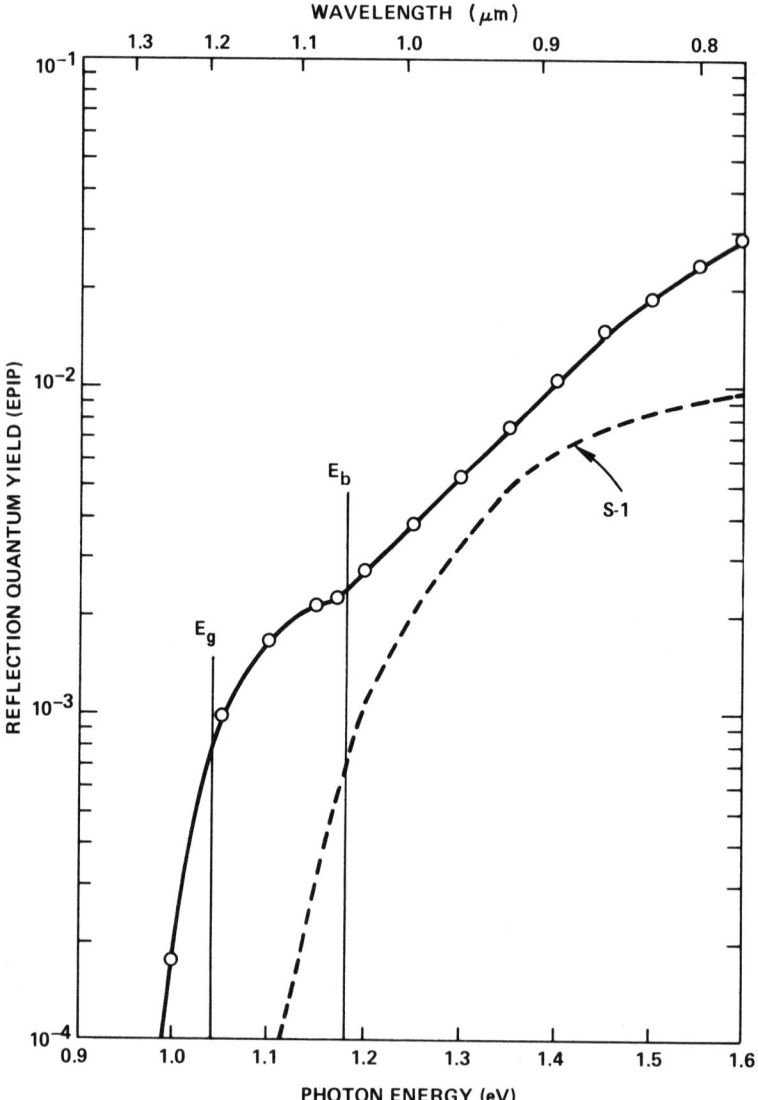

FIG. 33. Reflection-mode quantum yield curve from a 1.04-eV bandgap InAsP cathode activated with Cs + O. The inflection point in the yield is characteristic of the effect of an interfacial or work function barrier at the cathode surface. The work function in this case is about 1.0 eV. (From Bell et al., 1971.)

Fisher et al. (1972) developed a semiempirical surface escape probability model which incorporated the interfacial barrier concept, electron losses in the band-bending region, and thermal energy spread (Boltzmann) of the electrons. The various model parameters were adjusted to give a single best fit to InGaAs alloy cathodes of bandgap range 1.38 to 0.74 eV. The model of Fisher et al. assumes a fixed work function $\Phi \approx 1.0$ eV and allows the interfacial barrier to be semitransparent to electrons. The "best-fit" thickness (~ 8 Å) and energy height (~ 1.25 eV) are consistent with other estimates for the optimized Cs + O activation layer thickness and interfacial barrier height (Section VI.6). Furthermore, the model gives satisfactory agreement with the experimental work function versus escape probability data on GaAs of Korotkikh et al. (1977) and the long wavelength photoemission data from InGaAsP alloy cathodes of James et al. (1973). [See Figs. 34 and 35.]

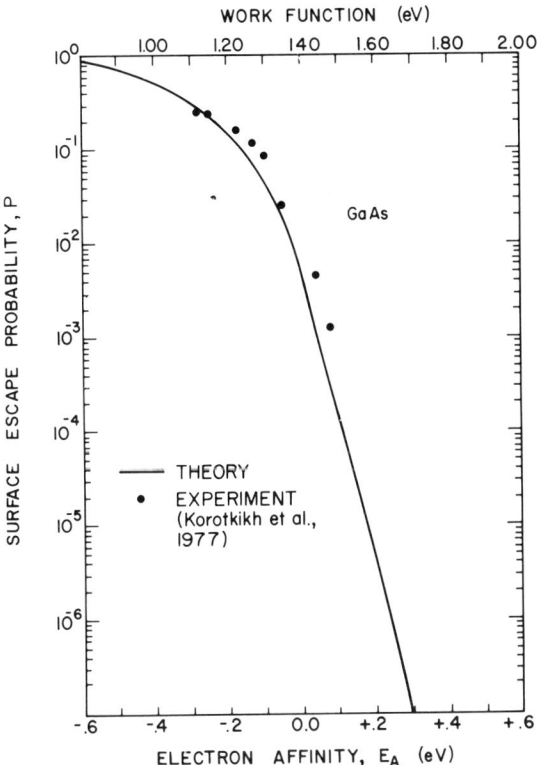

FIG. 34. Calculated and experimental surface escape probability versus work function and electron affinity (effective) for a NEA GaAs cathode. The parameters used in the calculation are unchanged from the original Fisher et al. (1972) model with $N_A = 1 \times 10^{19}/\text{cm}^3$. Experimental data from Korotkikh et al. (1977).

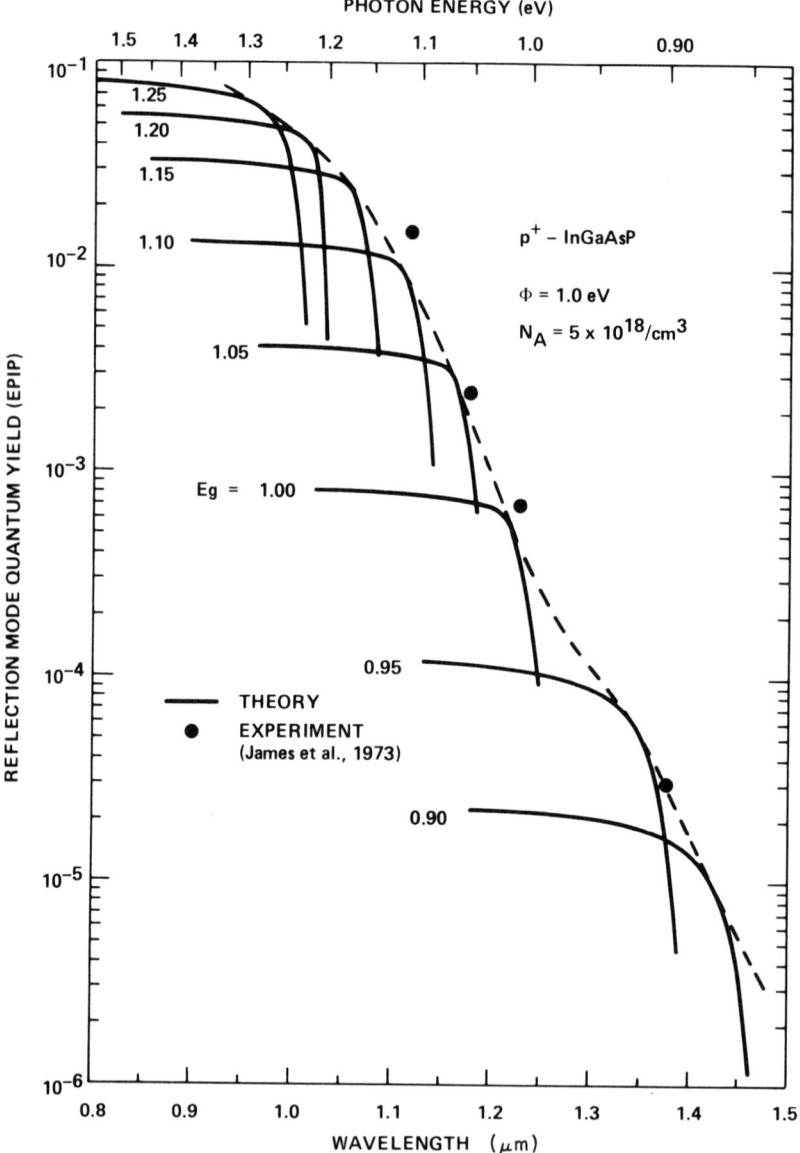

FIG. 35. Calculated and experimental reflection-mode quantum yield. Experimental data from p-InGaAsP cathodes (James et al., 1973). Calculation is based on the Fisher et al. (1972) surface escape probability model, Eq. (13), and Fig. 8.

From their EDC experiments and calculations on NEA VC GaAs (110) surfaces, James and Moll (1969) indicated that a number of electrons may be reflected at the cathode–vacuum interface. Not all reflected electrons will be lost to recombination, however, since many will be reaccelerated toward the surface and be emitted. [See Fig. 36.] The Fisher et al. (1972) escape probability model does not take reflected electrons explicitly into account but is incorporated indirectly through the fitting parameters of the interfacial barrier (height and width). Calculated EDC curves for NEA GaAs have been given by Escher and Schade (1973) using the Fisher et al. model and are in reasonable agreement with most experimental measurements (Eden et al., 1967; James and Moll, 1969; Kressel et al., 1973). Typical calculated and

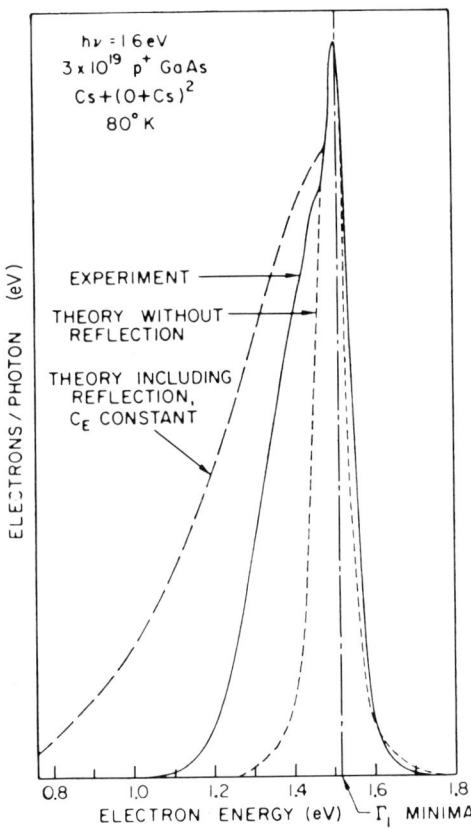

FIG. 36. Calculated and experimental EDCs from NEA, vacuum-cleaved GaAs (110). (From James and Moll, 1969.)

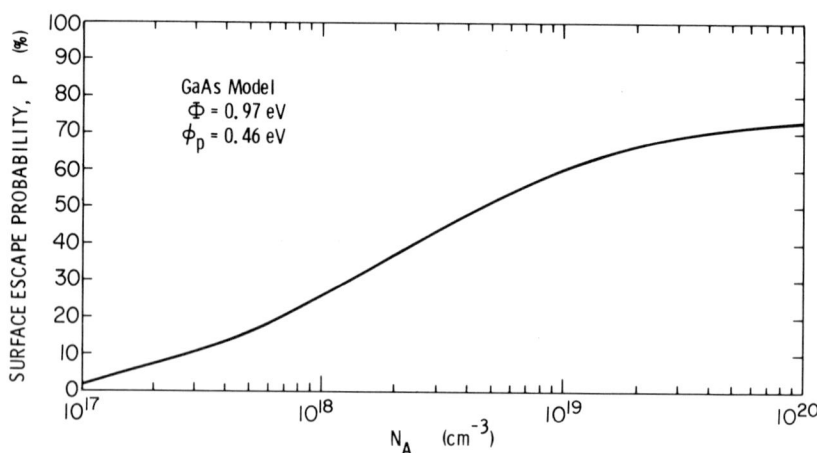

FIG. 37. Calculated surface escape probability P versus acceptor doping concentration N_A for NEA GaAs. Model parameters are unchanged from Fisher et al. (1972).

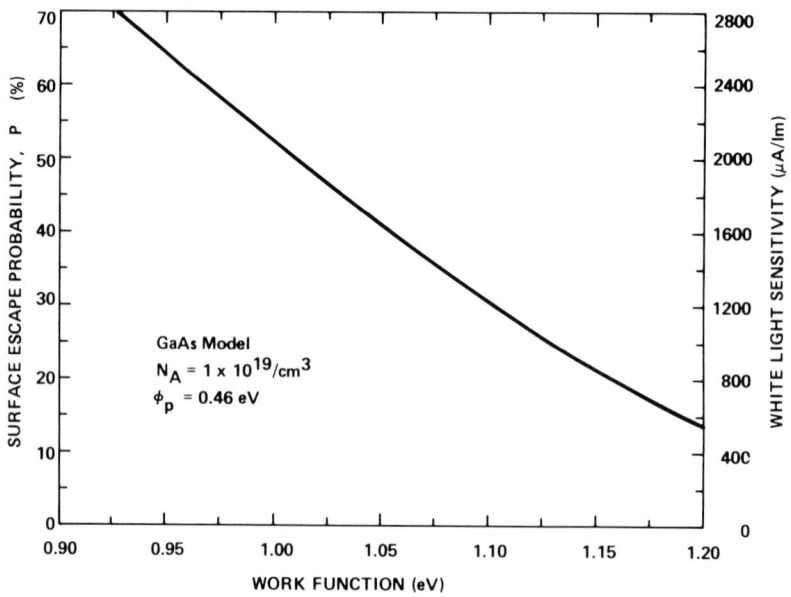

FIG. 38. Calculated surface escape probability P and integrated white light sensitivity in $\mu A/lm$ versus work function for NEA GaAs. The integrated sensitivity is calculated from P, Eq. (13), Fig. 8, and Eq. (21).

experimental half-widths are 100–150 meV. Other model calculations for NEA GaAs are shown in Figs. 37 and 38.

For $E_g > E_b$, the surface escape probability from III–V surfaces would be expected to increase slowly with increasing bandgap assuming all other factors constant. Experimental yield data of Escher and Antypas (1977) from GaAsP alloy cathodes (1.4–2.1 eV bandgap) show the anticipated trend in surface escape probability. [See Figs. 39–41.] The data are fitted to the model of Fisher *et al.* in Fig. 39. Note in Fig. 40 that the yield from the highest bandgap GaAsP cathodes is significantly less than from slightly lower bandgap cathodes. The reason for the drop in yield is likely associated with the direct to indirect transition whereby the escape probability for L electrons (indirect region) is about a factor of 4 less than that for Γ electrons (direct) (Burt, 1978). Earlier studies of NEA GaAsP have been made by Simon *et al.* (1969), Garbe (1969b) and Sommer (1973b).

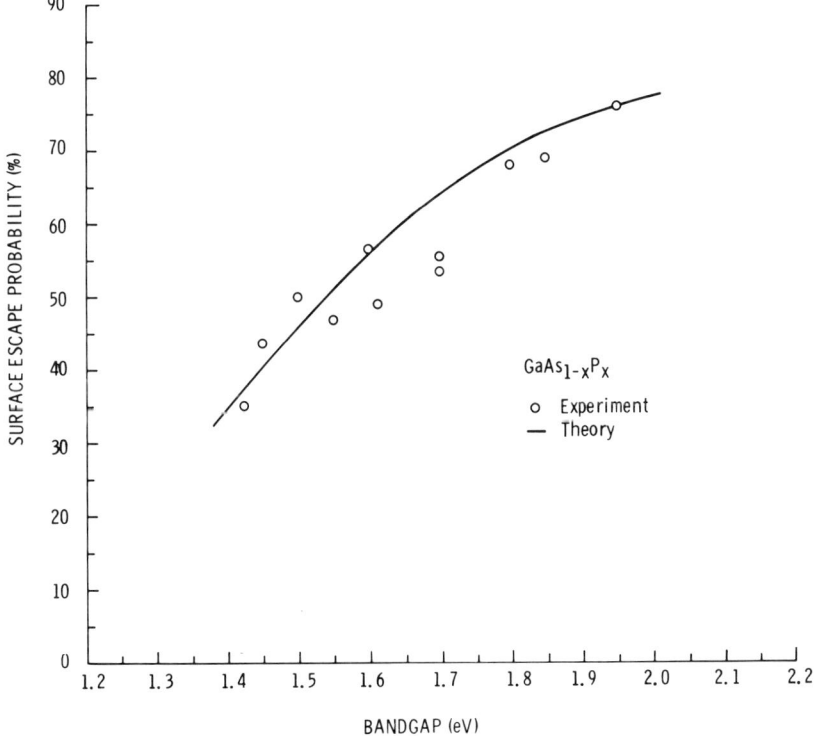

FIG. 39. Calculated and experimental surface escape probabilities from NEA GaAsP (111) B alloy cathodes. (From Escher and Antypas, 1977.)

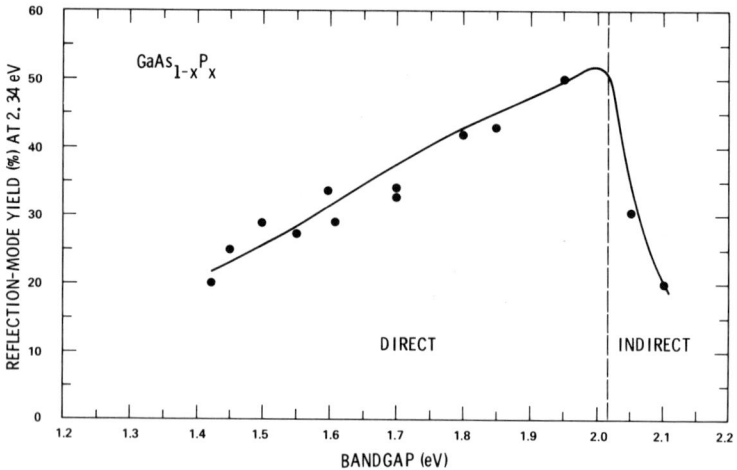

FIG. 40. Experimental reflection-mode quantum yield at 2.34 eV (0.53 μm) versus GaAsP bandgap.

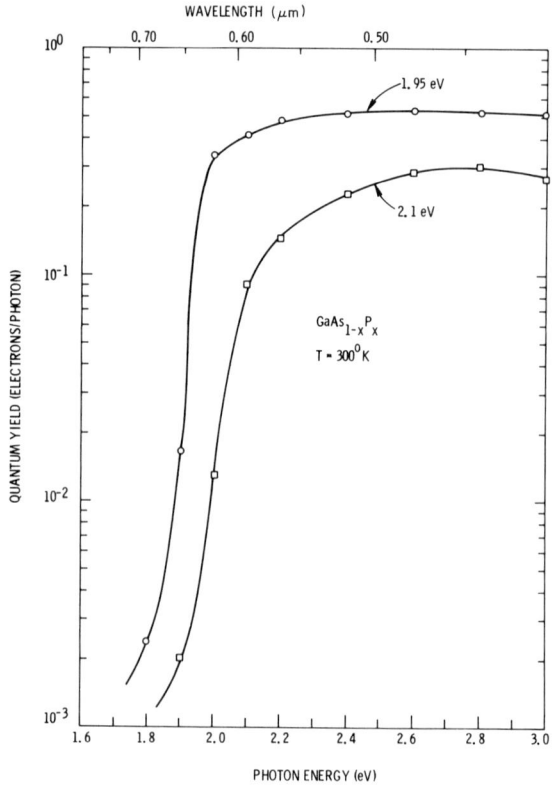

FIG. 41. Reflection-mode quantum yield from 1.95-eV and 2.1-eV bandgap GaAsP cathodes.

In contrast to the increasing escape probability with higher bandgap for direct-bandgap GaAsP alloy cathodes, Martinelli and Ettenberg (1974) found P to decrease with increasing bandgap from AlGaAs alloy cathodes (1.4–1.8 eV, (111) B orientation). The reason for this behavior may be the difficulty in completely removing the surface oxide from Al-containing alloys or perhaps to a change in surface composition from the HC cycle similar to that observed by Goldstein and Szotak (1975b) on GaInAs alloy cathodes (Martinelli, 1979).

e. Electric Field Enhancement

It is well known from work on proximity-focused image tubes with conventional cathodes that application of a high external electric field ($E \sim 10^4$ V/cm) can significantly enhance the quantum yield of the cathode (e.g., Garfield *et al.*, 1976). A similar field enhancement in quantum yield has been observed with NEA GaAs and Si surfaces by Howorth *et al.* (1973a, 1976b). The enhancement can be accounted for by a change in work function $\Delta\Phi$, which can be estimated from the Schottky effect, i.e.,

$$\Delta\Phi = (qE/4\pi\varepsilon_0)^{1/2}, \quad (19)$$

where q is the electronic charge and ε_0 the permittivity of vacuum (Sze, 1969, p. 365). For 1×10^4 and 1×10^5 V/cm fields, $q\Delta\Phi = 38$ and 120 meV, respectively. The model calculations shown in Fig. 34 suggest that the largest gain would be for E_A (effective) >0. For GaAs and $\Phi = 1.20$ eV, an increase in escape probability from $P \approx 15\%$ to $P \approx 35\%$ would be expected for $E = 1 \times 10^5$ V/cm. [See Fig. 38.] The experimental field-enhancement gains measured by Howorth *et al.* (1973a) are typically 1.5–1.7X for $E = 5$–7×10^3 V/cm. Similar gains would be expected for both RM and TM photoemission, secondary electron gain, and dark current emission. Dark current field-enhanced gain from NEA Si has also been observed by Howorth *et al.* (1975b, 1976b).

f. Temperature

Cooling an activated NEA surface generally results in a slightly higher surface escape probability and a small shift in threshold response to higher photon energy (e.g., James *et al.*, 1973; Cole and Ryer, 1972; Martin, 1976). Since threshold response is bandgap limited, the change in threshold with temperature is simply due to the bulk bandgap variation with temperature, e.g., $\Delta E_g/\Delta T \approx -2.4 \times 10^{-4}$, -4.3×10^{-4} eV/°K for Si and GaAs, respectively, at $T \approx 200$–$350°$K (Sze, 1969, p. 24). Cooling an NEA GaAs cathode to 200°K, for example, would cut approximately 250 Å off the long wavelength threshold response. The increase in escape probability can largely be accounted for by the slightly larger bandgap, i.e., "more NEA" and perhaps slightly reduced electron losses in the band-bending region due to the increase in optical–phonon mean free path. The gain in escape probability on cooling would be expected to be larger for surfaces with E_A (effective) >0 than for E_A (effective) <0 and most evident for thermalized emission near threshold. The temperature data of Martin (1976) on commercial NEA photomultiplier tubes (dynodes shorted together) indicate a cathode gain of approximately $1-3\%/°C$ near 8000 Å for most GaAs and InGaAsP ($E_g \approx 1.14$ eV, 300°K) tubes. Much smaller percentage gains ($0.1-0.2\%/°C$) would be expected from the highest sensitivity GaAs or higher bandgap cathodes, e.g., GaAsP. Note that cooling III–V cathodes would not be expected to result in a large increase in cathode electrical resistivity since the active cathode is essentially degeneratively doped and more than 1 μm thick. $\rho(300°K) \approx 10^{-2}$ Ω cm for $10^{19}/cm^3$ p-Si or p-GaAs. Heating an activated NEA surface in UHV to 50–100°C will desorb Cs from the surface and the work function will rise. Within a properly processed sealed-off tube, however, the partial pressure of Cs within the tube is often sufficient to maintain a high escape probability above room temeprature. Above 200°C, Cs begins to rapidly desorb and maintenance of a low work function even under a high Cs partial pressure would likely be difficult. (See Section XI.)

g. Strain

It has been observed by James (1973) that mechancial strain applied to the surface of an activated GaAs cathode reduces the surface escape probability. The effect is reversible only for small strains applied for a short time and may be associated with the piezoelectric effect. The change in escape probability observed by James was as much as 50%. A similar "strain effect" was suggested by Liu et al. (1973) to explain their generally lower escape probabilities on thinned GaAs/AlGaAs/GaAs structures which showed bowing.

VIII. Photocathode Sensitivity Measurements

The quantum yield curve of a photocathode, e.g., Fig. 12, is one of the most useful measures of photosensitivity. A closely related measure is the absolute sensitivity in mA/W, σ. The quantity σ and quantum yield, $Y(EPIP)$, are related by

$$Y(\text{EPIP}) = \frac{hv \text{ (eV)} \, \sigma \text{ (mA/W)}}{10^3}$$

$$= \frac{1.2395 \, \sigma \text{ (mA/W)}}{\lambda \text{ (nm)}}, \tag{20}$$

TABLE V

TYPICAL AND BEST REPORTED μA/LM SENSITIVITIES FROM
NEA REFLECTION-MODE PHOTOCATHODES

Cathode type	S_R (μA/lm)	
	Typical	Best
GaP	10–20	35
		(Piaget et al., 1977a)
GaAsP	200–300	375
($E_g \approx 1.9$ eV)		(Escher and Antypas, 1977)
GaAs	800–1400	2150
		(Olsen et al., 1977)
InGaAsP	200–500	1640
($E_g \approx 1.15$ eV)		(Escher et al., 1976b)
Si	300–600	1500
		(Howorth et al., 1975b)

where λ (nm) is the incident wavelength measured in nanometers and $hc/q = 1.2395$ nm W/mA. Since the 1940s it has been traditional to measure the response of a photocathode to a standard tungsten light source—the so-called lumen source. The present standard is an operating color temperature of 2856°K. The integrated response is measured in microamps per lumen, μA/lm (Biberman, 1971). Experimentally it is convenient to activate cathodes with a standard lumen source of 10^{-2} to 10^{-3} lm intensity since it involves a single measurement and changes in cathode sensitivity can be easily followed. However, the spectral output of a lumen source is very low in the blue region of the spectrum and is highest in the near IR, peaking ~ 1.01 μm (Engstrom, 1955). Therefore, using μA/lm as a comparative measure of photocathode sensitivity is best reserved for comparisons between photocathodes of the same type. Since the term "lumen" is a photometric term (i.e. related to the response of the human eye), it is logically inconsistent to use μA/lm as a measure of sensitivity for photocathodes with significant near-IR response (Biberman, 1967). However, the tradition has held on and most device studies continue to express senstivities in these units. Typical and peak μA/lm values, S_R, for various reflection-mode NEA surfaces are summarized in Table V.

Planck's equation for blackbody radiation from a $T = 2856°K$ source can be used to calculate S_R (μA/lm) from the quantum yield curve (Engstrom, 1955; Engstrom and Morehead, 1967):

$$S_R (\mu A/lm) = \int_0^\infty Y(E)\chi(E)\,dE,$$

where (21)

$$\chi(E) \approx \frac{CE^2}{\exp(E/k_B T) - 1},$$

$E = h\nu$ (eV), $k_B T = 0.246$ eV, and $C \approx 2.5 \times 10^6$.

For passive night viewing applications using semitransparent imaging cathodes, the quantum yield curve can be integrated with the night sky radiance to gain a convenient measure of a cathode's sensitivity to the night sky. Calculations of this nature for conventional and NEA cathodes are given by Richards (1969), Lamport (1975), Richard et al. (1977), and Howorth (1979).

IX. NEA Devices

There have been a large number of developmental NEA devices designed for photoemission, secondary electron emission, and cold-cathode emission. Most of these have been fully described in previous reviews (e.g., Bell, 1973; Martinelli and Fisher, 1974). RM NEA GaAs photomultiplier tubes and Cs/p-GaP dynodes have been available commercially for over ten years. More recent developments include the demonstration of a semitransparent (TM) GaAs cathode suitable for low-light-level applications, commercial availabity of an end-on GaAs PMT tube, RM III–V cathodes with a long wavelength threshold to ~ 1.1 μm, and RM GaAsP cathodes with high quantum efficiency (20–40%) throughout the visible. Most recently a TM silicon photocathode (with cooler) coupled to a low-light-level TV camera tube has been announced (Hopkins et al., 1979). The NEA GaAs surface has also been found to be a new source of spin-polarized electrons suitable for a wide range of basic physics experiments (Pierce et al., 1975a,b). Developmental work is continuing on NEA Si devices and a number of III–V cold-cathode structures are being investigated.

11. RM NEA Photomultiplier Tubes

There are currently a number of photomultiplier tubes available with RM (opaque) NEA cathodes from several manufacturers (EMR Schlumberger, Hamamatsu TV, Co., LTD, RCA Corporation, Varian Associates, Inc., at the time of this writing). The NEA GaAs photomultiplier tube (PMT) was the first NEA PMT commercially introduced (RCA) and is the most commonly available. Advantages of the GaAs PMT over conventional tubes are the high quantum efficiency ($>10\%$) nearly flat to 1.45 eV (855 nm) and low cathode dark current emission at 300°K. Other RM cathodes which are commercially available in PMT tubes include GaAsP alloys for the highest cathode quantum efficiency in the visible region and low dark current, and InGaAsP or InGaAs alloy cathodes for broad response out to approximately 1.1 μm. [See Fig. 42.]

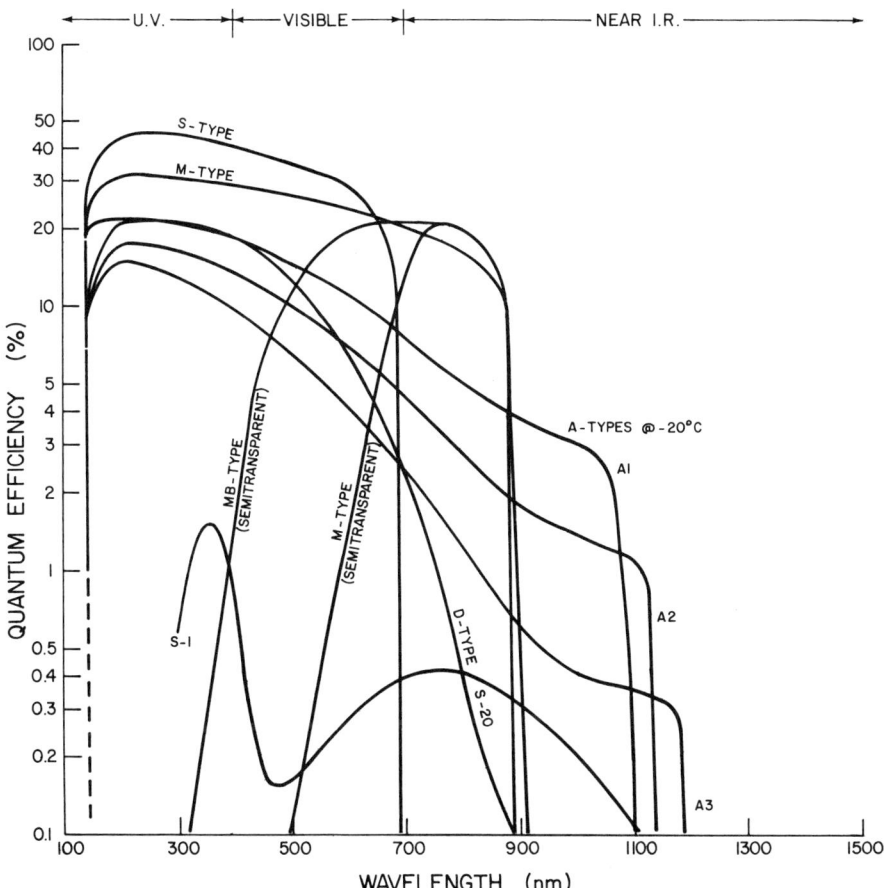

FIG. 42. Reflection-mode quantum yield curves from commerically available NEA cathodes (Varian Associates, Inc.). S-type cathodes are GaAsP, M-type GaAs, and A-type InGaAsP. Conventional S-1 and S-20 cathode yield curves are shown for comparison.

Recent years have also seen the development of high-speed PMTs with 120-psec rise times for dc–3 GHz detection (Persky and Crawshaw, 1973; Wilcox et al., 1979). These are commercially available with III–V cathodes. Application of computerized electron optics design for these tubes has been especially helpful (e.g., Krall and Persyk, 1972). A cross section of a static cross-field high-speed PMT is shown in Fig. 43. NEA Si PMTs have not been developed commercially, to date, due in part to the necessity of cooling for most applications (to reduce cathode dark current emission), the relatively more difficult activation, and the weak long wavelength optical absorption of silicon compared with a similar bandgap III–V (recall Fig. 8).

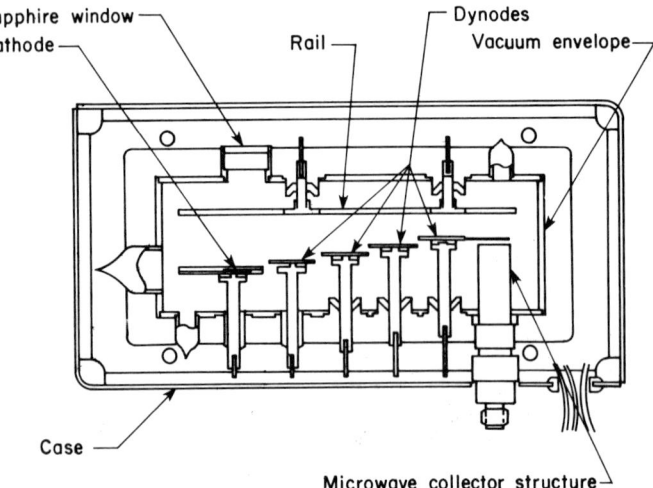

FIG. 43. Cross section of a static crossed-field photomultiplier tube (PMT). The line focusing properties of crossed electric and magnetic static fields are used in this design to achieve electron beam focusing. Tubes with four dynodes have typically achieved 120-psec rise and fall times at 5×10^3 gain. (From Wilcox et al., 1979.)

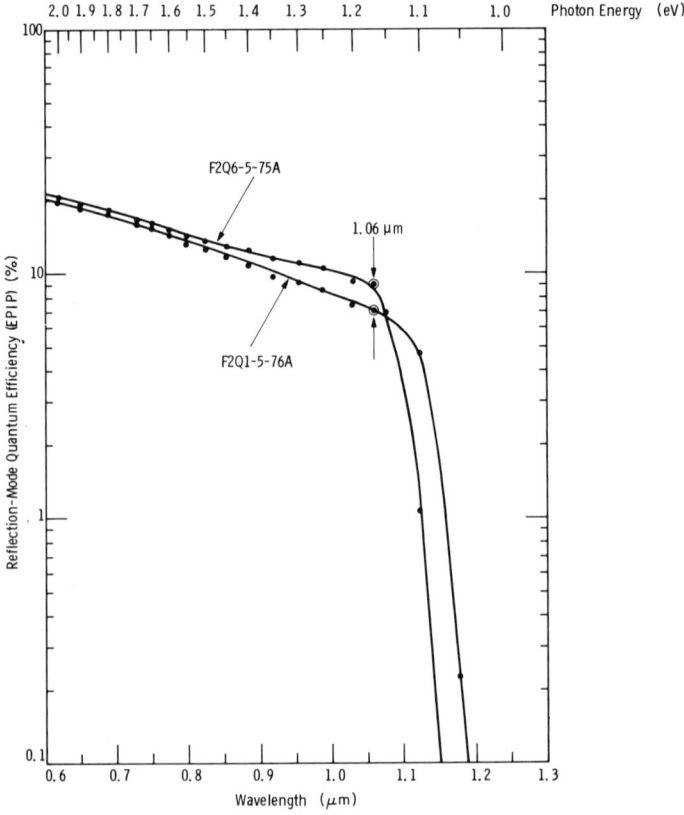

FIG. 44. Experimental RM quantum yield from NEA p-InGaAsP (111) B cathodes. 300°-K shelf life in sealed-off tubes has been a problem with these cathodes and has been circumvented only by cooling of the tube to approximately $-20°C$. (From Escher et al., 1976b.)

A number of developmental RM and TM long wavelength III–V cathodes (i.e. to about 1.1 μm) have been reported with quantum efficiencies considerably better than the commercially available S-1 cathode (Antypas et al., 1970, 1972; Antypas and James, 1970; Enstrom et al., 1971; James et al., 1971b,c, 1973; Jackson and Yee, 1971; Fisher et al., 1971, 1972, 1974; Escher et al., 1976b). [See Fig. 44.] However, shelf-life stability of III–V cathodes with $E_g \gtrsim 1.3$ eV in sealed-off tubes at room temperature is a problem that can only be circumvented, at present, by cooling of the tube (see Section IX).

12. THE SEMITRANSPARENT (TM) GaAs PHOTOCATHODE

A major goal of NEA developmental work for approximately ten years has been the fabrication of a viable semitransparent GaAs photocathode. [Some early work was done on a reflectronics image tube which utilized a RM NEA GaAs cathode (e.g., Palmer et al., 1973).] The very earliest attempts were in situ evaporation techniques onto glass or sapphire substrates (Steinberg, 1968; Syms, 1969; Yasar and Steinberg, 1970; Andrew et al., 1970; Hyder, 1971; Antonova et al., 1976). Although the sensitivities achieved in some cases were comparable to conventional cathodes, it was clear that such relatively crude (compared with MBE) fabrication schemes generally suffered from poor diffusion lengths, poor control over the dopant, and a high surface recombination velocity, $S \gg 1$, between the active cathode and the substrate. In order to overcome these limitations, epitaxial techniques on higher bandgap III–V substrates were investigated, e.g., GaAs/GaP, GaAs/GaAsP/GaP, GaAs/InGaP/GaP, and GaAs/AlGaAs/GaP (Gutierrez and Pommerrenig, 1973; Bell, 1973, p. 45; Fisher and Martinelli, 1974; Hughes et al., 1974; van Oirschot et al., 1975, 1977). To date, only GaAs/AlGaAs and GaAs/InGaP "lattice-matched" heterostructures have demonstrated good diffusion lengths within the active GaAs cathode layer and a reasonably low interfacial recombination velocity at the cathode-"substrate" interface (e.g., Allenson et al., 1972; Liu et al., 1973; Frank and Garbe, 1974; Gutierrez et al., 1974; Fisher et al., 1974; Antypas and Edgecumbe. 1975; Enstrom and Fisher, 1975; Fisher and Olsen, 1979). Nearly ideal semitransparent response can be achieved from these structures with transmission quantum efficiencies generally higher than those in reflection over the entire active photon energy range of the cathode (with an antireflection coating on the back of the cathode). [Recall Fig. 29.] The high-energy cutoff of semitransparent response is essentially the bandgap of the AlGaAs or InGaP substrate, ~ 1.8–2.1 eV.

A major technological development has been the so-called "glass-bonded" or "inverted" cathode first described (and developed) by Antypas and Edgecumbe (1975; see also Piaget et al., 1977c). By suitable epitaxial growth and etching techniques, it was demonstrated that a GaAs/AlGaAs/glass structure could be fabricated with good photoelectric properties and the

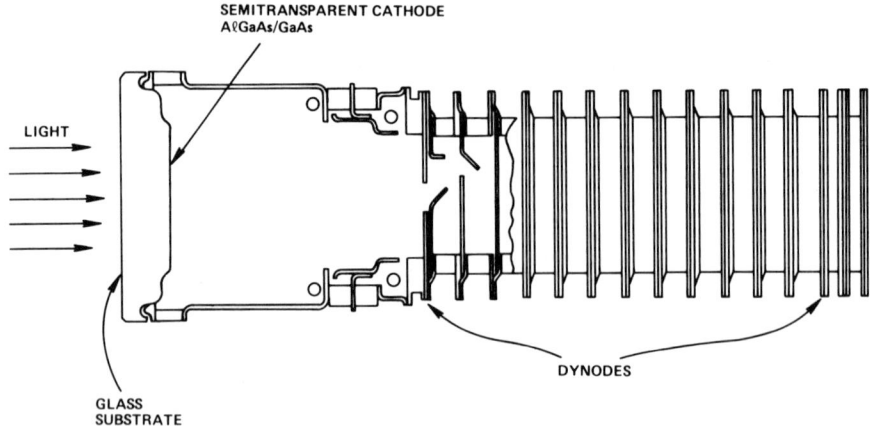

FIG. 45. Glass-bonded AlGaAs/GaAs semitransparent photomultiplier. The active photocathode diameter is 18 mm and the dynode gain is typically 10^6 (VPM-192 Series, Varian Associates, Inc.). The quantum yield curve for this type cathode is shown in Fig. 46.

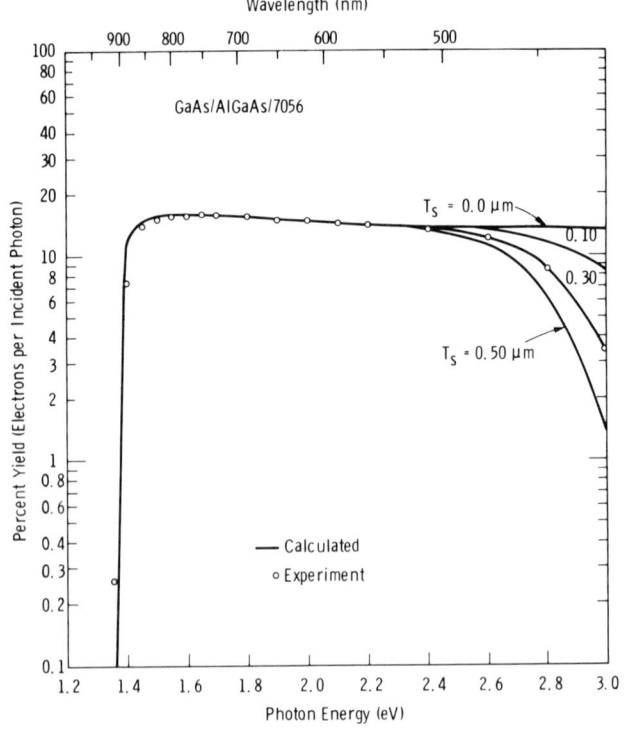

FIG. 46. Semitransparent (TM) quantum yield from a GaAs/AlGaAs glass PMT tube. The AlGaAs is grown relatively thin so that it becomes optically semitransparent but still provides a low recombination interface to the GaAs. (From Antypas et al., 1978.)

significant practical feature of mechanical integrity. An "end-on" PMT tube employing the glass-bonded GaAs cathode with a broadband (1.4–3.0 eV) response achieved with an optically thin AlGaAs layer (~ 0.3 μm) is described by Antypas et al. (1978) and is available commercially (Varian Associates, Inc.). [See Figs. 45 and 46.] Experimental demonstration of the feasibility of x-ray imaging (~ 0.8 to ~ 3 keV) with a NEA III–V cathode has been established by Van Speybroeck et al. (1974) and Bardas et al. (1978).

13. THE SEMITRANSPARENT (TM) SI PHOTOCATHODE

Development of a TM Si photocathode has been relatively straightforward. The technology for producing large area, uniformly thin (~ 10 μm), mechanically stable silicon wafers is well developed from Si vidicon TV camera work. Techniques have been developed to effectively reduce the back surface recombination velocity by means of a thin p^{++} layer at the back surface (Howorth and Pool, 1978) and semitransparent sensitivities as high as 1000 μA/lm have been reported from optimized cathodes (Howorth et al., 1975a, 1976a). Dark current emission at room temperature from NEA Si is similar to that from the S-1 cathode and high sensitivity III–V cathodes of similar bandgap, $\sim 10^{-11}$–10^{-12} A/cm^2. For low-light-level applications some modest cooling, perhaps to $-20°C$, will be needed during operation. A TM Si photocathode coupled to an Isocon low-light-level TV system has been reported (Hopkins et al., 1979).

14. RM NEA SECONDARY ELECTRON EMITTERS (RSE)

In most low-light-level applications the photoelectron current from the cathode alone is very low, e.g., $< 10^{-12}$ A. The electrons from the cathode, however, can be readily accelerated within the vacuum tube to several hundred volts and appropriately focused onto a surface (dynode) which exhibits high secondary electron gain. In the case of photomultiplier tubes (PMTs) there is a series of 4–20 dynodes each of which amplifies the accelerated electron current (3–50X) from the previous dynode in a cascading fashion. Total gains of 10^6–10^7 are typical. Electron "multiplication" in this fashion is generally superior to external means of amplification (e.g., Zwicker, 1977). This is the primary reason why under low-light conditions vacuum electron emission devices are often preferred to all-solid-state light detection devices.

In general, high quantum efficiency RM NEA photocathodes are also high-gain RM secondary electron emitters. The essential difference between the semiconductor RM NEA photocathode and secondary electron emitter is the generation mechanism of minority carriers. In the case of photon generation, the optical bandgap of the semiconductor is important and the depth of generation is on the order of a micron. (Recall Fig. 8.) For significant secondary electron emission, however, the energy of the incident electrons must be several orders of magnitude higher than the semiconductor bandgap, yet the depth of secondary electron generation is usually less than a micron. (Section VII.8.) In the case of most PMTs, the accelerating potential between dynodes may only be 150–600 V, which implies very shallow primary electron beam penetration into the bulk of the dynode, ~100 Å (Table III). Bulk material properties are not critical at these low energies as long as the semiconductor is p-type, a high number of secondary electrons are generated, and a low work function is achieved at the emitting surface. A high percentage of the emitted secondary electrons, however, would be expected to remain as hot electrons due to the shallow penetration depth. As a consequence, their finite-energy spread must be taken into account in the design of high-speed PMTs (Piaget et al., 1975).

Much higher secondary electron gains were measured from NEA GaP surfaces than conventional dynode materials early in the development of NEA devices (Simon and Williams, 1968; Simon et al., 1968). Their practical utilization as a high-gain first dynode in PMTs came relatively easily. This was due in part to the ability to grow polycrystalline p-type GaP on curved

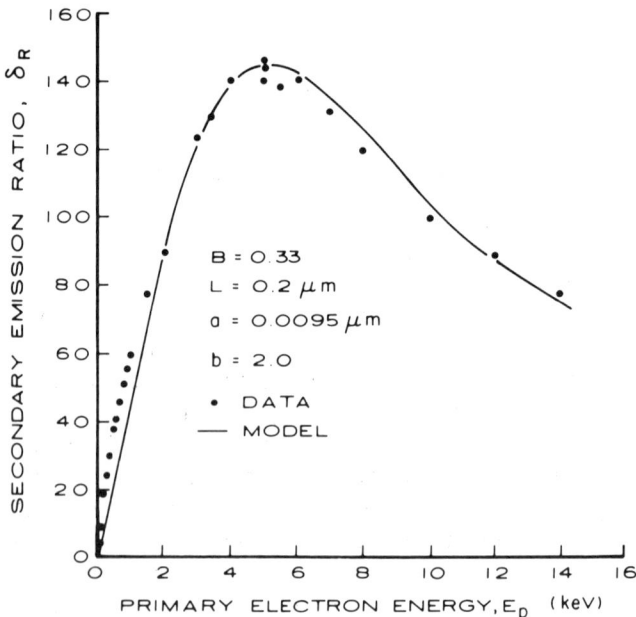

FIG. 47. RM secondary electron gain versus primary electron energy from a Cs/p-GaP/metal dynode. Model calculations based on Eqs. (9) and (18). B is the surface escape probability, L the diffusion length, and a and b the range parameters (Table III). The excess gain at low primary beam energies relative to the model calculation is from hot-electron emission. (From Martinelli, 1974b.)

metal substrates and the relative ease of activating GaP to NEA with Cs-only improved photoelectron statistics for low-light-level scintillation experiments was anticipated and demonstrated (Morton et al., 1968, 1969; Krall et al., 1970). Conventional dynodes (e.g., Be–Cu) are limited to gains of 4–10 at 400–600 V primary electron energy while single-crystal or polycrystalline GaP dynodes are capable of gains of 30–50 in this energy range (Piaget et al., 1975).

A strict thermalized electron diffusion model analysis of secondary electron emission from NEA surfaces for the relatively low energies (<1 kV) used with PMTs is inadequate due to the emission of a high percentage of hot electrons and uncertainty in the variation in backscattering factor $(1-f)$ in this range (Coates, 1970; Afonina et al., 1971; Martinelli, 1974b; Afonina and Stuchinski, 1973; Piaget et al., 1977b). At higher energies, however, the diffusion model can be used to describe the data quite well. Figure 47 shows the secondary electron gain versus primary electron energy from an NEA Cs/p-GaP dynode (Martinelli, 1974b). The dynode in this case was polycrystalline GaP deposited onto a metal substrate. This is the fabrication scheme often used for PMT dynodes.

High RM secondary electron gains have also been measured from single-crystal NEA GaAs surfaces (Martinelli et al., 1972b; Gutierrez et al., 1972), and polycrystalline NEA GaAs (Andronov et al., 1976). The best gains reported are approximately 400 and 1500 for GaAs and Si, respectively, at 15 kV. The higher gain from Si is due in part to the generally longer diffusion length and lower energy required to create an electron–hole pair than GaAs (ε_p, Table III). Due to the early success of the GaP dynode, neither GaAs nor Si has been used commercially for dynodes.

15. TM Secondary Electron Emitters (TSE)

The high gains achieved from NEA GaAs and Si surfaces suggests the possibility of fabricating transmission secondary electron emitters (TSE). These could be used as compact electron multipliers perhaps reducing transit time fluctuations in certain PMTs or as a multiplier for the TM NEA photocathode for low-light-level imaging applications. Some work along these lines has been reported for TSE GaAs (Martinelli et al., 1972b; Gutierrez et al., 1972; Martinelli and Olsen, 1976) and TSE Si (Martinelli, 1970b; Howorth et al., 1975a, 1976a). The shallow penetration depth of the primary electrons dictates very thin self-supporting structures (2–4 μm) for reasonable TSE gains.

Just as with the TM NEA photocathode, optimal TSE gain requires a low back surface recombination velocity. Presumably the use of a p^{++} layer at the back surface of the Si TSE dynode would be effective (Howorth et al., 1976a). Thin InGaP/GaAs TSE dynodes have been shown by Martinelli and

Olsen (1976) to be superior to GaAs dynodes alone. TSE gains of approximately 300 at 15 kV were reported for optimized InGaP/GaAs dynodes and gains of 300–500 at 15 kV from unpassivated Si dynodes have been measured (Martinelli and Fisher, 1974; Howorth et al., 1976a). NEA TSE devices have not found practical application to date, however, due to the difficulties in reliable fabrication and activation of such thin, self-supporting structures.

16. COLD-CATHODE EMITTERS

Following the experimental photoemission results of Scheer and van Laar (1965) with Cs/GaAs cathodes, it was soon realized that the NEA surface could be effectively employed with a forward-biased p–n junction to achieve a cold-cathode structure (e.g., Geppert, 1966). See Fig. 48. The forward-biased junction injects electrons from the n region into the electron emitting p region which is activated to NEA. The p region is made sufficiently thin so that injected electrons will have a reasonable probability of diffusing to the surface. Initial applications for such a device would likely be low-light-level TV camera tubes where improved resolution and faster response to illumina-

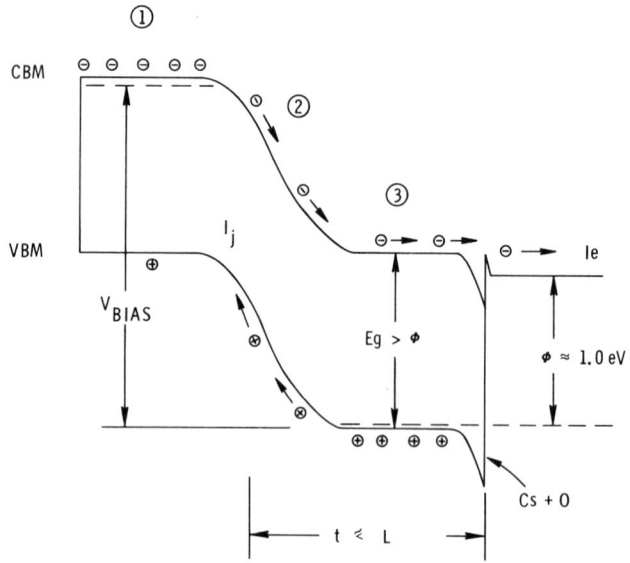

FIG. 48. Simple p–n homojunction cold-cathode energy-band diagram under bias conditions. The substrate (1) is n-type and provides the source of electrons. Under forward bias, electrons are injected (2) from the substrate into the thin p-type emitter region whose surface is activated to NEA (3). If the p region is sufficiently thin, i.e., approximately a diffusion length L, a large fraction of the injected electrons will diffuse-drift to the emitting surface and a fraction P of those will escape into vacuum. The cold-cathode efficiency in % is $(I_e/I_j) \times 100$ where I_e is the emitted electron current into vacuum and I_j the total internal junction current.

tion changes could be anticipated due to the lower-energy spread anticipated from a NEA cold cathode (e.g., Deasley and Faulkner, 1972). Instant start and low power consumption are other potential advantages.

A variety of developmental NEA-based cold-cathode devices has been reported. These include *p-n* homojunction GaAs (Williams and Simon, 1969), GaAsP (Deasley and Faulkner, 1972; Faulkner *et al.*, 1973; Howorth *et al.*, 1973b, 1976c), GaP (Stupp *et al.*, 1977) and Si (Kohn, 1971a, 1973). Heterojunction GaAs–AlGaAs (Schade *et al.*, 1972; Kressel *et al.*, 1973) and GaP-GaAlP (Kan *et al.*, 1979) cold cathodes have been described as well as an optoelectronic GaAs–AlGaAs cold cathode (Kressel *et al.*, 1970; Schade *et al.*, 1971). Peak cold-cathode efficiencies (electrons emitted/internal junction current) are on the order of 1.0% under dc operating conditions with emission currents well above 10 μA possible. An important feature of most cold-cathode designs is a fabrication scheme which forces injected electrons from the *n* region to pass into a confined emitting *p* region (e.g., Kohn, 1971b; Schade *et al.*, 1972).

Experimental vidicon results have been reported for Si (Cope *et al.*, 1973), and GaAsP (Howorth *et al.*, 1973b, 1976c). Generally improved performance is observed due to the narrower electron beam energy spread compared with a conventional hot cathode. Further work is needed, however, to improve operating life, tube design, and tube activation procedures.

17. THE NEA GaAs SPIN-POLARIZED ELECTRON SOURCE

Garwin *et al.* (1974) suggested that an NEA GaAs surface may be a viable source of low-energy spin-polarized electrons. A similar proposal was made by Lampel and Weisbuch (1975) indicating that GaAs (and possibly other semiconductors) would likely be a very bright, intense polarized electron source suitable for a wide range of scattering experiments in atomic and molecular physics. Experimental confirmation of these proposals shortly followed with the work of Pierce *et al.* (1975a,b). Successful application of the GaAs polarized electron source has been reported with high-energy particle physics experiments (e.g., Prescott *et al.*, 1978) and a number of other physics applications, e.g., spin-polarized LEED, have been proposed and demonstrated (e.g., Pierce and Meier, 1976; Campagna *et al.*, 1976; Wang *et al.*, 1979; Celotta *et al.*, 1979; Pierce *et al.*, 1980).

Photoemission of spin-polarized electrons follows essentially from the bulk band structure of GaAs and the relatively high surface escape probability which can be achieved with proper activation of *p*-GaAs surfaces. For GaAs the valence band maximum at Γ is split into a fourfold degenerate $P_{3/2}$ level and a twofold degenerate $P_{1/2}$ level by the spin–orbit interaction. These valence-band levels are separated in energy by $\Delta E = 0.34$ eV. Recall Fig. 9. For circularly polarized light of photon energy, $E_g < h\nu < E_g + \Delta E$, it can be shown that three times as many electrons are excited into $S = -\frac{1}{2}$ states as $S = +\frac{1}{2}$ states (Pierce and Meier, 1976). For higher photon energies, excitation from the $P_{1/2}$ valence-band mix with those from the $P_{3/2}$ and reduce the overall polarization of the emitted electrons. The maximum polarization for electrons thermalized in the band at Γ is $P = 50\%$ and values near this have been experimentally measured (Pierce et al., 1975a,b; Pierce and Meier, 1976; Erbudak and Reihl, 1978; Pierce et al., 1979). Most experiments are made with the GaAs cooled to 100°K to reduce thermally induced depolarization effects. Measurements at 300°K indicate a polarization reduction by a factor of 1.3 (Erbudak and Reihl, 1978).

Pierce et al. (1979) have noted from their results and those of Erbudak and Reihl (1978) that experimentally one can achieve a significantly higher spin polarization for NEA GaAs (100) surfaces than (110), e.g., 43% versus 21%, respectively. Pierce et al. suggest that this difference may be associated with the much higher surface recombination velocity for the (100) face than the (110) as discussed by Burt and Inkson (1977). The lower polarization for the (110) case may be due to multiple escape attempts and depolarization within the band-bending region and the Cs–O activation layer. Erbudak and Reihl (1978) have demonstrated, however, that the lower polarization associated with the NEA (110) surface can be partially overcome by activating the surface to slightly positive electron affinity. In this case thermalized electron emission is primarily from the L minima. Useful experimental details associated with the design, activation, and performance of a NEA GaAs spin-polarized source are given by Pierce et al. (1980).

X. Dark Current Emission from NEA Cathodes

Most photocathode applications are for detection of very low light level signals. The quality of the measured signal is dependent on the photoemission current $[I(\text{PE}) = qI_0 Y(\text{EPIP})$, where q is the electronic charge and I_0 is the incident light intensity] and the noise level. In many practical devices, the noise level is dominated by dark current emission J_{dark} from the cathode. Therefore, the signal-to-noise ratio is generally a function of both $Y(\text{EPIP})$ and J_{dark}. An approximate measure of the minimum input signal level necessary for PMT detection is the equivalent noise input (ENI). If the dominant tube noise is from dark current emission, ENI can be calculated from

$$\text{ENI} = [2qI_{\text{dark}}(\Delta f/G)]^{1/2}/\sigma, \tag{22}$$

where $q = 1.6 \times 10^{-19}$ C, I_{dark} is the dark current in A, Δf is the bandwidth (usually 1.0 Hz), G is the total dynode chain gain, and σ is the cathode sensitivity in A/W [(Eq. (20)].

Typical quantum yields from conventional and NEA cathodes are 10^{-1}–10^{-3} (EPIP). Cathode dark current densities, however, show considerable

TABLE VI

TYPICAL PHOTOCATHODE DARK
CURRENT EMISSION CHARACTERISTICS

Cathode	E_g (eV) $T = +25°C$	$J_{dark}\left(\dfrac{A}{cm^2}\right)$ $T = +25°C$	$\dfrac{J(-20°C)}{J(25°C)}$	$\dfrac{J(-75°C)}{J(25°C)}$
GaAsP	~1.8	$<10^{-14}$	10^{-2}	$<10^{-3}$
GaAs	1.425	10^{-14}–10^{-16}	10^{-3}	10^{-5}
InGaAsP	~1.1	10^{-11}–10^{-13}	10^{-3}	10^{-5}
Si	1.12	10^{-9}–10^{-13}	10^{-2}	no data
S-1	<1.0	10^{-11}–10^{-13}	10^{-3}	10^{-5}
S-20	~1.0	10^{-15}–10^{-16}	10^{-2}	10^{-2}

variation depending on the cathode type and operating temperature. [See Table VI.] Cooling the cathode from +25°C to −20°C decreases the dark current by a factor of 10^2–10^3 suggesting thermally activated electron emission over an energy barrier E_b of the form $J_{dark} \sim \exp(-E_b/K_BT)$, where $E_b = 0.8$–1.2 eV, and $k_BT = 0.0259$ eV at +25°C and 0.0220 eV at −20°C. For $I_{dark} = 10^{-15}$ A, $G = 10^5$, and $\sigma = 0.10$ A/W [i.e., Y(EPIP) = 15% at 826 nm] ENI = 5.6×10^{-19} W.

Figure 49 shows a schematic diagram of the activation of a NEA cathode showing work function, photoemission, and dark current emission versus

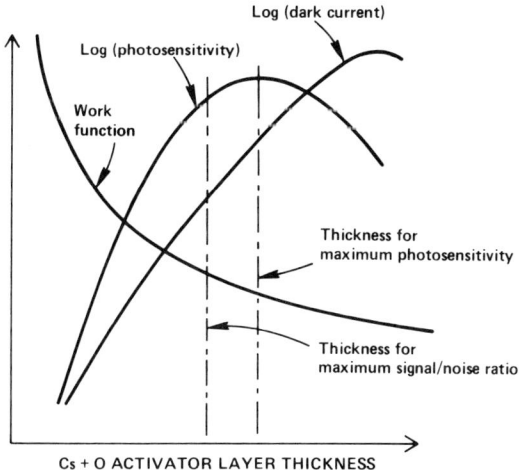

FIG. 49. Schematic diagram of the photoemission and dark current emission characteristics of a NEA III–V surface as a function of the Cs + O activator thickness. Activation of NEA Si (100) is similar except the dark current generally peaks before the photoemission.

Cs + O activator thickness. In general, the peak photoemission point does not coincide with the peak dark current emission point. The behavior sketched in Fig. 49 is generally what is observed with NEA III–V cathodes (Bell, 1973, p. 103), while NEA Si activations show the dark current emission peaking before the photoemission peak (Howorth *et al.*, 1975b; Martinelli, 1974a). In either case, it is apparent that there can be a relatively large variation in dark current emission for cathodes of nearly equal quantum efficiency.

The most extensive theoretical study of thermionic dark current emission from NEA cathodes is the work of Bell (1969, 1970). Bell considers five potential sources. These are (1) surface-state and trapping center generation within the Cs + O activation layer—the "activation layer" component, (2) generation via surface states at the cathode-activator interface—the "surface-state" component, (3) generation via traps within the band-bending region of the cathode—the "generation" component, (4) thermal generation of electrons in the bulk of the cathode—the "diffusion" component, and (5) generation by the ambient infrared photons—the "radiative" component. Order of magnitude calculations can be made of each source component for NEA GaAs (Bell, 1969) and NEA Si cathodes (Bell, 1970; Howorth *et al.*, 1975b). These estimates are summarized in Table VII. Note that the diffusion and radiative components are exponentially dependent on the cathode bandgap while the others are exponentially dependent on the work function (ignoring the interfacial barrier). There is considerable uncertainty in these estimates, since many of the materials parameters needed in the calculation are not accurately known.

There are no detailed, published experimental dark current data available on NEA III–V surfaces. The III–V data for Table VI are based on dark cur-

TABLE VII

CALCULATED DARK CURRENT COMPONENTS FROM
NEA GaAs AND Si AFTER BELL (1969, 1970)
AND HOWORTH *et al.* (1975b)

Dark Current Component	J_{dark} in A/cm², 300°K	
	GaAs	Si
Cs + O activation layer	10^{-12}–10^{-15}	10^{-9}–10^{-12}
Surface states	$\sim 10^{-17}$	10^{-10}–10^{-13}
Trap generation	10^{-15}–10^{-20}	$\sim 10^{-14}$
Bulk diffusion	$\sim 10^{-19}$	$\sim 10^{-15}$
300°K Radiative component	$\sim 10^{-21}$	$\sim 10^{-17}$
Bandgap (eV)	1.425	1.122
Work function (eV)	~ 1.0	~ 0.85
Escape probability (%)	~ 50	~ 30
Bulk doping (cm^{-3})	$\sim 10^{19}$	$\sim 10^{19}$

rent measurements from commercially available PMT tubes which could be in error due to non-cathode-related dark current sources within the tube. However, the activation behavior of the dark current outlined in Fig. 49 suggests, in agreement with Table VII, that the Cs + O activation layer component may be dominant for activations near peak photoemission. Furthermore, in the case of GaAs, a 10^{-3} reduction in dark current on cooling from 300°K to 255°K is consistent with thermionic emission over a 1.0-eV barrier rather than generation across the bandgap. Dark current densities at 300°K from GaAs and higher bandgap III–V cathodes are generally extremely low, $\sim 10^{-16}$ A/cm$^2 \approx 600$ electrons/sec/cm^2, making these cathodes suitable for a wide range of applications.

Detailed dark current experiments have been made on NEA Si cathodes (Martinelli, 1974a; Howorth et al., 1975b, 1976b). These experiments, as mentioned above, generally show the dark current emission peaking before the photoemission peak. Martinelli (1974a) demonstrated that n- and p-type Si (100) surfaces activated with Cs + O yield nearly indentical dark current emission. Howorth et al. (1975b) showed that the externally applied electric field enhancement characteristics of photoemission and dark current emission are not the same. Furthermore, the temperature dependence of the dark current emission is consistent with thermionic emission over the work function rather than bandgap generation. Hence the surface-state and activation-layer components fit both the experimental observations and the order of magnitude calculations of Table VI. The experiments are also consistent with the "trap generation" component, but the calculations suggest that this will be relatively low. The work of Clark et al. (1976) on thick Cs + O activation layers on NEA Si surfaces does not give information on the dark current emission versus activator thickness which would help decide on which component is dominant in the case of optimally activated Si surfaces. The experimental studies on Si also show that the dark current decreases as much as a factor of 10 within several hours after initial activation, while the photoemission remains relatively unchanged. These changes are similar to the decrease in surface-state emission observed by Wagner and Spicer (1972) on VC Si (111) surfaces maintained under different UHV conditions. Martinelli (1973a) has seen a slight shoulder in the photoemission yield curve of some NEA Si cathodes in the threshold region near 1.0 eV photo energy which is not in the optical absorption data. This "extra" yield near threshold may well be from surface-state emission. Overall, the evidence seems to favor surface-state generation as the dominant source of dark current emission from NEA Si surfaces.

Unfortunately the dark current densities at 300°K reported from NEA Si surfaces are generally too high for most applications, 10^{-10}–10^{-12} A/cm^2. This has been the most serious limitation, to date, to an otherwise highly efficient, stable electron emitting surface. A viable processing or fabrication technique that could overcome this problem would make NEA Si devices extremely attractive.

A summary of dark current characteristics and temperature dependence of conventional cathodes, including the S-1 and S-20, is given by Rome (1971). Dark current measurements on NEA GaAs and InGaAsP PMT tubes have been published by Cole and Ryer (1972) and Martin (1976). Information in regard to a specific tube type should, however, best be obtained directly from the manufacturer.

XI. Shelf Life and Operating Life of NEA Devices

The development of Cs/p–GaP dynodes and their incorporation into photomultiplier tubes came relatively easily and was the first successful use of a III–V NEA surface. Practical experience with PMT tubes having GaP dynodes has been generally good with shelf-life and operating-life characteristics at least as good as conventional dynode materials (Sommer, 1973b,c,d). The stability of GaP is due in part to the large bandgap (2.22 eV) which gives a large effective NEA and to the fact that a relatively large fraction of the secondary electrons are emitted as hot electrons. For these reasons, the mean energy of the secondaries is well above the vacuum level (Piaget et al., 1977b). Small changes in work function, which may happen over a long period of time within a sealed tube, have little effect on the overall dynode gain and stability of the tube.

Of the III–V NEA photocathodes, GaAsP and GaAs cathodes have demonstrated good shelf life in properly processed tubes (i.e. comparable to a high sensitivity conventional S-20 tube, 300°K storage). Operating life is equally good *provided* the tube has not been subject to excessive cathode currents from high-light-level usage. Most manufacturers suggest a maximum continuous cathode current of 10^{-11}–10^{-9}A. The reason for this is thought to arise from the fact that the emitted electrons from the cathode are accelerated and focused onto the first dynode surface. Depending on the accelerating voltage and cathode current density, there can be an electron "scrubbing" effect of the first dynode resulting in positive ions being liberated and accelerated back onto the NEA surface. Neutrals can also be liberated. The effect is a continuous increase in the cathode work function under these conditions (Schade et al., 1972; Kressel et al., 1973). Another factor which may accelerate a cathode's degradation under high current operating conditions is electron stimulated desorption of the Cs + O activation layer (Schade, 1976). This effect was studied by Schade (1976) with a GaAs cold-cathode structure where the electron emission current densities are quite high relative to typical photocathode current densities.

Developmental NEA Si tubes seem to have good shelf-life characteristics (Martinelli, 1974a; Howorth et al., 1975b) while operating-life characteristics of Si photocathodes have not been reported to date. The good shelf life of NEA Si tubes is somewhat surprising considering that small changes in work function (± 0.1 eV) can have a very large effect on the surface escape probability. Similar bandgap III–V photocathodes ($E_g \approx 1.1$ eV) are generally not stable at room temperature for this reason. (See Fig. 50.) At the present time, the only known method of achieving satisfactory shelf-life stability in tubes with III–V cathodes of bandgap less than about 1.3 eV is to maintain the tube at or below about $-20°$C. The improved stability of NEA Si relative to a comparable bandgap III–V cathode may be due to the favor-

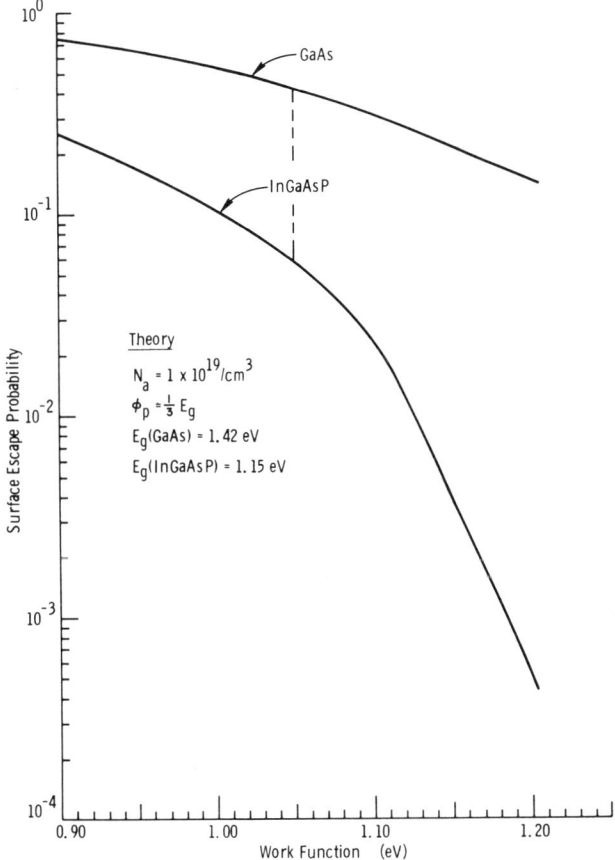

FIG. 50. Calculated surface escape probability versus work function for NEA GaAs and InGaAsP ($E_g = 1.15$ eV) cathodes. A relatively small work function increase from 1.00 to 1.10 eV, which could happen over an extended time within a sealed-off tube, changes the GaAs escape probability from 50% to 30% and the InGaAsP escape probability from 10% to 2.5%. Calculations based on the escape probability model of Fisher *et al.* (1972).

able "pedestal" and "cave" activation sites and the apparent lack of a significant Cs + O interfacial barrier.

NEA tube fabrication technology is still new. All of the factors which are important in achieving tube stability in conventional tubes carry over to NEA tubes (Sommer, 1973c,d). In general, a III–V NEA surface is not stable in ultrahigh vacuum (e.g., Yee and Jackson, 1972). At 300°K, Cs will slowly desorb from the surface in UHV at a rate dependent, in part, on the partial pressure of Cs within the vacuum system. Additional Cs will restore a fraction (70–90%, perhaps) of the original surface escape probability. However, the optimal Cs + O stoichiometry evidently is destroyed by this procedure. In a sealed tube, however, there is a finite partial pressure of Cs (10^{-12}–10^{-13} Torr) which is sufficient to stabilize the Cs + O activation layer. Present NEA photocathodes are not as "rugged" or "forgiving" as conventional photocathodes; however, the technology will likely improve with further production processing experience.

XII. Bias-Assisted Photoemitters

From Section VII.10.d, it is apparent that the NEA cathode is limited to a useful long wavelength threshold of about 1.1 μm. In an effort to extend photoemission thresholds beyond this limit a number of externally biased (field-assisted) photocathode designs have been proposed and studied over the years (e.g., Schagen and Turnbull, 1969; Milnes and Feucht, 1971; Dalal, 1972; Bell et al., 1974; Sahai et al., 1975). The early work of Burton (1957) and Simon and Spier (1960a,b) on reverse-biased silicon and germanium p–n junctions led to a number of edge- and plane-parallel junction field-assisted photoemission studies (Bartelink et al., 1963; Thornton and Northrup, 1965; Davies and Thornton, 1967; Ward et al., 1969; Escher, 1973; D'Haenens et al., 1974; Shahriary et al., 1979). MOS (Foss, 1971; Miller and Jones, 1980) and thin Schottky barrier devices, mostly on Si (Itch et al., 1967, 1970; Musatov and Shulepov, 1971), have been investigated. The field-emission photocathode based on arrays of Si-based point emitters has also been developed (Thornton, 1973; Schroder et al., 1974).

18. TE Photoemission

A more recent approach to longer wavelength photoemission is the transferred-electron (TE) cathode first proposed and demonstrated (photo- and cold-cathode emission) by Bell et al. (1974). TE photoemission is based on the fact that for certain III–V semiconductors, such as InP, InGaAsP alloys, and GaAs, electrons can be promoted to upper conduction-band valleys for electric fields on the order of 10^4 V/cm (e.g., Sze, 1969, p. 743). Figure 51 is a schematic energy-band diagram for a simple p-InP TE cathode. Photoexcited electrons are generated in the bulk of the InP by >1.35 eV photons incident

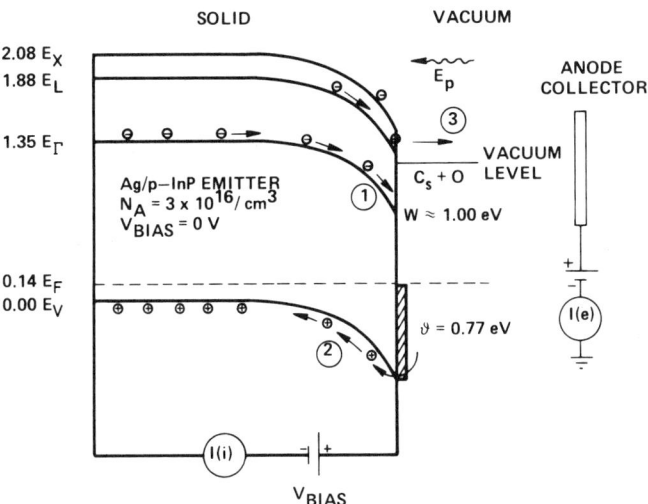

FIG. 51. Schematic energy-band diagram for a Ag/p-InP transferred-electron (TE) cathode.

upon a thin semitransparent Schottky barrier. Some of these minority carriers will diffuse to the InP/Ag interface. Most of these will be collected in the surface Schottky contact; process "1" in Fig. 51. If, however, a reverse bias is applied to the Schottky barrier, a small fraction of the photogenerated electrons will escape into vacuum; process "3" in Fig. 51.

The TE photoemission process consists of six steps. They are (1) the optical transmission of incident photons through the thin Ag film, (2) photogeneration of electron–hole pairs within the bulk of the p-InP, (3) diffusion of thermalized electrons in the conduction band to the surface depletion region, (4) acceleration of these electrons within the reverse-biased depletion region to higher energies in the Γ, L, or X valleys, (5) hot-electron transport through the Ag film, and (6) electron emission into vacuum over the Cs + O-activated metal surface work function. The term transferred-electron photocathode comes from step (4) whereby a significant fraction of the photoexcited carriers will transfer into a higher mass valley (or heat up within the Γ conduction band). It is convenient to define steps (1)–(4) as the internal collection efficiency or yield, Y(Int.). Experimental reflection-mode Y(Int.) and photoemission quantum yield from a Ag/p-InP cathode are shown in Figs. 52 and 53.

It is known from the work of Uebbing and James (1970) that the work function of Ag can be lowered to ~1.0 eV with Cs + O. (Recall Fig. 10.) For efficient photoemission, therefore, electrons must reach the Ag/InP interface with more than 1 eV energy relative to the pinned Fermi level at the surface. The Schottky barrier height for Ag/InP can be deduced from internal photoemission measurements (process "2" in Fig. 51) and is approximately 0.75

eV (Escher et al., 1976a). (See Fig. 54.) The direct bandgap of InP is 1.35 eV (300°K). Therefore the L and X conduction valleys at the surface will be slightly above the vacuum level, while the Γ conduction-band edge will be 0.42 eV below (James et al., 1970). (Recall Fig. 51.) It follows then that electrons which successfully transfer to the upper mass valleys or ~ 0.5 eV hot Γ electrons will have a chance of escaping into vacuum.

Monte Carlo computer simulation techniques have been used successfully to model the transferred-electron effect in III–V semiconductors (e.g., Maloney and Frey, 1977; Ridley, 1977; Littlejohn et al., 1977). Similar calculations have been performed for the Ag/p-InP TE cathode (Maloney et al., 1980). Calculated electron energy distributions of electrons incident at the Ag/InP interface within the Γ and L valleys for a 6-V bias and $2 \times 10^{15}/\text{cm}^3$ doping are shown in Fig. 55. Note that the hot Γ electrons tend to bunch up just below the transfer energy. The effect of applied bias on the percentage of electrons in each valley and the mean energy in each valley is shown in Figs. 56 and 57. Experimental EDC measurements would likely be helpful in testing these calculations but have not been performed to date.

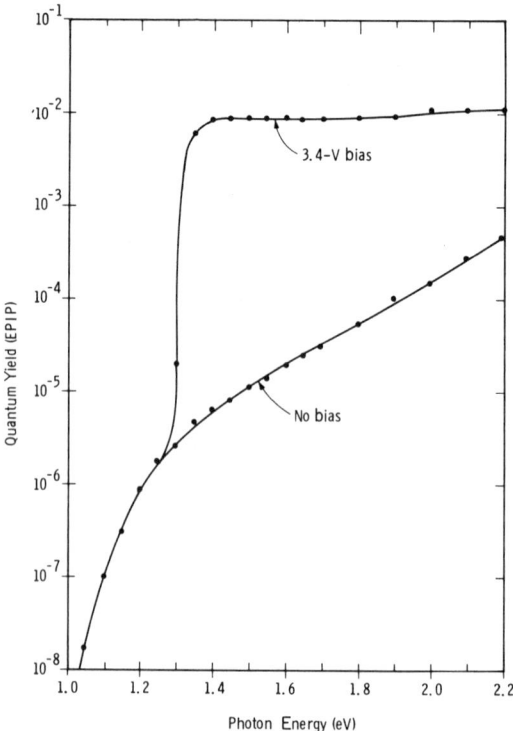

FIG. 52. Experimental reflection-mode quantum yield from a Ag/p-InP transferred-electron photocathode.

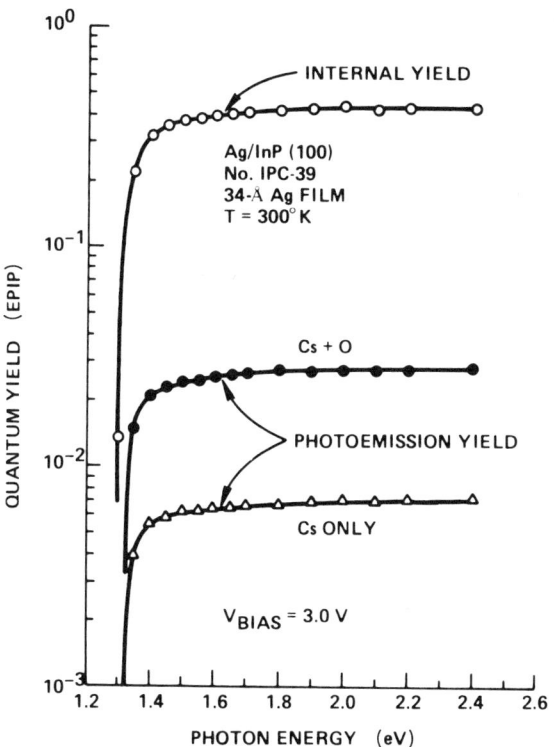

FIG. 53. Internal quantum yield $Y(\text{int.})$ and photoemission quantum yield $Y(\text{PE})$ for Cs-only and Cs + O activations. (From Escher et al., 1979.)

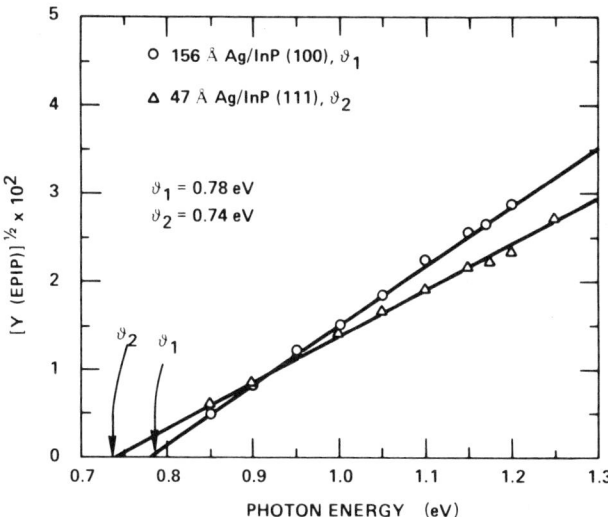

FIG. 54. Fowler plot of the internal Schottky barrier yield (process "2" in Fig. 51) from Ag/p-InP (100) and Ag/p-InP (111) **B** surfaces.

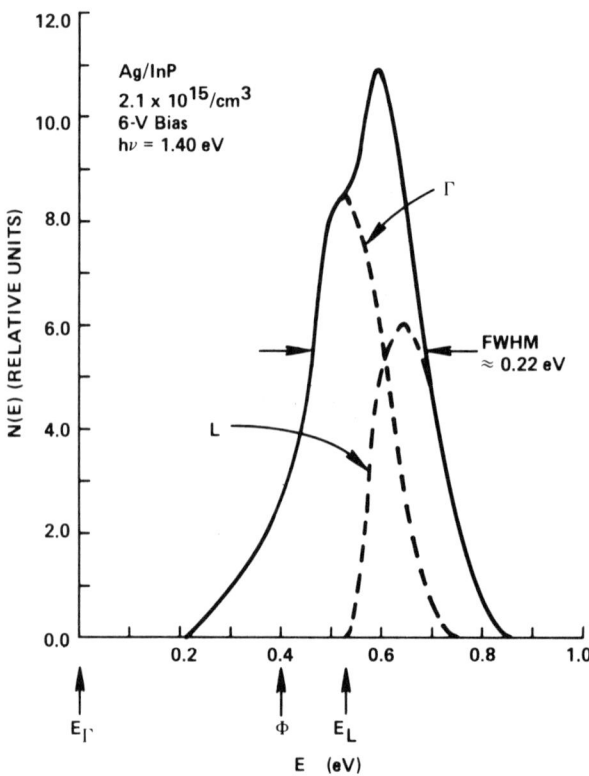

FIG. 55. Calculated electron energy distribution in the Γ and L valleys of a Ag/p-InP TE cathode for $h\nu = 1.40$ eV, $N_A = 2 \times 10^{15}/\text{cm}^3$, and 6.0-V bias. The full width at half maximum FWHM is about 220 meV. (From Maloney et al., 1980.)

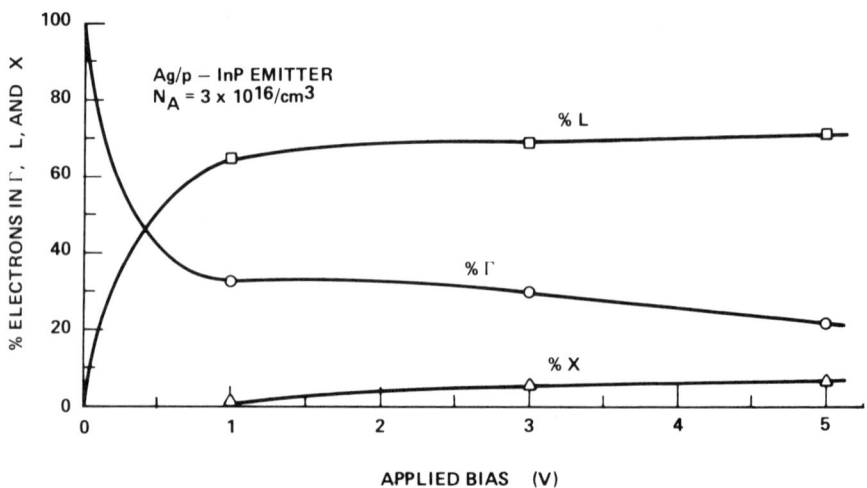

FIG. 56. Monte Carlo calculation of the percentage of electrons in the upper mass conduction-band valleys at the surface of InP versus applied bias for a fixed $3 \times 10^{16}/\text{cm}^3$ acceptor doping concentration. (From Escher et al., 1979.)

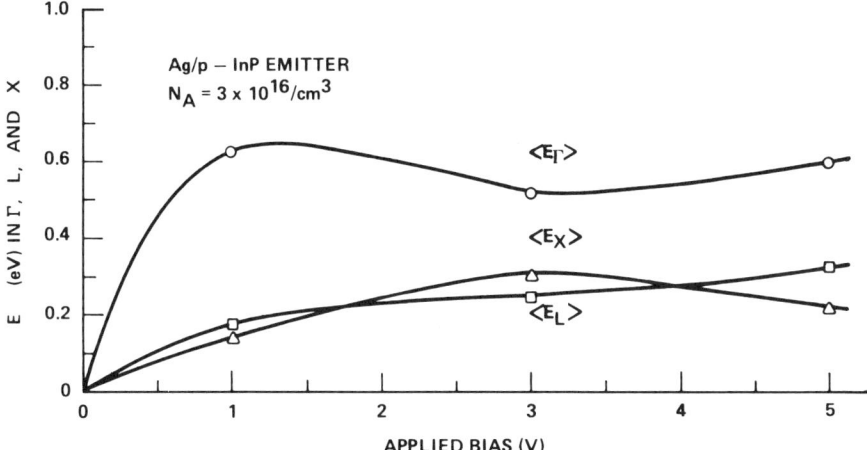

FIG. 57. Monte Carlo calculation of the mean electron energy in the upper mass conduction-band valleys as measured from the respective band minima at the surface of InP versus applied bias for a fixed acceptor doping concentration of $3 \times 10^{16}/\text{cm}^3$. (From Escher et al., 1979.)

A simple analysis of data similar to that in Fig. 53 can be made to estimate the transmission of electrons through the Ag film and the escape probability for electrons from the activated Ag surface (Escher et al., 1979). The transmission of hot electrons through the film can be approximated as being proportional to $\exp[-T(\text{Ag})/L]$, where $T(\text{Ag})$ is the Ag film thickness and L is the mean electron attenuation length within the Ag film (e.g., Crowell and Sze, 1967; Mead, 1962). The vacuum photoemission quantum yield, $Y(\text{PE})$, can then be written as

$$Y(\text{PE}) = PY(\text{Int.})\exp[-T(\text{Ag})/L], \qquad (23)$$

where P is defined as the surface escape probability from the activated Ag film surface into vacuum, step (6). Note that both the surface reflectance and optical transmission losses in the Ag film are incorporated into $Y(\text{Int.})$. Experimental data for the ratio of $Y(\text{PE})/Y(\text{Int.}) = I(e)/I(i)$ versus $T(\text{Ag})$ is shown in Fig. 58. The attenuation length L from the slope of these data is ~ 50 Å which is quite low for ~ 1.9-eV hot electrons in Ag. (Recall Fig. 4). The reason for this is not known but is thought to be due to enhanced electron scattering from Ag–InP intermixing or alloying (e.g., Williams et al., 1977; Williams, 1979; Escher et al., 1979). The escape probability P is a strong function of applied bias increasing from $<0.1\%$ at zero bias to a maximum at ~ 5.0-V bias. Experimentally the peak escape probability from Ag/p-InP (100) and Ag/p-InP (111) B cathodes is similar even though theoretically one might expect differences in emission probability (Burt, 1979).

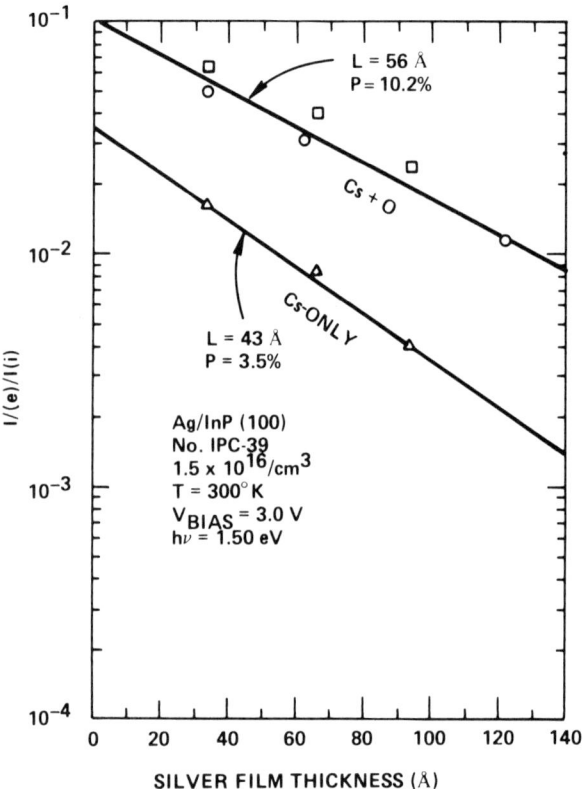

FIG. 58. Experimental ratio of $I(e)/I(i) = Y(PE)/Y(Int.)$ versus Ag film thickness for a constant 3.0-V applied bias and fixed photon energy of 1.50 eV. L is the attenuation length of ~ 1.9-eV hot electrons in Ag and P is the surface escape probability from the activated Ag surface. (From Escher et al., 1979.)

19. DIRECT EMITTER CATHODES

Although the TE photoemission threshold from InP extends essentially to the bandgap, 1.35 eV (~ 0.92 μm), the measured yields are generally no better than from a NEA p-InP cathode (Bell and Uebbing, 1968; James et al., 1971b). Therefore TE photoemission from InP itself is only of interest as an efficient emitter of "transferred" electrons. However, it was soon discovered that TE photoemission could be achieved directly from Ag/InGaAsP cathodes over the entire bandgap range of quaternary alloys (1.35–0.75 eV) which are lattice matched to InP substrates (Escher and Sankaran, 1976; Sankaran et al., 1976a; Escher et al., 1978b). (See Figs. 59–61.) The lower bandgap quaternary TE cathodes ($E_g < 1.0$ eV) generally improve in quantum efficiency upon cooling. The reason for the higher yield is not understood at this time but may be associated with reduced thermionic emission

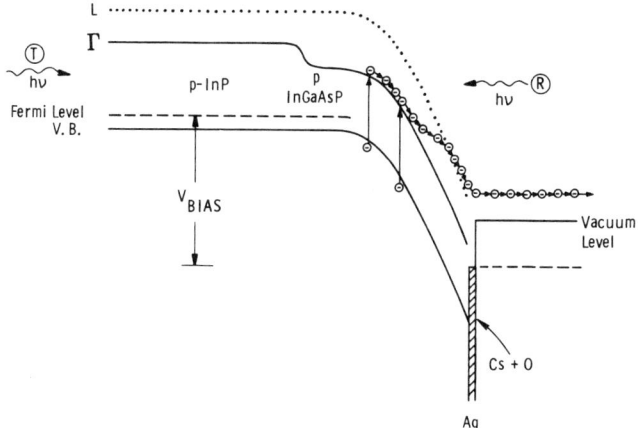

FIG. 59. Schematic energy-band diagram for a Ag/p-InGaAsP direct emitter TE cathode under bias conditions.

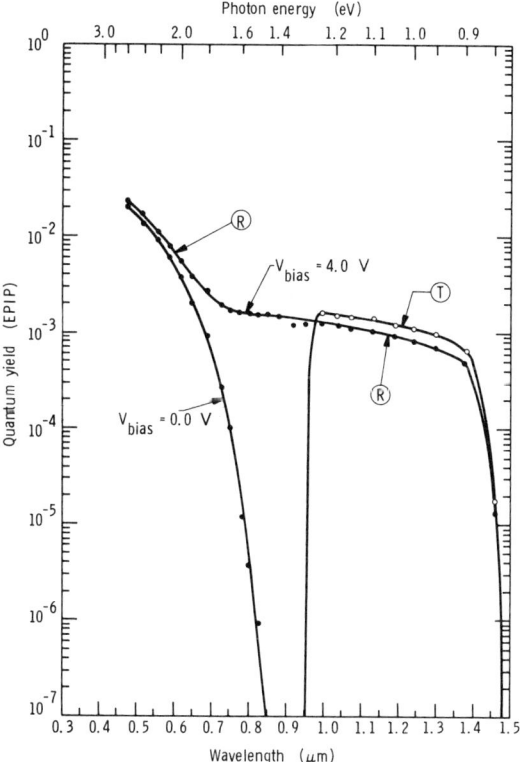

FIG. 60. Experimental RM and TM quantum yield from a Ag/p-InGaAsP ($E_g = 0.85$ eV, 300°K) direct emitter TE cathode. (From Escher and Sankaran, 1976.)

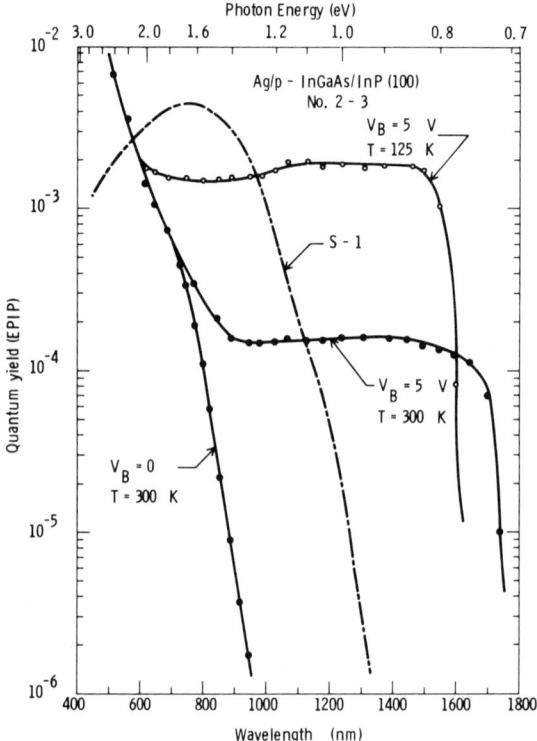

FIG. 61. Experimental reflection-mode quantum yield curves for bias and no-bias conditions. A typical S-1 cathode yield is included for reference. (From Escher et al., 1978b.)

of holes over the Schottky barrier, and a decrease in ionized acceptor density within the bulk cathode associated with filling of acceptorlike traps (Escher et al., 1978b). TE photoemission from the higher bandgap quaternaries, including InP, does not change significantly upon cooling, except for a small threshold change due to the slight increase in bandgap. The long wavelength limit to efficient TE photoemission from a low bandgap direct emitter cathode will likely be set by the onset of electron impact ionization for electrons with energy $\sim 3E_g$ and the work function at the emitting surface ~ 1.0 eV. TE photoemission to 2.1-μm threshold and 0.2% RM yield at 125°K from a Ag/p-In$_{0.77}$Ga$_{0.23}$As cathode has been reported by Gregory et al. (1980).

20. HETEROJUNCTION TE CATHODES

The Ag/p-InGaAsP alloy cathodes of bandgap less than about 1.0 eV suffer from the fact that the functions of photogeneration and electron emission are preformed within the same active cathode layer. A higher efficiency TE cathode should be possible by separating these two functions into distinct photon absorbing and electron emitting layers—a heterojunction TE cath-

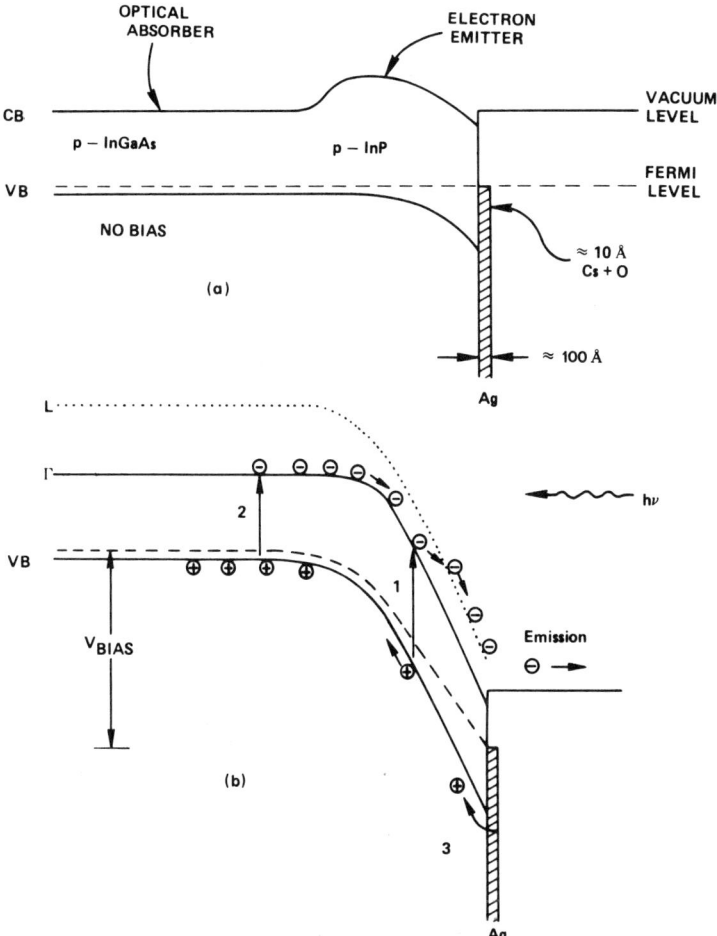

FIG. 62. Schematic energy-band diagram of a heterojunction TE cathode. The applied bias permits efficient minority-carrier transport of photogenerated electrons from the absorber into the emitter and also provides the internal field for TE photoemission from the p-InP emitter. For incident photon energies, $h\nu$, greater than E_g (InP), 1.35 eV, photoelectrons are generated within the InP emitter, process "1". For E_g (InP) > $h\nu$ > E_g (InGaAs), photoelectrons are generated within the InGaAs absorber, process "2". The reverse-bias Schottky leakage current consists of holes, process "3".

ode. Figure 62 is an energy-band diagram for such a p–p heterojunction cathode utilizing the lattice-match pair InP (emitter)/In$_{0.53}$Ga$_{0.47}$As (absorber). Under zero-bias conditions there is a ~0.6-eV conduction-band barrier to thermalized electron diffusion between the absorber and emitter. Under reverse bias of the Ag/p-InP Schottky barrier, however, the depletion field essentially eliminates the conduction-band barrier (Gregory et al., 1978).

For photon energies below the InP bandgap, electron–hole pairs are created in the absorber layer. The electrons then diffuse/drift across the heterojunction into the InP where they can be emitted into vacuum via TE photoemission. Similar heterojunction absorber-emitter concepts have been proposed by Milnes and Feucht (1971) and Sahai et al. (1975) with a NEA surface emitter. The first successful experimental demonstration of the heterojunction TE cathode was by Escher et al. (1975) with a Ag/p-InGaAsP (1.2-eV) cathode and Escher et al. (1978a) with a Ag/p-InGaAsP (1.28-eV)/p-InGaAs(0.75-eV) cathode.

Experimental reflection-mode yield from a Ag/p-InP/p-InGaAs (0.75-eV) cathode is shown in Fig. 63. The zero-bias yield is primarily from photoexcited electrons generated from the Cs + O-activated Ag film. For applied biases of about 1.0 V or more, the InP emitter becomes a TE photocathode. Bandgap-limited photoemission from the InP is clearly seen in Fig. 63 for the 2.0-V bias yield. For photon energies less than 1.35 eV and greater than 0.75 eV, the InP is optically transparent and photoelectrons are generated in the

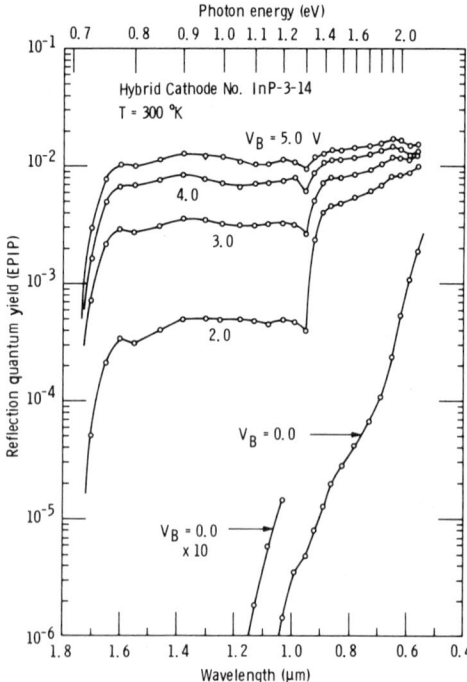

FIG. 63. Experimental RM quantum yield from a InP/In$_{0.53}$Ga$_{0.43}$As heterojunction TE cathode. The yield from 1.35 to 2.0 eV is from electrons photogenerated within the p-InP emitter. The yield from 0.75 to 1.35 eV is from electrons generated within the In$_{0.53}$Ga$_{0.47}$As absorber, transferred into the emitter, and then into vacuum. (From Escher et al., 1977.)

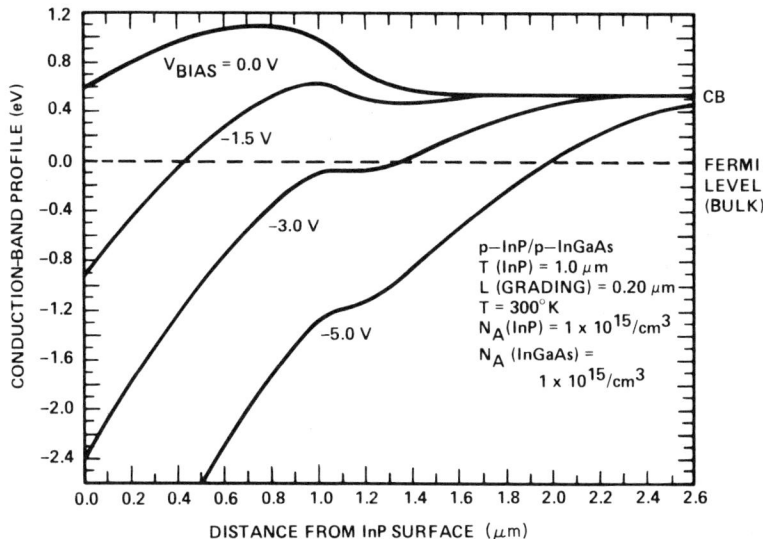

FIG. 64. Model calculation of the Γ conduction-band profile across a Ag/p-InP/p-In$_{0.53}$Ga$_{0.47}$As heterojunction under different applied reverse biases to the Schottky contact. Thermalized electron transport from the In$_{0.53}$Ga$_{0.47}$As (on the right) into the InP can occur, in this case, for biases of 3.0 V and higher. The grading distance over which the bands are assumed to change smoothly from In$_{0.53}$Ga$_{0.47}$As to InP is $L = 2000$ Å. (From Gregory et al., 1978.)

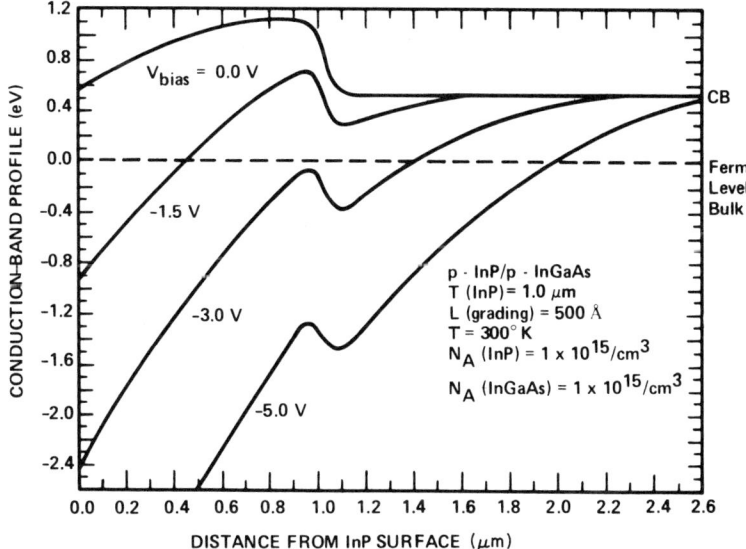

FIG. 65. Model calculation of the Γ conduction-band profile across a Ag/p-InP/p-In$_{0.53}$Ga$_{0.47}$As heterojunction under different applied reverse biases. Thermalized electron transport from the In$_{0.53}$Ga$_{0.47}$As into the InP is impossible in this case for any bias voltage shown due to the short bandgap grading distance, $L = 500$ Å, assumed in this calculation. (From Gregory et al., 1978.)

absorber (process "2" in Fig. 62). At 2.0-V bias approximately 10% of the electrons in the InGaAs are able to reach the emitter. With higher biases the heterojunction transfer efficiency for electrons to cross the absorber–emitter interface increases to nearly 100%. 5.0% RM quantum yield has been achieved from an experimental heterojunction cathode (Escher et al., 1980).

Model calculations of the conduction-band profile across the InP–InGaAs interface under bias conditions are shown in Figs. 64 and 65 (Gregory et al., 1978).

21. DARK CURRENT EMISSION

Due to the complex nature of the TE cathode, there are a number of potential sources of dark current emission. Of the potential sources of dark current

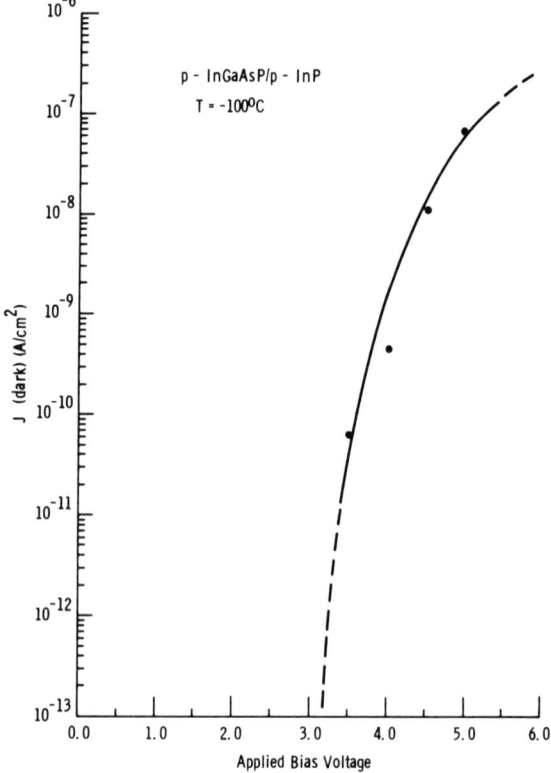

FIG. 66. Experimental dark current emission versus applied bias from a Ag/p-InGaAsP direct emitter cathode. The rapid increase in dark current emission with bias is thought to derive from impact ionization. (From Escher and Sankaran, 1976.)

from NEA cathodes discussed by Bell (1969, 1970), thermal generation of electrons ("diffusion" component) from the relatively low bandgap absorber will certainly be important. Another source of dark current emission considered by Bell is the "radiative" component from the 300°K background. For NEA cathodes this is not important; however, for high sensitivity, TE cathodes whose threshold reaches into the 1.8–2.0 μm range, this could become important. Dark current emission from electron generation via traps can only be estimated but could be significant. The earliest dark current measurements from Ag/p-InP and Ag/p-InGaAsP cathodes showed dark current emission that was a very strong function of applied bias. (See Fig. 66.) This observation suggested yet another source of dark current emission from TE cathodes: impact ionization of hot holes (Escher and Sankaran, 1976). The applied field which accelerates electrons to higher energy also accelerates holes from the Schottky barrier film. If these holes attain sufficient energy they can lead to a rapid generation of new electron–hole pairs and hence dark current. A simple calculation of the generation of electrons from the ionization rate data of Kao and Crowell (1978) gives good agreement with the higher bias dark current data from Ag/p-InP cathodes.

XIII. Summary

NEA photocathodes represent a significant new class of semiconductor electron emission devices. Their original conception and subsequent development has derived from a rather detailed knowledge of the surface physics of electron emission, widespread availability of ultrahigh vacuum equipment, and parallel development of semiconductor materials technology. A great deal of information on the potential and limitations of these devices has been gained. Although a number of fundamental questions still remain in regard to the atomic and electronic structure of the NEA surface, the newer surface physics tools give considerable promise of resolving many of these in the near future.

The engineering difficulties of developing viable tube fabrication techniques, adequate shelf and operating life stability, etc., are being met. GaP, GaAsP, GaAs, InGaAs, InGaAsP, and Si devices are commercially available at the present time. Their superior performance often comes, however, at a higher cost and more restricted operating range when compared with conventional cathodes. For these reasons, the NEA cathode will likely remain a "premium" specialized device for some years to come. Nevertheless, the proven potential coupled with a diversity of applications assures a firm future for the NEA semiconductor photocathode.

Acknowledgments

The author wishes to acknowledge his many colleagues at RCA and Varian who have contributed significantly to the work discussed here. He also wishes to give special thanks to Drs. R. L. Bell and A. H. Sommer, and Prof. W. E. Spicer for many helpful discussions, Mrs. N. Anderson for typing of the draft, and his family for their patience and understanding during the preparation of this chapter.

References

Afonina, L. F., and Stuchinski, G. B. (1973). *Fiz. Tverd. Tela (Leningrad)* **15**, 2179[Engl. transl., *Sov. Phys.—Solid State* **15**, 1448 (1974)].
Afonina, L. F., Lagodinskii, V. M., and Stuchinskii, G. V. (1971). *Izv. Akad. Nauk SSSR, Ser. Fiz.* **35**, 1046 [Engl. transl., *Bull. Acad. Sci. USSR, Phys. Ser.* **35**, 964].
Allen, F. G., and Gobeli, G. W. (1964). *J. Appl. Phys.* **35**, 597.
Allen, F. G., and Gobeli, G. W. (1966). *Phys. Rev.* **144**, 558.
Allen, G. A. (1971). *J. Phys. D* **4**, 308.
Allen, G. A. (1973). *Acta Electron.* **16**, 229.
Allenson, M., and Bass, S. J. (1976). *Appl. Phys. Lett.* **28**, 113.
Allenson, M. B., King, P. G., Rowland, M. C., Steward, G. J., and Syms, C. H. A. (1972). *J. Phys. D* **5**, 189.
Andrew, D., Gowers, J. P., Henderson, J. A., Plummer, M. J., Stocker, B. J., and Turnbull, A. A. (1970). *J. Phys. D* **3**, 320.
Andronov, A. N., Maslevtsov, A. V., Gin, B. V., and Lepeshinskaya, V. N. (1976). *Izv. Akad. Nauk SSSR, Ser. Fiz.* **40**, 1627 [Engl. transl., *Bull. Acad. Sci. USSR, Phys. Ser.* **40**, No. 8, 64].
Antonova, L. I., Klimin, A. I., Nemchenok, R. L., Pal'ts, T. N., and Strakovskaya, S. E. (1976). *Izv. Akad. Nauk SSSR, Ser. Fiz.* **40**, 1633 [Engl. transl., *Bull. Acad. Sci. USSR, Ser. 40*, No. 8, 68].
Antypas, G. A., and Edgecumbe, J. (1975). *Appl. Phys. Lett.* **26**, 371.
Antypas, G. A., and James, L. W. (1970). *J. Appl. Phys.* **41**, 2165.
Antypas, G. A., James, L. W., and Uebbing, J. J. (1970). *J. Appl. Phys.* **41**, 2888.
Antypas, G. A., Moon, R. L., James, L. W., Edgecumbe, J., and Bell, R. L. (1972). *Gallium Arsenide Relat. Comp., Inst. Phys. Conf. Ser.* No. 17, p. 48.
Antypas, G. A., Escher, J. S., Edgecumbe, J., and Enck, R. S. (1978). *J. Appl. Phys.* **49**, 4301.
Apker, L., Taft, E., and Dickey, J. (1948). *Phys. Rev.* **74**, 1462.
Appelbaum, J. A., and Hamann, D. R. (1978). *Surf. Sci.* **74**, 21.
Arthur, J. R. (1974). *Surf. Sci.* **43**, 449.
Aspnes, D. E., and Studna, A. A. (1973). *Phys. Rev. B* **7**, 4605.
Aspnes, D. E., Olson, C. G., and Lynch, D. W. (1976). *Phys. Rev. Lett.* **37**, 766.
Bardas, D., Kellogg, E., Murray, S., and Enck, R. (1978). *Rev. Sci. Instrum.* **49**, 1273.
Bartelink, D. J., Moll, J. L., and Mayer, N. J. (1963). *Phys. Rev.* **130**, 972.
Bayliss, C. R., and Kirk, D. L. (1975). *Thin Solid Films* **29**, 97.
Bayliss, C. R., and Kirk, D. L. (1976). *J. Phys. D* **9**, 233.
Bell, R. L. (1969). *Solid-State Electron.* **12**, 475.
Bell, R. L. (1970). *Solid-State Electron.* **13**, 397.
Bell, R. L. (1973). "Negative Electron Affinity Devices." Oxford Univ. Press (Clarendon), London and New York.
Bell, R. L. (1975). *J. Phys. D* **8**, L118.
Bell, R. L., and Spicer, W. E. (1970). *Proc. IEEE* **58**, 1788.

Bell, R. L., and Uebbing, J. J. (1968). *Appl. Phys. Lett.* **12**, 76.
Bell, R. L., James, L. W., Antypas, G. A., Edgecumbe, J., and Moon, R. L. (1971). *Appl. Phys. Lett.* **19**, 513.
Bell, R. L., James, L. W., and Moon, R. L. (1974). *Appl. Phys. Lett.* **25**, 645.
Berglund, C. N., and Spicer, W. E. (1964). *Phys. Rev.* **136**, A1030.
Bertoni, C. M., Bisi, O., Manghi, F., and Calandra, C. (1978). *J. Vac. Sci. Technol.* **15**, 1256.
Biberman, L. M. (1967). *Appl. Opt.* **6**, 1127.
Biberman, L. M. (1971). In "Photoelectric Imaging Devices" (L. M. Bibermand and S. Nudelman, eds.), Vol. 1, p. 9. Plenum, New York.
Bradley, D. J., Allenson, M. B., and Holeman, B. R. (1977). *J. Phys. D* **10**, 111.
Burstein, E. (1954). *Phys. Rev.* **93**, 632.
Burt, M. G. (1977). *J. Phys. D* **10**, 1161.
Burt, M. G. (1978). *J. Phys. D* **11**, 1189.
Burt, M. G. (1979). Personal communication.
Burt, M. G., and Heine, V. (1978). *J. Phys. C* **11**, 961.
Burt, M. G., and Inkson, J. C. (1975). *J. Phys. D* **8**, L3.
Burt, M. G., and Inkson, J. C. (1976a). *Appl. Phys. Lett.* **28**, 5.
Burt, M. G., and Inkson, J. C. (1976b). *J. Phys. D* **9**, 43.
Burt, M. G., and Inkson, J. C. (1976c). *J. Phys. D* **9**, L5.
Burt, M. G., and Inkson, J. C. (1977). *J. Phys. D* **10**, 721.
Burton, J. A. (1957). *Phys. Rev.* **108**, 1342.
Callcott, T. A. (1967). *Phys. Rev.* **161**, 746.
Campagna, M., Pierce, D. T., Meier, F., Sattler, K., and Siegmann, H. C. (1976). *Adv. Electron. Electron Phys.* **41**, 113.
Cardillo, M. J., and Becker, G. E. (1978). *Phys. Rev. Lett.* **40**, 1148.
Cardona, M., and Ley, L., eds. (1978). "Photoemission in Solids I." Springer-Verlag, Berlin and New York.
Carriere, B., and Deville, J. P. (1979). *Surf. Sci.* **80**, 278.
Celotta, R. J., Pierce, D. T., and Wang, G. C. (1979). *Phys. Rev. Lett.* **43**, 728.
Chadi, D. J. (1978). *J. Vac. Sci. Technol.* **15**, 1244.
Chang, C. C. (1974). In "Characterization of Solid Surfaces" (P. F. Kane and G. B. Larabee, eds.), p. 509. Plenum, New York.
Chang, K. H., and Meijer, P. H. E. (1977). *J. Vac. Sci. Technol.* **14**, 789.
Chelikowsky, J. R. (1977). *Phys. Rev. B* **16**, 3618.
Chelikowsky, J. R., and Cohen, M. L. (1979). *Phys. Rev. B* **20**, 4150.
Chen, J. M. (1971). *Surf. Sci.* **25**, 305.
Cho, A. Y. (1971). *J. Appl. Phys.* **42**, 2074.
Cho, A. Y., and Arthur, J. R. (1975). *Prog. Solid State Chem.* **10**, 157.
Cho, A. Y., and Dernier, P. D. (1978). *J. Appl. Phys.* **49**, 3328.
Chye, P. W., Babalola, I. A., Sukegawa, T., and Spicer, W. E. (1975). *Phys. Rev. Lett.* **23**, 1602.
Chye, P. W., Babalola, I. A., Sukegawa, T., and Spicer, W. E. (1976). *Phys. Rev. B* **13**, 4439.
Chye, P. W., Su, C. Y., Lindau, I., Skeath, P., and Spicer, W. E. (1979). *J. Vac. Sci. Technol.* **16**, 1191.
Clark, J. (1975). *J. Phys. D* **8**, 535.
Clark, M. G. (1976). *J. Phys. D* **9**, 2139.
Clark, M. G., Howorth, J. R., and Holtom, R. (1976). *J. Phys. D* **9**, 2155.
Clemens, H., and Mönch, W. (1975). *Crit. Rev. Solid State Sci.* **5**, 273.
Clemens, H. J., Von Wienskowski, J., and Mönch, W. (1978). *Surf. Sci.* **78**, 648.
Coates, P. B. (1970). *J. Phys. D* **3**, L27.
Cole, M., and Ryer, D. (1972). *Electro-Opti. Syst. Des.* June, 16.
Cope, A. D., Luedicke, E., and Carroll, J. P. (1973). *RCA Rev.* **34**, 408.

Crowell, C. R., and Sze, S. M. (1967). *Phys. Thin Films* **4**, p. 325.
Dalal, V. (1972). *J. Appl. Phys.* **43**, 1160.
Dash, W. C., and Newman, R. (1955). *Phys. Rev.* **99**, 1151.
Davies, I. G., and Thornton, P. R. (1967). *Appl. Phys. Lett.* **10**, 249.
Deasley, P. J., and Faulkner, K. R. (1972). *Adv. Electron. Electron Phys.* **33A**, 459.
Derbenwick, G. F., Pierce, D. T., and Spicer, W. E. (1974). *Methods Exp. Phys.* **11**, 67.
Derrien, J., and Arnaud d'Avitaya, F. (1976). *Rev. Phys. Appl.* **11**, 377.
Derrien, J., and Arnaud d'Avitaya, F. (1977). *Surf. Sci.* **65**, 668.
Derrien, J., Arnaud d'Avitaya, F., and Bienfait, M. (1976). *Solid State Commun.* **20**, 557.
D'Haenens, I. J., Roth, J. A., and Anderson, C. L. (1974). Final Tech. Rep., Contract DAKO2-72-C-0442, Hughes Res. Lab., Malibu, California.
Drathen, P., Ranke, W., and Jacobi, K. (1978). *Surf. Sci.* **77**, L162.
Eastman, D. E. (1970). *Phys. Rev. B* **2**, 1.
Eastman, D. E. (1975). In "Vacuum Ultraviolet Radiation Physics" (E. Koch, R. Haensel, and C. Kunz, eds.) Vieweg, Braunschweig.
Eastman, D. E., and Grobman, W. D. (1972). *Phys. Rev. Lett.* **28**, 1378.
Eastman, D. E., Himpsel, F. J., Knapp, J. A., and Van Vechten, J. A. (1979). *Bull. Am. Phys. Soc.* **24**, 403.
Eden, R. C., Moll, J. L., and Spicer, W. E. (1967). *Phys. Rev. Lett.* **18**, 597.
Edwards, D., and Peria, W. T. (1978). *Appl. Surf. Sci.* **1**, 419.
Ehrenreich, H., and Phillip, H. R. (1962). *Phys. Rev.* **128**, 1622.
Einstein, A. (1905). *Ann. Phys. (Leipzig)* **17**, 145.
Engstrom, R. W. (1955). *RCA Rev.* **16**, 116.
Engstrom, R. W., and Morehead, A. L. (1967). *RCA Rev.* **28**, 419.
Enstrom, R. E., and Fisher, D. G. (1975). *J. Appl. Phys.* **46**, 1976.
Enstrom, R. E., Richman, D., Abrahams, M. S., Appert, J. R., Fisher, D. G., Sommer, A. H., and Williams, B. F. (1971). *Gallium Arsenide Relat. Compd., Inst. Phys. Conf. Ser.* No. 9, p. 30.
Erbudak, M., and Reihl, B. (1978). *Appl. Phys. Lett.* **33**, 584.
Escher, J. S. (1973). Final Tech. Rep. Contract DAAKO2-72-C-0412, RCA Lab., Princeton, N.J.
Escher, J. S. (1979). Unpublished work.
Escher, J. S., and Antypas, G. A. (1977). *Appl. Phys. Lett.* **30**, 314.
Escher, J. S., and Sankaran, R. (1976). *Appl. Phys. Lett.* **29**, 87.
Escher, J. S., and Schade, H. (1973). *J. Appl. Phys.* **44**, 5309.
Escher, J. S., Fairman, R. D., Antypas, G. A., Sankaran, R., James, L. W., and Bell, R. L. (1975). *Crit. Rev. Solid State Sci.* **5**, 577.
Escher, J. S., James, L. W., Sankaran, R., Antypas, G. A., Moon, R. L., and Bell, R. L. (1976a). *J. Vac. Sci. Technol.* **13**, 874.
Escher, J. S., Antypas, G. A., and Edgecumbe, J. (1976b). *Appl. Phys. Lett.* **29**, 153.
Escher, J. S., Gregory, P. E., Hyder, S. B., Houng, Y. M., and Antypas, G. A. (1977). *Tech. Dig. 1977, Int. Electron Devices Meet., Inst. Electr. Electron. Eng., New York* p. 460.
Escher, J. S., Gregory, P. E., Antypas, G. A., Sankaran, R., and Houng, Y. M. (1978a). *J. Appl. Phys.* **49**, 447.
Escher, J. S., Gregory, P. E., Hyder, S. B., and Sankaran, R. (1978b). *J. Appl. Phys.* **49**, 2591.
Escher, J. S., Maloney, T. J., Gregory, P. E., Hyder, S. B., and Houng, Y. M. (1978c). *IEEE Trans. Electron Devices* **25**, 1347.
Escher, J. S., Gregory, P. E., and Maloney, T. J. (1979). *J. Vac. Sci. Technol.* **16**, 1394.
Escher, J. S., Bell, R. L., Gregory, P. E., Hyder, S. B., Maloney, T. J., and Antypas, G. A. (1980). *IEEE Trans. Electron Devices* **ED-27**, 1244.
Ettenberg, M., Olsen, G. H., and Nuese, C. J. (1976). *Appl. Phys. Lett.* **29**, 141.

Evans, B. E., Swanson, L. W., and Bell, A. E. (1968). *Surf. Sci.* **11**, 1.
Fan, H. Y. (1945). *Phys. Rev.* **68**, 43.
Fan, H. Y. (1967). *In* "Semiconductors and Semimetals" (R. K. Willardson and A. C. Beer, eds.), Vol. 3, p. 405. Academic Press, New York.
Farrow, R. F. C. (1974). *J. Phys. D* **7**, 2436.
Farrow, R. F. C. (1975). *J. Phys. D* **8**, L87.
Faulkner, K. R., Astridge, R. A., Howorth, J. R., and Surridge, R. K. (1973). *Appl. Phys. Lett.* **23**, 298.
Feldman, C. (1960). *Phys. Rev.* **117**, 455.
Fischer, T. E. (1966a). *Phys. Rev.* **142**, 519.
Fischer, T. E. (1966b). *Phys. Rev.* **147**, 603.
Fischer, T. E. (1968). *Surf. Sci.* **10**, 399.
Fischer, T. E. (1969). *Surf. Sci.* **13**, 30.
Fischer, T. E., Allen, F. G., and Gobeli, G. W. (1967). *Phys. Rev.* **163**, 703.
Fisher, D. G. (1974). *IEEE Trans. Electron Devices* **21**, 541.
Fisher, D. G., and Fowler, G. O. (1973). *Conf. Photoelectr. Secondary Electron Emiss.*, Univ. Minnesota, Minneapolis.
Fisher, D. G., and Martinelli, R. U. (1974). *Adv. Image Pickup Disp.* **1**, 71.
Fisher, D. G., and Olsen, G. H. (1979). *J. Appl. Phys.* **50**, 2930.
Fisher, D. G., Enstrom, R. E., and Williams, B. F. (1971). *Appl. Phys. Lett.* **18**, 371.
Fisher, D. G., Enstrom, R. E., Escher, J. S., and Williams, B. F. (1972). *J. Appl. Phys.* **43**, 3815.
Fisher, D. G., Enstrom, R. E., Escher, J. S., Gossenberger, H. F., and Appert, J. R. (1974). *IEEE Trans. Electron Devices* **21**, 641.
Foss, N. A. (1971). *J. Appl. Phys.* **42**, 3762.
Fowler, R. H. (1931). *Phys. Rev.* **38**, 45.
Foxon, C. T., Harvey, J. A., and Joyce, B. A. (1973). *J. Phys. Chem. Solids* **34**, 1693.
Frank, G., and Garbe, S. (1973). *Acta Electron.* **16**, 237.
Frank, G., and Garbe, S. (1974). *Phys. Status Solidi A* **26**, K91.
Garbe, S. (1969a). *Solid-State Electron.* **12**, 893.
Garbe, S. (1969b). *Phys. Status Solidi* **33**, K87.
Garbe, S. (1970). *Phys. Status Solidi A* **2**, 497.
Garbe, S., and Frank, G. (1969). *Solid State Commun.* **7**, 615.
Garbe, S., and Frank, G. (1970). *Gallium Arsenide Relat. Compd., Inst. Phys. Conf. Ser.* No. 9, p. 208.
Garfield, B. R. C., Wilson, R. J. F., Goodson, J. H., and Butler, D. J. (1976). *Adv. Electron. Electron Phys.* **40A**, 11.
Garner, C. M., Lindau, I., Su, D. Y., Pianetta, P., Miller, J. N., and Spicer, W. E. (1978). *Phys. Rev. Lett.* **40**, 403.
Garwin, E. I., Pierce, D. T., and Siegmann, H. C. (1974). *Helv. Phys. Acta* **47**, 393.
Gatos, H. C. (1961). *J. Appl. Phys.* **32**, 1232.
Geppert, D. V. (1966). *Proc. IEEE* **54**, 61.
Gobeli, G. W., and Allen, F. G. (1962). *Phys. Rev.* **127**, 141.
Gobeli, G. W., and Allen, F. G. (1965). *Phys. Rev. A* **137**, 245.
Goldstein, B. (1973). *Surf. Sci.* **35**, 227.
Goldstein, B. (1975). *Surf. Sci.* **47**, 143.
Goldstein, B., and Martinelli, R. U. (1973). *J. Appl. Phys.* **44**, 4244.
Goldstein, B., and Szostak, D. (1975a). *Appl. Phys. Lett.* **26**, 111.
Goldstein, B., and Szostak, D. (1975b). *Appl. Phys. Lett.* **26**, 685.
Goldstein, B., Szostak, D. J., and Ban, V. S. (1976). *Surf. Sci.* **57**, 733.
Gregory, P. E., and Spicer, W. E. (1975). *Phys. Rev.* **12**, 2370.
Gregory, P. E., and Spicer, W. E. (1976). *J. Appl. Phys.* **47**, 510.

Gregory, P. E., Spicer, W. E., Ciraci, S., and Harrison, W. A. (1974). *Appl. Phys. Lett.* **25**, 511.
Gregory, P. E., Chye, P., Sunami, H., and Spicer, W. E. (1975). *J. Appl. Phys.* **46**, 3525.
Gregory, P. E., Escher, J. S., Hyder, S. B., Houng, Y. M., and Antypas, G. A. (1978). *J. Vac. Sci. Technol.* **15**, 1483.
Gregory, P. E., Escher, J. S., Saxena, R. R., and Hyder, S. B. (1980). *Appl. Phys. Lett.* **36**, 639.
Guichar, G. M., Sebenne, C. A., and Thuault, C. D. (1979). *J. Vac. Sci. Technol.* **16**, 1212.
Gundry, P. M., Holtom, R., and Leverett, V. (1974). *Surf. Sci.* **43**, 647.
Gutierrez, W. A., and Pommerrenig, H. D. (1973). *Appl. Phys. Lett.* **22**, 292.
Gutierrez, W. A., Pommerrenig, H. D., and Holt, S. L. (1972). *Appl. Phys. Lett.* **21**, 249.
Gutierrez, W. A., Wlson, H. L., and Yee, E. M. (1974). *Appl. Phys. Lett.* **25**, 482.
Harrison, W. A. (1970). "Solid State Theory," p. 330. McGraw-Hill, New York.
Henzler, M. (1967). *Phys. Status Solidi* **19**, 833.
Holtom, R., and Gundry, P. M. (1977). *Surf. Sci.* **63**, 263.
Hopkins, G. P., Howorth, J. R., Palmer, I. C., and Pettas, H. T. (1979). *Int. Conf. Low Light Therm. Imag., 2nd, Inst. Electr. Eng., London* p. 22.
Howorth, J. R. (1979). *Int. Conf. Low Light Therm. Imag., 2nd, Inst. Electr. Eng., London* p. 61.
Howorth, J. R., and Pool, P. J. (1978). U.S. Patent 4,099,198.
Howorth, J. R., Harmer, A. L., Trawny, E. W. L., Holtom, R., and Sheppard, C. J. R. (1973a). *Appl. Phys. Lett.* **23**, 123.
Howorth, J. R., Surridge, R. K., and Palmer, I. (1973b). *1973 Int. Electron Devices Meet. Tech. Dig., Inst. Electr. Electron. Eng., New York* p. 304.
Howorth, J. R., Sheppard, J. R., Holtom, R., and Trawny, E. W. L. (1975a). *Int. Conf. Low Light Therm. Imag. Syst., Inst. Electr. Eng., London* p. 229.
Howorth, J. R., Sheppard, C. J. R., Holtom, R., and Harmer, A. L. (1975b). *J. Appl. Phys.* **46**, 151.
Howorth, J. R., Folkes, J. R., Palmer, I. C., Holtom, R., Sheppard, C. J. R., and Trawny, E. W. L. (1976a). *J. Phys. D* **9**, 785.
Howorth, J. R., Holtom, R., Sheppard, C. J. R., and Trawny, E. W. L. (1976b). *Adv. Electron. Electron Phys.* **40A**, 387.
Howorth, J. R., Surridge, R. K., and Palmer, I. C. (1976c). *Adv. Electron. Electron Phys.* **40A**, 463.
Hughes, A. L., and DuBridge, L. A. (1932). "Photoelectric Phenomena." McGraw-Hill, New York.
Hughes, F. R., Savoye, E. D., and Thoman, D. L. (1974). *J. Electron. Mater.* **3**, 9.
Huijser, A., and van Laar, J. (1975). *Surf. Sci.* **52**, 202.
Huijser, A., van Laar, J., and van Rooy, T. L. (1977). *Surf. Sci.* **62**, 472.
Hyder, S. B. (1971). *J. Vac. Sci. Technol.* **8**, 228.
Itch, T., Matsuda, I., Hasegawa, K., and Umeaka, K. (1967). *J. Appl. Phys.* **38**, 3395.
Itch, T., Matsuda, I., and Hasegawa, K. (1970). *J. Appl. Phys.* **41**, 1945.
Jackson, D., and Yee, E. M. (1971). *Proc. IEEE* **59**, 90.
Jacobi, K. (1975). *Surf. Sci.* **51**, 29.
Jacobi, K., Muschwitz, C. V., and Ranke, W. (1979). *Surf. Sci.* **82**, 270.
James, L. W. (1973). *Conf. Photoelectr. Secondary Electron Emission, Univ. Minnesota, Minneapolis*.
James, L. W. (1974). *J. Appl. Phys.* **45**, 1326.
James, L. W., and Moll, J. L. (1969). *Phys. Rev.* **183**, 740.
James, L. W., and Uebbing, J. J. (1970). *Appl. Phys. Lett.* **16**, 370.
James, L. W., Moll, J. L., and Spicer, W. E. (1968). *Gallium Arsenide Relat. Compd., Inst. Phys., London* p. 230.
James, L. W., Van Dyke, J. P., Herman, F., and Chang, D. M. (1970). *Phys. Rev. B* **1**, 3998.

James, L. W., Antypas, G. A., Edgecumbe, J., Moon, R. L., and Bell, R. L. (1971a). *J. Appl. Phys.* **42**, 4976.
James, L. W., Antypas, G. A., Uebbing, J. J., Yep, T. O., and Bell, R. L. (1971b). *J. Appl. Phys.* **42**, 580.
James, L. W., Antypas, G. A., Uebbing, J. J., Edgecumbe, J., and Bell, R. L. (1971c). *Gallium Arsenide Relat. Compd., Inst. Phys. Conf. Ser.* No. 9, p. 195.
James, L. W., Antypas, G. A., Moon, R. L., Edgecumbe, J. J., and Bell, R. L. (1973). *Appl. Phys. Lett.* **22**, 270.
Jona, F. (1965a). *Appl. Phys. Lett.* **6**, 205.
Jona, F. (1965b). *IBM J. Res. Dev.* **9**, 375.
Jona, F., and Marcus, P. (1977). *Comments Solid State Phys.* **8**, 1.
Jona, F., Shih, H. D., Ignatiev, A., Jepsen, D. W., and Marcus, P. M. (1977). *J. Phys. C* **10**, L67.
Joyce, B. D., and Williams, E. W. (1970). *Gallium Arsenide Relat. Compd., Inst. Phys. Conf. Ser.* No. 9., p. 57.
Kahn, A., Cisneros, G., Bonn, M., Mark, P., and Duke, C. B. (1978a). *Surf. Sci.* **71**, 387.
Kahn, A., So, E., Mark, P., Duke, C. B., and Meyer, R. J. (1978b). *J. Vac. Sci. Technol.* **15**, 1223.
Kan, H., Nakamura, T., Katsuno, H., Hagino, M., and Sukegawa, T. (1979). *Appl. Phys. Lett.* **34**, 545.
Kane, E. O. (1962). Phys. Rev. **127**, 131.
Kao, C. W., and Crowell, C. R. (1978). *Workshop Hot-Electron Phenom. Semicond.*, Cornell Univ., Ithaca, N.Y.
Kapitsa, M. L., Telyatnikova, T. A., Pustyl'nik, I. A., Shemelev, V. I., and Kruglov, M. V. (1976). *Izv. Akad. Nauk SSSR, Ser. Fiz.* **40**, 1638 [Engl. transl., *Bull. Acad. Sci. USSR, Phys. Ser.* **40**, No. 8, 72].
Kingdon, K. H. (1924). *Phys. Rev.* **24**, 510.
Kingdon, K. H., and Langmuir, I. (1923). *Phys. Rev.* **21**, 380.
Klein, W. (1969). *J. Appl. Phys.* **40**, 4384.
Klein, W. (1972). *J. Appl. Phys.* **43**, 4832.
Kobayashi, T. (1972). *Appl. Phys. Lett.* **72**, 150.
Kohn, E. S. (1971a). *Appl. Phys. Lett.* **18**, 272.
Kohn, E. S. (1971b). *J. Appl. Phys.* **42**, 2493.
Kohn, E. S. (1973). *IEEE Trans. Electron Devices* **20**, 321.
Korotkikh, V. L., Korinfskii, A. D., Matyash, A. A., Musatov, A. L., Strel'chenko, S. S., and Titov, V. A. (1977). *Fiz. Tverd. Tela (Leningrad)* **19**, 2869 [Engl. transl., *Sov. Phys.—Solid State* **19**, 1681].
Koval', I. F., Kryn'ko, Yu. N., Mel'nik, P. V., Nakhodkin, N. G., and Goîsa, S. N. (1977). *Fiz. Tverd. Tela (Leningrad)* **19**, 2681 [Engl. transl., *Sov. Phys.—Solid State* **19**, 1570].
Koval', I. F., Mel'nik, P. V., Nakhodkin, N. G., and Goîsa, S. N. (1978). *Fiz. Tverd. Tela (Leningrad)* **20**, 2070 [Engl. transl., *Sov. Phys.—Solid State* **20**, 1769].
Krall, H. R., and Persyk, D. E. (1972). *IEEE Trans. Nucl. Sci.* **19**, 45.
Krall, H. R., Helvy, F. A., and Persyk, D. E. (1970). *IEEE Trans. Nucl. Sci.* **17**, 71.
Kressel, H., Kohn, E. S., Nelson, H., Tietjen, J. J., and Weisberg, L. R. (1970). *Appl. Phys. Lett.* **16**, 359.
Kressel, H., Schade, H., and Nelson, H. (1973). *J. Lumin.* **7**, 146.
Krolikowski, W. F., and Spicer, W. E. (1969). *Phys. Rev.* **185**, 882.
Kudman, I., and Seidel, T. (1962). *J. Appl. Phys.* **33**, 771.
Lampel, G., and Weisbuch, C. (1975). *Solid State Commun.* **16**, 877.
Lamport, D. L. (1975). *Int. Conf. Low Light Therm. Imag. Syst., Inst. Electr. Eng., London* p. 133.
Lander, J. J., and Morrison, J. (1962). *J. Chem. Phys.* **37**, 729.
Lang, N. D. (1971). *Phys. Rev. B* **4**, 4234.
Langmuir, I. (1932). *J. Am. Chem. Soc.* **54**, 2798.
Langmuir, I., and Kingdon, K. H. (1924). *Phys. Rev.* **23**, 112.

Langmuir, I., and Taylor, J. (1944). *Phys. Rev.* **44**, 423.
Lapeyre, G. J., and Anderson, J. (1975). *Phys. Rev. Lett.* **35**, 117.
Levine, J. (1973). *Surf. Sci.* **34**, 90.
Ley, L., and Cardona, M., eds. (1979). "Photoemission in Solids II." Springer-Verlag, Berlin and New York.
Lindau, I., and Spicer, W. E. (1974). *J. Electron Spectrosc. Relat. Phenom.* **3**, 409.
Lindau, I., Pianetta, P., Garner, C. M., Chye, P. W., Gregory, P. E., and Spicer, W. E. (1977). *Surf. Sci.* **63**, 45.
Lindau, I., Chye, P. W., Garner, C. M., Pianetta, P., Su, C. Y., and Spicer, W. E. (1978a). *J. Vac. Sci. Technol.* **15**, 1332.
Lindau, I., Pianetta, P., Spicer, W. E., Gregory, P. E., Garner, C. M., and Chye, P. W. (1978b). *J. Electron Spectrosc. Relat. Phenom.* **13**, 155.
Littlejohn, M. A., Hauser, J. R., and Gilsson, T. H. (1977). *Appl. Phys. Lett.* **30**, 242.
Liu, Y. Z., Moll, J. L., and Spicer, W. E. (1969). *Appl. Phys. Lett.* **14**, 275.
Liu, Y. Z., Moll, J. L., and Spicer, W. E. (1970). *Appl. Phys. Lett.* **17**, 60.
Liu, Y. Z., Hollish, C. D., Stein, W. W., Bolger, D. E., and Greene, P. D. (1973). *J. Appl. Phys.* **44**, 5619.
Lubinsky, A. R., Duke, C. B., Lee, B. W., and Mark, P. (1976). *Phys. Rev. Lett.* **36**, 1058.
Ludeke, R., and Esaki, L. (1974). *Phys. Rev. Lett.* **33**, 653.
Luscher, P. E. (1977). *Solid State Technol.* **20** (12), 43.
McCaldin, J. O., McGill, T. C., and Mead, C. A. (1976). *J. Vac. Sci. Technol.* **13**, 802.
McKelvey, J. P. (1966). "Solid State and Semiconductor Physics." Harper, New York.
MacRae, A. U. (1966). *Surf. Sci.* **4**, 247.
Madey, T. E., and Yates, J. T. (1971). *J. Vac. Sci. Technol.* **8**, 39.
Maloney, T. J., and Frey, J. (1977). *J. Appl. Phys.* **48**, 781.
Maloney, T. J., Burt, M. G., Escher, J. S., Gregory, P. E., Hyder, S. B., and Antypas, G. A. (1980). *J. Appl. Phys.* **51**, 2879.
Margaritondo, G., Rowe, J. E., and Christman, S. B. (1976). *Phys. Rev. B* 14, 5396.
Mark, P., Pianetta, P., Lindau, I., and Spicer, W. E. (1977). *Surf. Sci.* **69**, 735.
Martin, H. (1976). *Electro-Opt. Syst. Des. Nov.*, p. 16.
Martinelli, R. U. (1970a). *Appl. Phys. Lett.* **16**, 261.
Martinelli, R. U. (1970b). *Appl. Phys. Lett.* **17**, 313.
Martinelli, R. U. (1973a). *J. Appl. Phys.* **44**, 2566.
Martinelli, R. U. (1973b). *Appl. Opt.* **12**, 1841.
Martinelli, R. U. (1974a). *J. Appl. Phys.* **45**, 1183.
Martinelli, R. U. (1974b). *J. Appl. Phys.* **45**, 3203.
Martinelli, R. U. (1979). Personal communication.
Martinelli, R. U., and Ettenberg, M. (1974). *J. Appl. Phys.* **45**, 3896.
Martinelli, R. U., and Fisher, D. G. (1974). *Proc. IEEE* **62**, 1339.
Martinelli, R. U., and Olsen, G. H. (1976). *J. Appl. Phys.* **47**, 1332.
Martinelli, R. U., Schultz, M. L., Shrader, M. B., and Hughes, F. R. (1972a). Final Tech. Rep., Contract No. DAAKO2-71-C-0329, RCA Corp. Princeton, N.J.
Martinelli, R. U., Schultz, M. L., and Gossenberger, H. F. (1972b). *J. Appl. Phys.* **43**, 4803.
Massies, J., Devoldere, P., and Linh, N. T. (1979). *J. Vac. Sci. Technol.* **16**, 1244.
Mead, C. A. (1962). *Phys. Rev. Lett.* **8**, 56 [errata, *Phys. Rev. Lett.* **9**, 46 (1962)].
Mele, E. J., and Joannopoulos, J. D. (1978). *Phys. Rev. Lett.* **40**, 341.
Michaelson, H. B. (1977). *J. Appl. Phys.* **48**, 4729.
Michaelson, H. B. (1978). *IBM J. Res. Dev.* **22**, 72.
Miller, B., and Jones, T. (1980). *IEEE Trans. Elect. Dev.*, Nov.
Miller, D. J., and Haneman, D. (1979). *Surf. Sci.* **82**, 102.
Milnes, A. G., and Feucht, D. L. (1971). *Appl. Phys. Lett.* **19**, 383.

Milton, A. F., and Baer, A. D. (1971). *J. Appl. Phys.* **42**, 5095.
Mityagin, A. Y., Orlov, V. P., Panteleev, V. V., Khronopulo, K. A., and Cherevatski, N. Y. (1972). *Fiz. Tverd. Tela* **14**, 1870 [Engl. transl., *Sov. Phys.—Solid State* **14**, 1623].
Miyao, M., Goto, R., Sukegawa, T., and Hagino, M. (1978). *Surf. Sci.* **71**, 148.
Miyao, M., Tanaka, A., Sukegawa, T., and Hagino, M. (1979). *J. Appl. Phys.* **50**, 5966.
Mönch, W. (1970). *Phys. Status Solidi* **40**, 257.
Mönch, W. (1977). *Surf. Sci.* **63**, 79.
Mönch, W., and Clemens, H. J. (1979). *J. Vac. Sci. Technol.* **16**, 1238.
Mönch, W., Auer, P. P., and Fedder, R. (1979). *J. Vac. Sci. Technol.* **16**, 1286.
Morton, G. A., Smith, H. M., and Krall, H. R. (1968). *Appl. Phys. Lett.* **13**, 356.
Morton, G. A., Smith, H. M., and Krall, H. R. (1969). *IEEE Trans. Nucl. Sci.* **16**, 92.
Musatov, A. L., and Shulepov, L. N. (1971). *Fiz. Tverd. Tela* **12**, 3343 [Engl. transl., *Sov. Phys.—Solid State* **12**, 2711].
Musatov, A. L., Korotkikh, V. L., and Korinfskii, A. D. (1976). *Izv. Akad. Nauk SSSR Ser. Fiz.* **40**, 2523 [Engl. transl., *Bull. Acad. Sci. USSR* **40**, *No. 12*, 64].
Muscat, J. P., and Newns, D. M. (1974). *J. Phys. C* **7**, 2630.
Nethercot, A. H. (1974). *Phys. Rev. Lett.* **33**, 1088.
Norman, D., McGovern, I. T., and Norris, C. (1977). *Phys. Lett. A* **63**, 384.
Olsen, G. H., Szostak, D. J., Zamerowski, T. J., and Ettenberg, M. (1977). *J. Appl. Phys.* **48**, 1007.
Palmberg, P. W. (1967). *J. Appl. Phys.* **38**, 2137.
Palmer, H. L., Koda, N. J., and Dinan, J. H. (1973). *Proc. Soc. Photo-Opt. Instrum. Eng.* **42**, 31.
Pendry, J. B. (1974). "Low Energy Electron Diffraction." Academic Press, New York.
Persyk, D. E., and Crawshaw, D. D. (1973). *RCA Rev.* **34**, 344.
Petersen, H., and Hagstrom, S. B. M. (1978). *Phys. Rev. Lett.* **41**, 1314.
Petrov, N. N. (1972). *Zh. Tekh. Fiz.* **41**, 2473 [Engl. transl., *Sov. Phys. Tech. Phys.* **16**, 1965].
Piaget, C., Guittard, P., Andre, J. P., and Saget, P. (1975). *Gallium Arsenide Relat. Compd., 1974, Inst. Phys. Conf. Ser.* No. 24, p. 266.
Piaget, C., Polaert, R., and Richard, J. C. (1976). *Adv. Electron. Electron Phys.* **40A**, 377.
Piaget, C., Vannimenus, J., and Saget, P. (1977a). *J. Appl. Phys.* **48**, 3901.
Piaget, C., Saget, P., and Vannimenus, J. (1977b). *J. Appl. Phys.* **48**, 3907.
Piaget, C., Richard, J. C., Monnier, M., and Guittard, P. (1977c). *Acta Electron* **20**, 333.
Pianetta, P., Lindau, I., and Spicer, W. E. (1978a). *In* "Quantitative Surface Analysis of Materials" (N. S. McIntyre, ed.), ASTM STP 643, p. 105. Am. Soc. Test. Mater., Philadelphia, Pennyslvania.
Pianetta, P., Lindau, I., Gregory, P. E., Garner, C. M., and Spicer, W. E. (1978b). *Phys. Rev. B* **18**, 2797.
Pianetta, P., Lindau, I., Gregory, P. E., Garner, C. M., and Spicer, W. E. (1978c). *Surf. Sci.* **72**, 298.
Pierce, D. T., and Meier, F. (1976). *Phys. Rev. B* **13**, 5484.
Pierce, D. T., Meier, F., and Zurcher, P. (1975a). *Phys. Lett. A* **51**, 465.
Pierce, D. T., Meier, F., and Zucher, P. (1975b). *Appl. Phys. Lett.* **26**, 670.
Pierce, D. T., Wang, G. C., and Celotta, R. J. (1979). *Appl. Phys. Lett.* **35**, 220.
Pierce, D. T., Celotta, R. J., Wang, G. C., Unertl, W. N., Galejs, A., Kuyatt, C. E., and Mielczarek, S. R. (1980). *Rev. Sci. Instrum.* **51**, 478.
Powell, C. J. (1974). *Surf. Sci.* **44**, 29.
Powell, R. A., Spicer, W. E., Fisher, G. B., and Gregory, P. (1973). *Phys. Rev. B* **8**, 3987.
Prescott, C. Y., Atwood, W. B., Cottrell, R. L. A., DeStaebler, H., Garwin, E. L., Gonidec, A., Miller, R. H., Rochester L. S., Sato, T., Sherden, D. J., Sinclair, C. K., Stein, S., Taylor, R. E., Clenenin, J. E., Hughes, V. W., Sasao, N., Schuler, K. P., Borghini, M. G., Lubelsmeyer, K., and Jentschke, W. (1978). *Phys. Lett. B* **77**, 347.
Ranke, W., and Jacobi, K. (1977). *Surf. Sci.* **63**, 33.

Ranke, W., and Jacobi, K. (1979). *Surf. Sci.* **81**, 504.
Richard, J. C. (1973). *Acta Electron.* **16**, 245.
Richard, J. C., Lamport, D. L., Roaux, E., and Vannesto, C. (1977). *Acta Electron.* **20**, 353.
Richards, E. A. (1969). *Adv. Electron. Electron Phys.* **28B**, 661.
Ridley, B. K. (1977). *J. Appl. Phys.* **48**, 754.
Riviere, J. C. (1972). *Contemp. Phys.* **14**, 513.
Rome, M. (1971). *In* "Photoelectronic Imaging Devices" (L. Biberman and S. Nudelman, eds.), Vol. 1, p. 147. Plenium, New York.
Rougeot, H., and Baud, C. (1976). *Rev. Tech. Thomson.—CSF* **8**, 449.
Rougeot, H., and Baud, C. (1979). *Adv. Electron. Electron Phys.* **48**, 1.
Rowe, J. E. Christman, S. B., and Margaritondo, G. (1975). *Phys. Rev. Lett.* **35**, 1471.
Rowe, J. E., Margaritondo, G., and Christman, S. B. (1977). *Phys. Rev. B* **15**, 2195.
Sahai, R., Harriss, J. S., Eden, R. C., Bubulac, L. O., and Chu, J. C. (1975). *Crit. Rev. Solid State Sci.* **5**, 565.
Sankaran, R., Antypas, G. A., Moon, R. L., Escher, J. S., and James, L. W. (1976a). *J. Vac. Sci. Technol.* **13**, 932.
Sankaran, R., Moon, R. L., and Antypas, G. A. (1976b). *J. Cryst. Growth* **33**, 271.
Schade, H. (1976). *Surf. Sci.* **55**, 20.
Schade, H. (1979). *Appl. Phys.* **18**, 339.
Schade, H., Nelson, H., and Kressel, H. (1971). *Appl. Phys. Lett.* **18**, 413.
Schade, H., Nelson, H., and Kressel, H. (1972). *Appl. Phys. Lett.* **20**, 385.
Schagen, P., and Turnbull, A. A. (1969). *Adv. Electron. Electron Phys.* **28A**, 393.
Scheer, J. J. (1960). *Philips Res. Rep.* **15**, 584.
Scheer, J. J., and van Laar, J. (1965). *Solid State Commun.* **3**, 189.
Scheer, J. J., and van Laar, J. (1967). *Solid State Commun.* **5**, 303.
Scheer, J. J., and van Laar, J. (1969). *Surf. Sci.* **18**, 130.
Schlier, R. E., and Farnsworth, H. E. (1975). *In* "Semiconductor Surface Physics," p. 3. Univ. of Pennsylvania Press, Philadelphia.
Schlier, R. E., and Farnsworth, H. E. (1959). *J. Chem. Phys.* **30**, 917.
Schroder, D. K., Thomas, R. N., Vine, J., and Nathanson, H. C. (1974). *IEEE Trans. Electron Devices* **21**, 785.
Sebenne, C., Bolmont, D., Guichar, G., and Balkanski, M. (1975). *Phys. Rev. B* **12**, 3280.
Seraphin, B. O., and Bennett, H. E. (1967). *In* "Semiconductors and Semimetals" (R. K. Willardson and A. C. Beer, eds.), Vol. 3, p. 499. Academic Press, New York.
Shahriary, I., Schwank, J. R., and Allen, F. G. (1979). *J. Appl. Phys.* **50**, 1428.
Shockley, W. (1939). *Phys. Rev.* **56**, 317.
Shockley, W., and Pearson, G. L. (1948). *Phys. Rev.* **74**, 232.
Simon, R. E., and Spicer, W. E. (1960a). *Phys. Rev.* **119**, 621.
Simon, R. E., and Spicer, W. E. (1960b). *J. Appl. Phys.* **31**, 1505.
Simon, R. E., and Williams, B. F. (1968). *IEEE Trans. Nucl. Sci.* **15**, 167.
Simon, R. E., Sommer, A. H., Tietjen, J. J., and Williams, B. F. (1968). *Appl. Phys. Lett.* **13**, 355.
Simon, R. E., Sommer, A. H., Tietjen, J. J., and Williams, B. F. (1969). *Appl. Phys. Lett.* **15**, 43.
Skeath, P., Saperstein, W. A., Pianetta, P., Lindau, I., and Spicer, W. E. (1978). *J. Vac. Sci. Technol.* **15**, 1219.
Skeath, P., Su, C. Y., Chye, P. W., Pianetta, P., Lindau, I., and Spicer, W. E. (1979a). *J. Vac. Sci. Technol.*
Skeath, P., Su, C. Y., Chye, P. W., Pianetta, P., Lindau, I., and Spicer, W. E. (1979b). *J. Vac. Sci. Technol.*
Smith, D. L., and Huchital, D. A. (1972). *J. Appl. Phys.* **43**, 2624.
Sommer, A. H. (1943). *Nature (London)* **152**, 215.
Sommer. A. H. (1958). *J. Appl. Phys.* **29**, 1568.

Sommer, A. H. (1968). "Photoemissive Materials." Wiley, New York.
Sommer, A. H. (1973a). *J. Phys.* **C6**, 51.
Sommer, A. H. (1973b). *RCA Rev.* **34**, 95.
Sommer, A. H. (1973c). *Gallium Arsenide Relat. Compd., 1972, Inst. Phys. Conf. Ser.* No. 17, p. 143.
Sommer, A. H. (1973d). *Appl. Opt.* **12**, 90.
Sommer, A. H., and Spicer, W. E. (1965). *In* "Photoelectric Materials and Devices" (S. Larach, ed.), p. 175. Van Nostrand, Princeton, New Jersey.
Sommer, A. H., Whitaker, H. H., and Williams, B. F. (1970). *Appl. Phys. Lett.* **17**, 273.
Sonnenberg, H. (1969). *Appl. Phys. Lett.* **14**, 289.
Sonnenberg, H. (1970a). *IEEE J. Solid-State Circuits* **5**, 272.
Sonnenberg, H. (1970b). *Appl. Phys. Lett.* **16**, 245.
Sonnenberg, H. (1971). *Appl. Phys. Lett.* **19**, 431.
Sonnenberg, H. (1972). *Appl. Phys. Lett.* **21**, 103.
Spicer, W. E. (1958a). *Phys. Rev.* **112**, 114.
Spicer, W. E. (1958b). *RCA Rev.* **19**, 555.
Spicer, W. E. (1962). *Phys. Rev.* **125**, 220.
Spicer, W. E. (1977). *Appl. Phys.* **12**, 115.
Spicer, W. E. (1978). *In* "Electron and Ion Spectroscopy of Solids," (L. Fiermans, J. Vennik, and W. Dekeyser, eds.), p. 54. Plenum, New York.
Spicer, W. E. (1979). Personal communication.
Spicer, W. E., and Bell, R. L. (1972). *Publ. Astron. Soc. Pac.* **84**, 110.
Spicer, W. E., Lindau, I., Gregory, P. E., Garner, C. M., Pianetta, P., and Chye, P. W. (1976). *J. Vac. Sci. Technol.* **13**, 780.
Spicer, W. E., Lindau, I., Miller, J. N., Ling, D. T., Pianetta, P., Chye, P. W., and Garner, C. M. (1977). *Phys. Scr.* **16**, 388.
Spicer, W. E., Lindau, I., Su, C. Y., Chye, P. W., and Pianetta, P. (1978). *Appl. Phys. Lett.* **33**, 934.
Spicer, W. E., Lindau, I., Chye, P. W., Skeath, P., and Su, C. Y. (1979). *J. Vac. Sci. Technol.* **16**, 1422.
Steinberg, R. F. (1968). *Appl. Phys. Lett.* **12**, 63.
Stocker, B. J. (1975). *Surf. Sci.* **47**, 501.
Stohr, J. (1979). *J. Vac. Sci. Technol.* **16**, 37.
Stupp, E., Pelissier, A., Kidder, M., and Milch, A. (1977). *J. Appl. Phys.* **48**, 4741.
Sturge, M. D. (1962). *Phys. Rev.* **127**, 768.
Su, C. Y., Chye, P. W., Pianetta, P., Lindau, I., and Spicer, W. E. (1979). *Surf. Sci.* **86**, 894.
Syms, C. H. A. (1969). *Adv. Electron. Electron Phys.* **28A**, 399.
Sze, S. M. (1969). "Physics of Semiconductor Devices." Wiley, New York.
Tamm, I. (1933). *Phys. Z. Sowjetunion* **1**, 733.
Taylor, J. B., and Langmuir, I. (1933). *Phys. Rev.* **44**, 423.
Thomas, C. M. (1973). *Proc. Soc. Photo-Opt. Instrum. Eng.* **42**, 71.
Thornton, P. R. (1973). Final Tech. Rep., Contract DAAKO2-72-C-0149. Stanford Res. Inst., Stanford, California.
Thornton, P. R., and Northrup, D. C. (1965). *Solid-State Electron.* **8**, 437.
Thuault, C. D., Guichar, G. M., and Sebenne, C. A. (1979). *Surf. Sci.* **80**, 273.
Thurmond, C. D. (1965). *J. Phys. Chem. Solids* **26**, 785.
Tong, S. Y. (1978). *Comments Solid State Phys.* **9**, 1.
Tong, S. Y., and Maldonado, A. I. (1978). *Surf. Sci.* **78**, 459.
Trueba, A., Munoz, E., and Piqueras, J. (1974). *Solid State Commun.* **15**, 199.
Turnbull, A. A., and Evans, G. B. (1968). *J. Phys. D* **1**, 155.
Uebbing, J. J. (1969). *J. Vac. Sci. Technol.* **7**, 81.

Uebbing, J. J. (1970). *J. Appl. Phys.* **41**, 802.
Uebbing, J. J., and Bell, R. L. (1967). *Appl. Phys. Lett.* **11**, 357.
Uebbing, J. J., and Bell, R. L. (1968). *Appl. Phys. Lett.* **12**, 76.
Uebbing, J. J., and James, L. W. (1970). *J. Appl. Phys.* **41**, 4505.
van Bommel, A. J., and Crombeen, J. E. (1974). *Surf. Sci.* **45**, 308.
van Bommel, A. J., and Crombeen, J. E. (1976). *Surf. Sci.* **57**, 109.
van Bommel, A. J., and Crombeen, J. E. (1978). *Surf. Sci.* **76**, 499.
van Bommel, A. J., Crombeen, J. E. and Van Oirschott, T. G. J. (1978). *Surf. Sci.* **72**, 95.
van Laar, J. (1973). *Acta Electron.* **16**, 215.
van Laar, J., and Huijser, A. (1976). *J. Vac. Sci. Technol.* **13**, 769.
van Laar, J., and Scheer, J. J. (1962). *Philips Res. Rep.* **17**, 101.
van Laar, J., and Scheer, J. J. (1967). *Surf. Sci.* **8**, 342.
van Laar, J., and Scheer, J. J. (1968). *Philips Tech. Rev.* **29**, 54.
van Oirschot, T. G. J., Acket, G. A., and Bartels, W. J. (1975). *J. Appl. Phys.* **46**, 1893.
van Oirschot, T. G. T., Hallais, J. P. and André, J. P. (1977). *Acta Electron.* **20**, 323.
Van Speybroeck, L., Kellogg, E. and Murray, S. (1974). *IEEE Trans. Nucl. Sci.* **21**, 408.
Wagner, L. F., and Spicer, W. E. (1972). *Phys. Rev. Lett.* **28**, 1381.
Wagner, L. F., and Spicer, W. E. (1974). *Phys. Rev.* **9**, 1512.
Wagner, P., Muller, K., and Heinz, K. (1977). *Surf. Sci.* **68**, 189.
Wang, G. C., Dunlap, B. I., Celotta, R. J., and Pierce, D. T. (1979). *Phys. Rev. Lett.* **42**, 1349.
Ward, S. A., Fischer, T. E., and Nowak, W. B. (1969). Tech. Rep. AFAL-TR-69-161, Contract AF33(615)-3638. CBS Lab., Stanford, Conn.
Weber, R. E., and Peria, W. T. (1969). *Surf. Sci.* **14**, 13.
Wehner, G. K. (1975). *In* "Methods of Surface Analysis" (A. W. Czaanderna, ed.), p. 5. Elsevier, Amsterdam.
White, S. J., and Woodruff, D. P. (1977). *Surf. Sci.* **64**, 131.
Wilcox, D. A., Abraham, W. G., Bardas, D., Gwilliam, G. F., and Enck, R. S. (1979). *Electro-Opt. Syst. Des.* March, p. 41.
Williams, B. F. (1969). *Appl. Phys. Lett.* **14**, 273.
Williams, B. F. (1972). *IEEE Trans. Nucl. Sci.* **19**, 39.
Williams, B. F., and Simon, R. E. (1967). *Phys. Rev. Lett.* **13**, 485.
Williams, B. F., and Simon, R. E. (1969). *Appl. Phys. Lett.* **14**, 214.
Williams, B. F., and Tietjen, J. J. (1971). *Proc. IEEE* **59**, 1489.
Williams, R. H. (1978). *Contemp. Phys.* **19**, 389.
Williams, R. H. (1979). *J. Vac. Sci. Technol.*
Williams, R. H., Varma, R. R., and Montgomery, V. (1979). *J. Vac. Sci. Technol.* **16**, 1418.
Williams, R. H., Varma, R. R., and McKinley, A. (1977). *J. Phys. C* **10**, 4545.
Yasar, T., and Steinberg, R. F. (1970). *J. Vac. Sci. Technol.* **7**, 139.
Yee, E. M., and Jackson, D. A. (1972). *Solid-State Electron.* **15**, 245.
Zwicker, H. R. (1977). *In* "Optical and Infrared Detectors" (R. J. Keyes, ed.), p. 149. Springer-Verlag, Berlin and New York.

Index

A

Absolute sensitivity, photocathode, *see* Sensitivity measurements, photocathode
Absorption coefficient
 metal, 203
 NEA semiconductors, 208, 234, 239
  GaAs, 208, 239
  GaP, (InGa)As, (InGa) (AsP), InP, and Si, 208
AES (Auger electron spectroscopy), 199, 200, 216
 surface reconstruction studies, 216, 231
Affinity rule, 138, *see also* Electron affinity
 proposed modification, 140
AlAs
 bandgap, 241
 contact fabrication, 12–15
 lattice constant, 241
AlP, contact fabrication, 11, 14
α, *see* Absorption coefficient; Diode-characteristics slope
Ambipolar diffusivity, 107
Ambipolar mobility, 107
Applied bias, *see also* Injection; Photomitters, bias assisted; *PN* junctions
 effect on electrochemical potential, 60–63, 65, 94–100, 129, 130
 effect on junction current, 66, 67, 90, *see also* Current–voltage relation
 energy-band diagram, 62, 70, 71, 98
 heterojunction, 152–154, 284–288, *see also* Transferred-electron cathode
 nonequilibrium, steady-state conditions, 64
 PN junction, 59–72, 94–97, 98–100
  reduction of diffusion barrier, 63, 64
  requirement for low-level injection, 66
 transferred-electron cathode 276–289, *see also* Transferred-electron cathode

B

Barrier width, 5, 57, *see also* Space-charge region
 effect of doping, 5, 6
Band bending
 contact interface, 2, 6, 8, 170, 182, 183
 effect on escape probability, 247
 field favorable for electron emission, 211
 heterojunction, 137, 145, 146, 149, 151, 152, 155, 160, 161, 174
  i–i, 146
  PN, 137, 145, 151, 152, 155, 160, 161, 174
 NEA surface, 220–223
 PN junction, 49, 54, 62, 63, 68, 70, 71, 98
  biased (injection), 62, 63, 70, 71, 98
Band edges, discontinuity
 simple, 138
 spike–notch, 137, 140, 152, 161, 162
Bandgap, *see also* specific materials
 dependence of surface escape probability, 249
 variation with lattice constant, AlAs, GaAs, GaP, InAs, (InGa)As, (InGa) (AsP), 241
Band structure, 209, 241, 246–257, 276–288
 conduction-band profile, heterojunction TE cathode, 287, 288
 GaAs, 209
 GaP, 248
 surface escape probability, 246–257, *see also* Escape probability
  band bending, 247, 248
  Γ electrons in GaAs, 247, 248
  L electrons in GaAs, 248
  tabulated values, 248

301

transitions between minima, 247, 276–288
  GaAs, 247, 248
  GaP, 247–249
  InP, 278–288
  Si, 248, 249
  transferred-electron cathode, 276–288
Barrier height, 2, 3–8, 48, 182, see also
    Schottky barrier
  effect of surface states, 3
  reduction by applied bias, 63, 64
  tabulated values, 4
Barrier layer, 48, 49, see also Schottky
    barrier; Space-charge region
  effect on escape probability, 249, 251, see
      also Escape probability
  strong electric field, 56
  depletion approximation, 56
Barrier potential, 136, 138, 170
Boltzmann distribution, 47, see also
    Maxwell–Boltzmann statistics
Boundary conditions in *PN* junction, 90–104,
    see also Applied bias; Electrochemical
    potential; *PN* junctions
  electrochemical potential, 91, 93, 95–97,
      99, 100, 103, 104, 129–131, 134
  Fletcher, 91–93, 95, 158
    modified, 97
  Harrick, 93
  heterojunction, 142–155, 158—162,
      168–175, 180–189
  Shockley, 90–94, 189
Buffer layer, 16

C

Capacity vs. voltage, 71, 150
  InP/CdS heterojunction, 150–152
Carrier concentration, equilibrium
  Fermi–Dirac statistics, 176–178
  Maxwell–Boltzmann statistics, 42–44, 46,
      47, 176, 177
    intrinsic case, 43, 44
Carrier concentration, nonequilibrium, 60, 69
  at edges of space-charge region, 130, 131
  Fermi–Dirac statistics, 176–178
  gradient, 61
  in space-charge region, 53, 55, 58
  steady-state conditions, 64

Carrier generation, 45, 73, 233–236, see also
    Generation–recombination
  photocathode, 234–236
  rate, 73
Carrier lifetime, 73, 76
  capture lifetime, 73
  effect on I–V characteristics, 122, 123, 129
  injection-independent, 108
Carrier recombination, 7, 45, 65, 72–90, see
    also Generation–recombination;
    Recombination centers
  Hall–Shockley–Read theory, 73–76
  ohmic contacts, 7
  rates, 45, 65, 73
  surface, 240
  traps, 72–74, 79, 87, 88, see also Traps
CdS
  dielectric constant, electron affinity, energy
      gap, lattice constant, 151
  N-CdS/P-InP heterojunction, 150–153
Chemical potential, 44, 46–55, see also
    Potential
  constancy in space-charge region, 64
  electrons, 59, 64
  heterojunction, 145, 149, 179
  holes, 59, 64
  incremental deviation from equilibrium, 58,
      59
  nonequilibrium conditions, 59–66
  relative, 50
Chemisorption, 197, 216
Contact potential difference, 169, 205
  determination of work function, 205
Contacts, 1–32, 162—189, see also Schottky
    barrier; Schottky diode
  amorphous intermediate layer, 7
  buffer layer, 16
  current–voltage relation, 3, 6, see also
      Current–voltage relation
  degradation, 27, 30, 31
  energy-band diagram, 2, 6–8, see also Band
      bending; Energy-band diagram
  heavily doped intermediate layer, 5
  ohmic, 2, 3, 5–7, 172, see also Ohmic
      contacts
  rectifying, 2, 3, 5, 6, 173
  resistance, 19, 23, 24
    dependence on doping, 23, 24
  resistance measurement, 18–24
    differential method, 19, 20

extrapolation, 19, 22
four contacts, 19, 21
four-dot array and planar back contact, 19, 23
I–V characteristics, 19
three contacts, 19–21
twin contacts, 19
Schottky barrier, 162, *see also* Schottky barrier
Schottky model, 2
structural characterization, 24–31
Contacts, ohmic, 1–32, 172, *see also* Contacts; Ohmic contacts
amorphous intermediate layer, 7
choice of metal, 9–18
degradation, 27, 30, 31
energy-band diagram, 2, 6–8, *see also* Energy-band diagram
heavily doped intermediate layer, 5
heterojunction intermediate layer, 7
Continuity equation, 44, 45, 105
Cs activation, 211–233, 244–246, 271, *see also* Photoemission
Ag surface, 212, 229
coverage dependency
photoemission, 226
work function, 224, 225
effect of cathode bandgap, 230
escape probabilities, 248, *see also* Escape probability
GaAs, Ga(AsP), GaP, (InGa) (AsP), Si, 248
Fermi level pinning, 210, 221–224, 230, 231
GaAs, 221–224
Si, 230, 231
GaAs, 213–216, 222–230, 245, 246
Hi–Lo technique, 246
GaP, 230, 266
GaSb, 228
(InGa)As, 246
Hi–Lo technique, 246
InP, 228
interfacial barrier, 228, 233
NEA-cathode dark current, 270–274, *see also* Dark current, photocathodes
activator layer thickness dependence, 271
Si, 230–233, 244–246
thermal desorption, 227, 228, 231

work function
GaAs, 225
Si, 233
Current density, *see also* Current–voltage relation
α, 79–81, *see also* Diode-characteristics slope
as function of applied voltage in *PN* junction, 66, 67, 90, *see also PN* junctions
diffusion, 61, 62, 65, 77, 106
diffusion-limited, 172
drift, 61, 62, 65, 77, 106
generation–recombination, 76–85
abrupt Si diode, 82–84
compared to diffusion current, 78
Sah–Noyce–Shockley theory, 76
high injection, 109–111
individual carrier contribution, 60, 61
injected minority carrier, 66
low injection, 109–111
Ohm's law, 44
recombination, 77, 127, 129
van Roosbroeck model, 121–129, *see also* van Roosbroeck model
Current–voltage relation, *see also* Applied bias; Diode-characteristics slope
accumulation layer, 158, 159, 172
commercial diodes, 132, 133
diffusion limited, 172
Ge diode, 106
generation–recombination current, 78–84
Si diode, 83, 84
heterojunction, 152–154, 157
nonrectifying diode, 160, 161
P-GaAs–*N*-Ge, 154, 162
P-Ga(SbAs)/*N*-(InGa)As, 157
high injection, 107, 109, 110
InSb diode, 84
low injection, 67, 109, 110
ohmic contact, 6
PN junction (characteristic equation), 67, 107, *see also PN* junctions
high injection, 107, 109, 110
low injection, 67, 109, 110
recombination current, 77, 90, 127, 129
Schottky barrier, 3, 172
Si diode, 86
trapping levels, 90, *see also* Traps
van Roosbroeck model, 121–129

D

Dark current, photocathodes, 208, 270–274, 288, 289
  equivalent noise input, 270
  possible sources, 272, 288, 289
  transferred-electron cathode, 288, 289
  typical values, 271, 272, 288
    GaAs, Si, 271–273
    Ga(AsP), (InGa) (AsP), 271, 288
Debye length, 50, 59
  extrinsic, 59
  intrinsic, 50
Degradation
  NEA devices, 274–276
  ohmic contacts, 27, 30, 31
Density of states, 42–44, 176
Depletion approximation, 56–58, 70, 72, 81, 189
  applied to heterojunctions, 136
  asymmetric junction, 72
  band shapes, transition region, 70, 71
  capacitance, 57, 71, 72
  charge distribution, 56
Depletion layer, 171, *see also* Space-charge region
  width, NEA photoemitter, 210, 213, 246
Diffusion
  photoelectrons, 237–242
  reduction of barrier, applied bias, 63, *see also* Applied bias
  Schottky barrier, 3
Diffusion constant, 61
  ambipolar, 107
  injection-dependent, 109
Diffusion current density, 61, 78, 106, 172
  compared to drift current, 106
  compared to generation–recombination current, 78
Diffusion length, 65
  injection-dependent, 109
  minority-carrier, 213
  photoelectron, 213, 237–239
Diffusion potential, 136, 138, 139
  barrier, 138, 170, 182
  P-InP/N-CdS heterojunction, 150
Diode-characteristics slope, α, 79–81, 83–86, 89, 90, *see also* Current–voltage relation; van Roosbroeck model
  abrupt Si diode, 83, 85, 86
  alloyed Ge diode, 106
  commercial diodes, 132
  dependence on injection level, 122
Distribution function, *see* Fermi–Dirac statistics; Maxwell–Boltzmann statistics
Doping concentration
  effect on escape probability, 246
  opposite effect on diffusion length, 246
  heavily doped intermediate layer, 5
  NEA photocathode, 246
  rectifying or ohmic contact, 5, 6

E

EDC (Electron energy distribution curve), 200, 205, 207, 280
  determination of work function, 205
  InP, Γ and L valleys, 280
  NEA GaAs surface, 253
Einstein photoelectric equation, 181, 201, 202
Einstein relation, 61
  generalized, 108
Electrochemical potential, 44, 46–55, 60, 62, 63, 65, 71, 91, 93–100, 129–131, 164, 174, 178, 180, 188
  across space-charge region, 93–100
  boundary conditions in PN junction, 65, 91, 93, 95–97, 99, 100, 103, 104, 129–131, 134, 164, 188
  effect of applied bias, 60–63, 65, 94–100, 129, 130
  electrons, 60
  holes, 60
  majority-carrier injection effect, 99, 100, 189
  metals in contact, 188
  nonequilibrium case, 60–63
  nonzero slope, 62, 63
  variation along diode, 134
Electron affinity, 2, 135, 141, 210, 221, *see also* Affinity rule; specific materials
  heterojunction, 135, 137, 141
  electrostatic term, 141
  polarization term, 141
  surface dipole, 142
Electron energy-loss mechanisms, *see also* Electron scattering
  electron–electron collisions, 202, 203
  impurity scattering, 205

optical-phonon scattering, 205
pair production (impact ionization), 205
Electron scattering, 202, 203, 213
  defect scattering, 205
  electron–electron collisions, 202
  hot electrons in metal, 203
  optical-phonon scattering, 205, 213
Electrostatic potential, 44–49
  heterojunction, 140, 142, 158, 164, 175, 178
  continuity, 142, 175
ELS (Electron energy loss spectroscopy), 199, 200
Emission face, surface escape probabilities, 246–248
Energy-band diagram
  applied bias, 62, 70, 71, 98
  asymmetric PN junction, abrupt, 54
  cold-cathode PN homojunction, 268
  contacts, 2, 6–8
  Cs + O/P-GaAs NEA photoemitter, 214, 268
  heterojunction, 137, 145–147, 151, 155, 160, 161, 168, 170, 174, 184, 186
    Schottky–Spenke approach, 186
  homojunction, 49, 54, 63, 68, 71, 98, 173, 174
  i-Ge/i-GaAs heterojunction, 146, 147
    continuous intrinsic level, 147
    continuous vacuum level, 146
  isolated metal and N-type semiconductor, 182
    Spenke model, 182
  metal surface photoemission, 201
  metal–semiconductor, 2, 6, 8, 170, 182, 183
    after contact, 170
    before contact, 170, 182
  metals
    in contact, 168
    separated, 167, 178–180
    Spenke model, 178, 180
  P-GaAs/N-Ge heterojunction, 152, 155
    continuous intrinsic level, 155
  P-Ga(SbAs)/N-(InGa)As heterojunction, 160, 161
    spike–notch discontinuity, 161, 162
  P-Ge/N-GaAs heterojunction
    continuous intrinsic level, 145
    continuous vacuum level, 137

photoelectric process, 181
photoemission, 206, 207, 214, 268, 277, 283, 285
P-InP/N-CdS heterojunction, 150, 151
P-InP/P-(InGa)As heterojunction, 285
PN junction in equilibrium, 49, 54, 174, see also Energy-band diagram, homojunction
PN junction under injection, 63, see also Energy-band diagram, homojunction
  band shape in transition region, 70, 71
  Shockley low-level theory, 68, see also Shockley low-level theory
  Schottky diode, 183, see also Schottky diode
  semiconductor, 42, 206
    effect of surface, 206
    n-type, 206
    p-type, 206
  symmetric PN junction, abrupt, 49
  thermostatic potentials, 44
  transferred-electron cathode, 277, 283, 285
Energy levels, equilibrium statistics, 42–44
Escape depth, 199, 205, 213
  GaAs photoelectrons, 199
  NEA-semiconductor photoelectrons, 213
Escape probability, 228, 238, 241–258, 275–281, see also specific materials
  bandgap dependence, 255
    Ga(AsP), 255
  dependence on band minima, 247–249
    transferred-electron cathode, 276–281
  dependence on surface face, 247
  dependence on work function, 251, 254, 275
    GaAs, 251, 254, 275
    (InGa) (AsP), 275
  doping dependence, 246, 254
  effect of reflections, 247, 253
  experimental and theoretical values, 228, 248, 251, 254, 255
    GaAs, 228, 248, 251, 254, 275
    Ga(AsP), 248, 255
    GaP, 248
    (InGa) (AsP)—248, 275
    InP, 228
    Si, 248
  influential factors, 228, 242–258
    band structure and emission face, 246–257

bulk doping, 246, 254
  Cs + O activation 228, 244–246
  electric-field enhancement, 257
  strain, 258
  surface cleaning, 243, 244
  temperature, 257, 258
  interfacial barrier, 228, 233, 249, 251
  NEA surfaces, *see* Escape probability, experimental and theoretical values

F

Fermi–Dirac distribution, 47, 176–178
Fermi level (energy), 2, 6, 8, 42–44, 46, 47, 59, 60, 63, 71, 98, 137, 145, 151, 155, 160, 161, 170, 174, 178, 180, 182, 183, 186, 208, 210, 221
  at surface, 208, 210, 221
  Cs activation, 222
  equivalence to electrochemical potential, 47
  nonequilibrium conditions, 59
  pinning, 210, 221–224, 230, 231
    GaAs, 221–224
    Si, 230, 231
  quasi, 60
  related to chemical potential and intrinsic level, 46, 47, 71
Fick's law, 61
Field emission, 5
  Schottky barrier, 5

G

(GaAl)As
  bandgap, 241
  contact fabrication, 17
  (GaAl)As/GaAs NEA cathode, 198
    cold-cathode emitter, 269
    quantum yield, 242, 264
    TM photomultiplier tube, 263, 264
  lattice constant, 241
(GaAl)P
  contact fabrication, 17
  (GaAl)P/GaP heterojunction cold-cathode emitter, 269
GaAs
  (AlGa)As/GaAs photocathode
    cold-cathode emitter, 269

quantum yield, 242, 264
TM photomultiplier tube, 263, 264
bandgap, 135, 241
band structure, 209
  transitions between minima, 247, 248
cold-cathode emitter, 268, 269
commercial NEA device, 289
contact resistance, 23, 24
Cs activation, 213–216, 222–230, 245, 246, 271
  escape probability, 228, 248
  Fermi level pinning, 221–224, *see also* Fermi level
  Hi–Lo technique, 246
  thermal desorption, 227, 228
  work function, 225
Cs + O/P-GaAs NEA photoemitter, 213–215, 268, 269
dielectric coefficient, 135
electron affinity, 4, 135
i-GaAs/i-Ge heterojunction, 146, 147
lattice constant, 135, 241
NEA photoemitter, 198, 208, 213–215, 242, 247, 248, 251, 254, 255, 259–265, 267–276
  dark current, 271–273
  device life, 274–276
  energy distribution curve, 253
  quantum yield, 215, 261, *see also* Quantum yield
  surface Fermi level, 221–224
N-GaAs/P-Ge heterojunction, 135–137, 145–147
ohmic contacts, 6–17
optical absorption coefficient, 208
P-GaAs/N-Ge heterojunction, 152–154, 157
photoelectron escape depth, 199, 213
Schottky barrier height, 4, *see also* Schottky barrier
secondary-electron emitter, 267
  gain, 268
  (InGa)P/GaAs, 268
secondary electrons, 235, 236
  primary beam range, 236
spin-polarized electron source, 269, 270
surface cleaning, 243, 244
surface escape probability, 228, 247, 248, 251, 254, 275, *see also* Escape probability
  dependence on work function, 275

effect of strain, 258
effect of temperature, 257, 258
electric-field enhancement, 257
surface reconstruction, 216–220
(GaAs)P
cold-cathode emitter, 269
commercial NEA device, 289
contact fabrication, 17, 18
escape probability, NEA surface, 248, 255
effect of temperature, 258
quantum yield, NEA surface, 256, 261
sensitivity, NEA surface, 259
dark current, 271
device life, 274
Ga(AsSb)
contact fabrication, 18
P-Ga(AsSb)/N-(InGa)As heterojunction, 157–162
(GaIn)As
commercial NEA device, 289
contact fabrication, 18
Cs activation, 246
Hi–Lo technique, 246
InP/(InGa)As heterojunction TE cathode, 284–288
conduction-band profile vs. bias, 287, 288
energy-band diagram, 285
quantum yield, 286
N-(InGa)As/PGa(SbAs) heterojunction, 157–162
optical absorption coefficient, 208
transferred-electron direct emitter cathode, quantum yield, 284
(GaIn)Sb, contact fabrication, 18
Galvani potential, 169, 180
Spenke model, 179
GaN, contact fabrication, 14
GaP
bandgap, 241
band structure, 247, 248
transition between minima, 247
commercial NEA devices, 289
contact fabrication, 11–15
Cs activation, 230, 248
Cs/GaP dynode
device life, 274
gain, 266
dynode, 198
electron affinity, 4

escape probability, NEA surface, 247–249
lattice constant, 241
optical absorption coefficient, 208
Schottky barrier height, 4
secondary electrons, 235, 236
primary beam range, 236
sensitivity, NEA surface, 259
surface cleaning, 243
surface reconstruction, 230
GaSb
contact fabrication, 13, 14
Cs activation, 228
interfacial barrier, 228
Ge (germanium)
bandgap, dielectric coefficient, electron affinity, lattice constant, 135
i-Ge/i-GaAs heterojunction, 146, 147
I–V characteristic, alloyed diode, 106
N-Ge/P-GaAs heterojunction, 152–154, 157
P-Ge/N-GaAs heterojunction, 135–137, 145–147
recombination via traps, 72
Generation, see Carrier generation
Generation–recombination, 72–90, see also Carrier generation; Carrier recombination; Recombination centers
current, 76–85
abrupt Si diode, 82–84
effect on device behavior, 72–90
Sah–Noyce–Shockley theory, 76–85
via traps, 74, 79, 87, 88

H

Hall–Shockley–Read theory, 73–76, see also Carrier recombination
Heterojunctions, 7–9, 135–189, 190, 197, 198, see also Boundary conditions in PN junction; specific materials
accumulation, 158, 159, 172
anisotype, 138
band-edge behavior, see Band bending; Band edges
current–voltage relation, 152–154, 157, 160—162, see also Current–voltage relation
depletion approximation, 136
energy-band diagram, see Energy-band diagram

enhancement of injection efficiency, 135
exclusion, 159, 160
extraction, 159, 160
isotype, 138
light–heavy junction, 159
materials, *see* Heterojunction materials
ohmic contact fabrication, 7–9
Te cathode, 284–288
Heterojunction materials
  i-Ge/i-GaAs, 146, 147
  P-GaAs/N-Ge, 152–154, 157
  P-Ga(SbAs)/N-(InGa)As, 157–162
  P-Ge/N-GaAs, 135–137, 145–147
  P-InP/N-CdS, 150–153
  P-InP/P-(InGa)As, 285–288

I

InAs
  bandgap, 241
  contact fabrication, 11, 12–15
  electron affinity, 4
  lattice constant, 241
  Schottky barrier height, 4
In(AsP)
  contact fabrication, 18
  photocathode, 250
In(AsSb)
  contact fabrication, 18
In(GaAs)
  bandgap, 241
  contact fabrication, 18
  In(GaAs)/InP heterojunction Te cathode, 284–288
    conduction-band profile vs. bias, 287, 288
    energy-gand diagram, 285
    quantum yield, 286
  lattice constant, 241
(InGa)(AsP)
  bandgap, 241
  commercial NEA device, 289
  dark current, photocathode, 271, 274, 288, 289
  escape probability, NEA surface, 248, 275
    dependence on work function, 275
    effect of temperature, 258
  lattice constant, 241
  optical absorption coefficient, 208

quantum yield, NEA surface, 252, 261, 262
  transferred-electron cathode, 283
sensitivity, NEA surface, 259
  device life, 275
transferred-electron direct emitter cathode, 282–284
  dark current, 288, 289
  energy-band diagram, 283
  heterojunction, 286
  quantum yield, 283, 284
(InGa)P/GaAs
  NEA cathode, 198
  TM secondary-electron emitter, 267, 268
  gain, 268
Injection, *see also* Applied bias; *PN* junctions; van Roosbroeck model
  asymmetric junction, 106
  currents
    high-level, 106, 109–111
    low-level, 106, 109–111
  high-level, 93, 101, 104–135
    current, 106, 109–111
    current–voltage characteristic, 106–111
  low-level, 59–68, 93, 104, 106, 109, *see also* Shockley low-level theory
    current, 106, 109–111
    current–voltage characteristic, 67, 106, 109–111
    necessary condition, 66
    space-charge region, 68–72
  PN junction, characteristic equation, 67, 93, 107
  law of the junction, 93
  reduction of diffusion barrier, 63, 64
  majority-carrier, 99, 100, 104, 189
    effect on electrochemical potential, 99, 100, 104
  Shockley low-level theory, *see also* Shockley low-level theory
    assumptions, 68, 90–94
InP
  bandgap, dielectric coefficient, lattice constant, 151
  contact fabrication, 12–15
  Cs activation, 228
  electron affinity, 4, 151
  optical absorption coefficient, 208
  P-InP/N-CdS heterojunction, 150–153
  P-InP/P-(InGa)As heterojunction TE cathode, 284–288

conduction-band profile vs. bias, 287, 288
energy-band diagram, 285
quantum yield, 286
Schottky-barrier height, 4
surface cleaning, 243, 245
transferred-electron cathode, 277–282, 284–289
  carrier fraction in upper minimum, 280
  dark current, 289
  energy-band diagram, 277, 285
  heterojunction, 284–289
  mean electron energy in upper minimum, 280, 281
  quantum yield, 278, 279
InSb
  contact fabrication, 15
  diode current–voltage characteristics, 84
  electron affinity, 4
  Schottky barrier height, 4
Intrinsic carrier concentration, 43, 44, see also Carrier concentration, equilibrium
Intrinsic level, 43, 44, 46, 50, 72
  continuity, 140, 142, 143, 145, 155, 174, 186
  discontinuity, 137, 142, 175
  heterojunction, 137, 140, 142, 143, 148, 155, 174, 175, 186

J

Junctions, see Contacts; Heterojunctions; *PN* junctions, Semiconducting junctions, theory

L

LEED (low-energy electron diffraction), 199, 200
  ordering of Cs, 226, 227
  surface reconstruction studies, 217–220, 231
Lifetime, see Carrier lifetime

M

Macropotential, 175, 178, 180, 183, 184, 186–188
  homojunction, 184

Maxwell–Boltzmann statistics, 42, 176, see also Boltzmann distribution
carrier equilibrium density, 42, see also Carrier concentration, equilibrium; Carrier concentration, nonequilibrium
limiting value of Fermi–Dirac integral, 176, 177
Metals for ohmic contacts, 9–18
  eutectic-temperature considerations, 9
  tabulation, 11–15, 17, 18

N

NEA (Negative electron affinity), 195–289, see also NEA devices; NEA semiconductor photoemitter; NEA surface; Photoemission; specific materials
  early developments, 198
  photocathode, 197, 240–242, 250–252, 255, 256, 258–289, see also Photocathode
  semiconductor photoemitter, 195–289
  surface, 197, 216–233, see also NEA surface
NEA devices, 260–270, 289, see also NEA; NEA semiconductor photoemitter; NEA surface; Transferred-electron cathode
  bias-assisted photoemitter, 276–289, see also Photoemitters, bias assisted
  cold-cathode emitter, 268, 269
  device life, 274–276
    improvement in NEA Si over III–V's, 275
    operating life, 274–276
    shelf life, 274–276
  GaAs as spin-polarized electron source, 269, 270
  materials, commercial devices, 289
  RM photomultiplier tube, 260–263, 270
    cross section, 262
    equivalent noise input, 270
    GaAs, 260
    Ga(AsP), 260
    (InGa)As, 260
    (InGa)(AsP), 260, 262
    quantum yield, 261–263
  RM secondary-electron emitters, 265–267
    Cs/GaP dynode, 266
    GaAs, 267
    gain, 268

(InGa)P/GaAs, 267
Si, 267
TM GaAs photocathode, 263–265
  (AlGa)As/GaAs, 263, 264
  cross section, 264
TM secondary-electron emitters, 267, 268
TM Si photocathode, 265
NEA semiconductor photoemitter, 198, 208, 213–216, 242, 246–289, *see also* NEA; NEA devices; NEA surface; Photocathode; Photoemission; Quantum yield; specific materials
  bias-assisted, 276–289, *see also* Photoemitters, bias-assisted
  cold-cathode, 268, 269
  dark current, 208, 270–274, *see also* Dark current, photocathodes
  depletion width, 213
  devices, *see* NEA devices
  energy-band diagram, 206, 207, 214, 268, 277, 283, 285
  energy distribution curve, 253
  extension of threshold, *see* Photoemitters, bias-assisted
  materials, commercial devices, 289
  p-type material favored, 211
  quantum yield, 215, 242, 250, 252, 256, *see also* Quantum yield
    GaAs, 215, 261
    GaAs/(AlGa)As, 242, 264
    Ga(AsP), 256, 261
    (InAs)P, 250
    (InGa)(AsP), 252, 261, 262
  reflection mode (RM), 213, 215, 234, 242, 250, 252, 256, 260–263, 265–267
  effect of Cs coverage, 226
  secondary-electron gain, 241
  sensitivity, *see* Sensitivity measurements, photocathode
  surface escape probability, *see* Escape probability
  TE cathode, 276–289, *see also* Transferred-electron cathode
  transmission mode (TM), 214, 234, 242, 263–265, 267
NEA surface, 216–233, *see also* NEA; NEA devices; Surface photoemission
  atomic structure, 216–220, *see also* Surface atomic structure
  chemisorption, 197, 216

cleanliness achievement, 216, 217
cleaving, 216, 219
epitaxial growth, 216, 219
ion bombardment, 216, 219
vacuum heat treatment, 217, 219
Cs activation studies, *see also* Cs activation
  Si, 230–233
  III–V materials, 224–230
dark-current emission, 208, 270–274, *see also* Dark current, photocathodes
decomposition, 219
electron escape probability, *see* Escape probability
electronic structure, 220–224, *see also* Surface electronic structure
reconstruction, 217–220, 231–233
  GaAs, 217, 218
  Si, 231–233
Schottky barrier formation, 216
sensitivity, 258–260, *see also* Sensitivity measurements, photocathode

O

Ohmic contacts, 1–32, 172, *see also* Contacts; Contacts, ohmic
  buffer layer, 16
    epitaxial growth, 16
    ion implantation, 16
    laser annealing, 16
    molecular-beam epitaxy, 16
  choice of metal, 9–18
  degradation, 27, 30, 31
    interdiffusion, 31
    migration, 30
  fabrication, 9–18
    balling up, 28
    evaporation technique, 10
    loss of As, 29
    plating technique, 10
    sputtering, 10
    surface cleanliness, 16, 28
    surface damage, 10
  general concepts, 2–9, 172
  resistance measurement, 18–24, *see also* Contacts
  structural characterization, 24–31
Ohm's law, 44, 105
  drift and diffusion terms, 61, 62
  individual carrier contributions, 60

P

PES (Photoelectron spectroscopy), 200, 201
Photocathode, 197–289, see also NEA; NEA devices; Photoemission; Photoemitters, bias-assisted
  Cs/GaP dynode, 198
  Cs + O activation, 214–216, see also Cs activation
  Cs + O/P-GaAs NEA surface, 214, 215
  equivalent noise input, 270
  (GaAl)As–GaAs, 198, 241
  GaAs, 198, 240, 241
    reflection-mode photomultiplier, 198
  (InGa)(AsP)–InP, 198, 241,
  (InGa)P–GaAs, 198, 241
  quantum yield, 242, 250, 252, 256, see also Quantum yield
  reflection mode, see NEA semiconductor photoemitter
  sensitivity measurements, 258–260, see also Sensitivity measurements, photocathode
  surface escape probability, see Escape probability
  transmission mode, see NEA semiconductor photoemitter
Photocathode sensitivity measurements, see Sensitivity measurements, photocathode
Photoelectric equation, 181
  energy-level diagram, 181
Photoelectric threshold, 202, 210, 211
  semiconductor
    n-type, 211
    p-type, 211
  temperature dependence, 205
Photoemission, see also NEA; NEA semiconductor photoemitter; NEA surface; Photoemission: three-step model; Photoemitters, bias-assisted
  conduction-band states, 206
  Cs coverage dependency, 226, see also Cs activation
  dark current, 208, 270–274, 288, 289, see also Dark current, photocathodes
  escape depth, 199, 205
  favored in p-type material, 211
  from metals, 201, 205
  from semiconductors, 205–211
    optical absorption coefficients, 208, 234, 239
  three-step process, 205, 209, 233–258, see also Photoemission: three-step model
  internal, 205
  space-charge surface region, 206
  surface states, 207–209
  threshold, 210, 211
  valence states, 206, 207
Photoemission: three-step model, 205, 209, 233–258
  band-structure considerations, 238–241
  diffusion length, 238
  diffusion-model calculations, 239–242
    RM and TM quantum yields, 242
  electron transport to surface, 237–242
    diffusion, 237–242
  optical absorption and electron generation, 234–236
  beam range, 236
  p-type material favored, 211
  quantum yield, see also Quantum yield
    internal, 237, 238
    photoemission, 238
  secondary-electron emission, 235, 236
    primary beam range, 236
  secondary-electron gain, 241
  solution of diffusion equation, 239, 240
  surface escape probability, 242–258, see also Escape probability
  surface recombination, 240
  threshold, 210, 211
  vacuum emission, 242–258, see also Escape probability
Photoemitters, bias assisted, 276–289
  TE (Transferred-electron) photoemission, 276–289, see also Transferred-electron cathode
Photomultiplier tube, 260–263, see also NEA devices
P–N junctions, see PN junctions
PN junctions, see also Applied bias; Current–voltage relation, Diode-characteristics slope; Energy-band diagram; Injection; Semiconducting junctions; specific materials; van Roosbroeck model
  abrupt, 49–57, 82–84
  asymmetric, abrupt
    band shape, transition region, 70, 71
    chemical potential, 53

depletion approximation, 72
electric field, 53
energy-band diagram, 54
injection current, 106
total charge, 53
barrier layer, 48, see also Barrier layer
basic concepts, 41–48
   electromagnetic theory, 44, 45
   energy-band theory, 41
   equilibrium statistics, 42–44
   thermostatics, 45–48
boundary conditions, 90–104, see also
   Boundary conditions in *PN* junction
characteristic equation, 67
cold-cathode emitter, 268
equilibrium conditions, 48–59, 174
   constancy of chemical potential, 51
   electrochemical, 48
   energy-band diagram, 49, see also
      Energy-band diagram
   mechanical, 48
   thermal, 48
heterojunction, see Heterojunctions;
   Heterojunction materials
high-level injection, 93, 101, 104–135, see
   also Injection
linearly graded, 57–59
   chemical potential, 58
   electric field, 58
   total charge, 58
low-level injection, 59–68, see also
   Injection; Shockley low-level theory
   energy-band diagram, 63
   space-charge region, 68–72
macropotential, 184, see also
   Macropotential
Shockley low-level theory assumptions, 68,
   see also Shockley low-level theory
symmetric, abrupt
   chemical potential, 55
   electric field, 55
   energy-band diagram, 49
   total charge, 55
van Roosbroeck model, 105, 111–135, see
   also van Roosbroeck model
Poisson–Boltzmann equation, 50, 58, 69,
   147
Poisson's equation, 45, 56–58, 105, 210
Potential, see also Thermostatic potentials
   barrier, diffusion, or built-in, 49

chemical, 44, 46–55, see also Chemical
   potential
electrochemical, 44, 46–49, see also
   Electrochemical potential
electrostatic, 44–49, see also Electrostatic
   potential
surface, 186, 187

Q

Quantum efficiency, 202, see also Quantum
   yield
Quantum yield, 200, 202, 212, 215, 242, 250,
   252, 256, 261, 264, 278, 279, 283, 284,
   286
   (AlGa)As–GaAs photocathode, 242, 264
   Cs + O activated Ag, 212
   electric-field enhancement, 257, see also
      Applied bias
   GaAs photocathode, 215, 261
   GaAs–(AlGa)As photocathode, 242
   Ga(AsP) photocathode, 256, 261
   In(AsP) photocathode, 250
   (InGa)(AsP) photocathode, 252, 261, 262
   internal, 237, 238, 279
   photoemission, 238
   RM commercial cathodes, 261
   RM NEA GaAs surface, 215, 261
   transferred-electron cathode, 278, 279, 284,
      286
   direct-emitter cathode, 282–284
   effect of bias, 278, 284
   heterojunction, 286
   typical values, 270
   dark currents, 270–274, see also Dark
      current, photocathodes

R

Recombination, see Carrier recombination;
   Generation–recombination
Recombination centers, 72, 79, 87, 88, see
   also Generation–recombination
contrasted to traps, 72, 87, 88
Rectifying contact, see Contacts
Reflection mode (RM), see also NEA devices;
   NEA semiconductor photoemitter
definition, 213

S

Sah–Noyce–Shockley theory, 76–85, *see also* Generation–recombination
   functional relationship, 81
   generation–recombination current, 76
Schottky barrier, 2, 3–8, 162, 170, 205, 210, 216, *see also* Barrier height; Barrier layer; Schottky diode
   current–voltage relation, 3, 172
   diffusion, 3
   formation, 197
   heights (tabulated), 4
   internal photoemission studies, 205
   thermionic emission, 3
Schottky diode, 148, 171, *see also* Schottky barrier
   accumulation layer, 172
   energy-band diagram, 183
Schottky model, 2, 5
Secondary-electron emitters, 265–268, *see also* NEA devices
Semiconducting junctions, theory, 39–189, *see also* Contacts; Heterojunctions; *PN* junctions
   basic concepts, 41–48
   boundary conditions, 90–104, *see also* Boundary conditions in *PN* junctions
   equilibrium conditions, 48–59
   generation–recombination processes, 72–90, *see also* Generation–recombination
   heterojunctions, 135–189, *see also* Heterojunctions
   high-level injection, 104–135, *see also* Injection
   low-level injection, 59–68, *see also* Injection
   space-charge region, 68–71, *see also* Space-charge region
   van Roosbroeck model, 111–135, *see also* van Roosbroeck model
Sensitivity measurements, photocathode, 258–260
   absolute sensitivity, 258–260
   experimental data
      Ga(AsP), 259
      GaP, 259
      (InGa)(AsP), 259
      Si, 259
   quantum-yield curve, 258–260

Shockley low-level theory, 59–68, 90–94, 104, 189, *see also* Injection, *PN* junctions
   applied to heterojunction, 162
   assumptions, 68, 90–94
   boundary conditions, 90–94, *see also* Boundary conditions in *PN* junction
   law of the junction, 93
   energy bands for *PN* junction, 68, *see also* Energy-band diagram
Si
   abrupt diode, 82–84
   cold-cathode emitter, 269
   Cs activation, 230–233, *see also* Cs activation
      Fermi-level pinning, 230, 231
      thermal desorption, 231
      work function dependence on oxygen, 232
      work function, optimal activation, 233
   diode impurity profile, 85
      I–V characteristics, 86
      two trapping levels, 89
   diode I–V characteristics, 86, 90
   electron affinity, 221
   escape probability, NEA surface, 248, 249
      effect of temperature, 257
      electric-field enhancement, 257
   NEA surface, 217–219, 230–233, 259, 265, 268, 271–276, 289
      dark current, 271–273
      device life, 274–276
      sensitivity, 259
      TM photocathode, 265
      TM secondary-electron emitter, 268
   optical absorption coefficient, 208
   recombination current, 82–84
   recombination via traps, 72
   secondary electrons, 235, 236
      primary beam range, 236
   surface cleaning, 244
   surface reconstruction, 217–219, 231–233
   work function, 221, 233
Space-charge neutrality, 91, 92, 94, 101, 116
Space-charge region, 48, 64, *see also* Barrier layer; Depletion layer
   carrier densities, 53, 55, 58, 130, 131
   behavior of electrochemical potential, 95–97, *see also* Electrochemical potential
   majority-carrier injection, 99, 100

generation–recombination currents, 77–85, see also Current density
low-level injection, 68–72
band shape, transition region, 70, 71
NEA photoemitter, 213, 246
photoemission origin, 206
width, 57, 210, 213, 246
Spenke model, 175, 178–182, 185–188
definition of potentials, 185
energy-band structure of heterojunction, 186
States density, see Density of states
Surface atomic structure, 197, 216–220, see also NEA surface
decomposition, 219
reconstruction, 217–220, 230—233
Surface charge, 167, see also Surface dipoles; Surface states
dipole potentials, 186, 187
double layer, 167, 180
Surface cleaning, 243, 244
GaAs, 243, 244
GaP, 243
InP, 243, 245
Si, 244
Surface dipoles, 167, 180, 184, 186, 187, see also Surface charge; Surface states
Surface electronic structure, 197, 198, see also NEA surface; Surface states
band bending, 220–223
Fermi level pinning, 208, 210, 221–224
Surface escape probability, see Escape probability
Surface photoemission, 199, see also Escape probability; PES; Photoemission; Photoemitters, bias assisted; Surface research techniques
Cs coverage dependency, 226, see also Cs activation
metals, 201–205
copper, 202
energy-band diagram, 201
semiconductors, 205–211, see also NEA semiconductor photoemitter
Surface reconstruction, 217–220, 231–233
GaAs, 217, 218
Si, 231–233
Surface research techniques, 199–201, 221, see also NEA surfaces
AES, 199, 200
cleanliness achievement, 216, 217

EDC, 200
ELS, 199, 200
LEED, 199, 200
PES, 200, 201
photoemission, 200
Surface states, 3, 162, 200, 208, 209, 216, 220–224, see also Surface charge; Surface dipoles
band bending, 220–223
dark-current emission, 208
Fermi level pinning, 208, 210, 221-224
NEA photoemitter, 216, 220–224
PES, 200
photoemission, 207, 208

T

Thermionic emission, 3, 5, 6, 171, 172, 179, 205
determination of work function, 205
ohmic contact, 6
Schottky barrier, 3, 5, 171, 172
Thermodynamic relationships, 45–48
chemical potential, 44, 46–55
electrochemical potential, 44, 46–49
electrostatic potential, 44–49
Thermostatic potentials, 44–48, see also Potential
barrier, diffusion, built-in, 49
chemical, 44, 46–55, see also Chemical potential
electrochemical, 44, 46–49, see also Electrochemical potential
electrostatic, 44–49, see also Electrostatic potential
Transferred-electron (TE) cathode, 276–289
carrier fraction in upper minimum, 280
dark-current emission, 288, 289
direct-emitter cathode, 282–284
energy-band diagram, 283
quantum yield, 283, 284
quaternary alloy (InGa)(AsP), 282–284
energy-band diagram, 277, 283, 285
heterojunction cathode, 284–288
conduction-band profile vs. bias, 287, 288
energy-band diagram, 285
photon-absorbing and photon-emitting layers, 284
quantum yield, 286

mean electron energy in upper minimum, 281
quantum yield, 278, 279, 284, 286
  effect of bias, 278, 284, 286
TE effect, 276–282
  transitions to upper-band minima, 276–278, 280, 281
Transmission mode (TM), *see also* NEA device; NEA semiconductor photoemitter
  definition, 214
Traps, 72–74, 79, 87, 88, 90, *see also* Carrier recombination; Generation–recombination
  dark-current emission, 208
Tunneling, 3, 5–7, 41
  heterojunctions, 162
  ohmic contact, 6, 7, 19
  Schottky barrier, 3, 5

V

Vacuum level, 135, 137, 163, 167, 188
  continuity, 137, 175
  discontinuity, 141, 150
    surface dipole, 141
van Roosbroeck model, 105, 111–135, 190
  comparison with experiment, 131–133
  diode parameters, 117–119
  I–V characteristics, 121–129
    bulk-region length, 122
    doping, 125
    lifetime changes, 122, 123, 129
    recombination current, 127, 129
    reverse current, 126
    trap position, 124
  numerical analysis, 111–135
    comparison with recombination–diffusion solution, 128

P-region low-level injection
  implications, 115, 116
  recombination–diffusion solution, 117–121, 127–129
    comparison with numerical solution, 128
  Stone's method, 112–117, 131–133
  three-part diode, 112–117
    abrupt, 114
    linear-field, 120
Volta potential, 170, 180–183, 188, 189

W

Work function, 2, 164–166, 179, 189, 201–205, 210, 221, 225, 232, 233, 251, 254, 275
  component of photoelectric threshold, 210, 211
  Cs coverage dependency, 224, 225
    effect of oxygen, 232
    GaAs, 225
    Si, 233
  effect on surface escape probability, 251, 254, 275
  effective, 182
  electric-field variation, 257
  lowering with Cs + oxygen, 211–213
    crystal-face orientation, 213
  optical vs. Kelvin, 229
  photoelectric, 164, 165, 201–205
    determination, 203–205
  related to electronegativity, 205
  related to sublimation entropies, 205
  semiconductor, *n*- vs. *p*-type, 210
  tabulated values, 4, 165, 204
  thermionic, 165, 210
  true, 164